ENGINEERING WITH RIGID PVC

PLASTICS ENGINEERING

Series Editor

Donald E. Hudgin
Princeton Polymer Laboratories
Plainsboro, New Jersey

1. Plastics Waste: Recovery of Economic Value, *Jacob Leidner*
2. Polyester Molding Compounds, *Robert Burns*
3. Carbon Black-Polymer Composites: The Physics of Electrically Conducting Composites, *edited by Enid Keil Sichel*
4. The Strength and Stiffness of Polymers, *edited by Anagnostis E. Zachariades and Roger S. Porter*
5. Selecting Thermoplastics for Engineering Applications, *Charles P. MacDermott*
6. Engineering with Rigid PVC: Processability and Applications, *edited by I. Luis Gomez*

Other Volumes in Preparation

ENGINEERING WITH RIGID PVC

Processability and Applications

Edited by I. LUIS GOMEZ
Monsanto Company
Springfield, Massachusetts

CRC Press
Taylor & Francis Group
Boca Raton London New York

CRC Press is an imprint of the
Taylor & Francis Group, an **informa** business

First published 1984 by Marcel Dekkar, Inc.

Published 2019 by CRC Press
Taylor & Francis Group
6000 Broken Sound Parkway NW, Suite 300
Boca Raton, FL 33487-2742

© 1984 by Taylor & Francis Group, LLC
CRC Press is an imprint of Taylor & Francis Group, an Informa business

First issued in paperback 2019

No claim to original U.S. Government works

ISBN 13: 978-0-367-45179-0 (pbk)
ISBN 13: 978-0-8247-7080-8 (hbk)

Visit the Taylor & Francis Web site at
http://www.taylorandfrancis.com

and the CRC Press Web site at
http://www.crcpress.com

LIBRARY OF CONGRESS CATALOGING IN PUBLICATION DATA

Main entry under title:

Engineering with rigid PVC.

(Plastics engineering ; 6)
    Bibliography: p.
    Includes index.
    1. Polyvinyl chloride. I. Gomez, I. Luis.
II. Series: Plastics engineering (Marcel Dekker,
Inc.) ; 6
TA455.P58E54   620.1'923   84-12161
ISBN 0-8247-7080-3

Foreword

There are so many technical publications offered to the plastics industry each year that when one comes along which truly fills a *real need*, special notice must be taken. *Engineering with Rigid PVC: Processability and Applications*, in my opinion, is such a publication. Although much has been published about the science and technology of PVC (the polymer itself, additives, processing, and testing), nowhere to my knowledge has the specific topic of rigid PVC been presented in a single volume in such a complete, practical, and useful manner.

Complete, since it covers all aspects of PVC compounding (ingredient selection, matching formulation to the process equipment, and blending techniques), major melt-processing methods (single and multiscrew extrusion, injection molding, blow-molding, and calendering), and postforming or downstream cooling and handling techniques. Rheological properties of the melt and quality control requirements during production are also included.

Practical and useful, because this volume is not merely an academic discussion, but is directed to the real-world concerns of those who formulate, compound, and process rigid PVC, and who are also responsible for product quality and production efficiency. The unique properties and problems of rigid PVC, combined with its tremendous growth in commercial importance over the past 20 years, make a publication such as this long overdue.

In comparison with other large-volume thermoplastic resins (polyolefins and styrenics) and even some of the engineering thermoplastic resins, compounding and processing rigid PVC presents quite a challenge: unlike polyethylene, one cannot take a coffee break during a plant power failure with machines sitting idle containing molten PVC.

Reflect for a moment on what has transpired (and continues to do so) with rigid PVC: a polymer that by itself is brittle and horny, has very high melt viscosity, easily degrades under heat and shear to

evolve corrosive HCl, tends to stick to hot metal surfaces, nevertheless
has exhibited remarkable growth in such applications as pipe, conduit,
fittings, siding, window profiles, sheet, and bottles, to the point that
rigid PVC products now consume about 40% of all PVC resin produced
in the United States, which totaled around 5.4 billion pounds in 1982.
Why? The versatility of PVC—its ability to respond so well to a host
of additive ingredients (modifying resins, process aids, lubricants,
stabilizers, fillers, blowing agents, and so on) to achieve such a wide
range of attractive physical and visual properties at reasonable cost—
this versatility is unmatched among the large-volume, if not all, thermoplastic resins.

The opportunity to review several chapters of this volume has
prompted my recollection of over 20 years' involvement with so many
facets of the rigid PVC industry—its fascination, and both its reward-
ing and frustrating moments in laboratories and plants here and overseas. Rigid PVC growth has been healthy, at times vigorous, and
occasionally controversial, but never dull. There has always been
room for honest differences of opinion, such as

Pellet versus powder blends (for extrusion, injection, and blow-
   molding).
Vented, long L/D versus nonvented, shorter L/D extruders.
Single-screw versus multiscrew extruders.
Parallel versus conical twin versus four-screw extruders.
Lead versus tin stabilization of pipe.
"One-pack" additive system versus separate components.
Lubrication theories: Is calcium stearate internal?
Extrusion versus injection blow-molding.
Tin versus calcium-zinc stabilization of food-grade bottles.
High-intensity blending: single, double, or triple batching versus
   ribbon-blended cold mix.
Profile versus postformed siding extrusion.
Weathering: laboratory results versus outdoors—best test sites.
Use of TiO_2—European versus U.S. formulation practices.
Impact testing: which method?
Torque rheometer: correlation with real-world production, or not.

In light of all the above, this volume does indeed provide an excellent
view of the current state of the art for rigid PVC, along with a little
historical perspective.

To I. Luis Gomez, W. J. Fudakowski, J. A. Baclawski, J. L. Murrey,
Robert DeLong, A. W. M. Coaker, and R. A. McCarthy, my compli-
ments and appreciation to you for your efforts in compiling a very
valuable contribution to the literature on rigid PVC. For those of us
intimately involved with this dynamic industry, this book should
gather no dust on our shelves.

George A. Thacker, Jr.
Anderson & Co., Inc.

Preface

The last few decades have seen the development and growth of a significant industry—the rigid PVC industry—which converts rigid PVC
into a wide range of valuable products as diverse as pipes, house
siding, bottles, window frames, and packaging films. Perhaps the most
versatile and useful of all the thermoplastics, rigid PVC possesses a
unique combination of much sought after properties—nonflammability,
good weatherability, excellent chemical resistance, low gas permeability,
and when properly modified, a rather wide processability range. More-
over, rigid PVC is relatively low cost and has cost stability. These
attractive features have made this polymer of interest not only to polymer manufacturers but also to supporting industries: the manufacturers of additives/modifiers which enhance heat stability, flow and release properties, impact strength, and weatherability, and the manufacturers of processing equipment. Undoubtedly, this multi-industry
attention is a considerable factor in bringing the rigid PVC industry
to its present advanced stage.

In spite of all the progress achieved by this industry, the thermal
and/or shear sensitivity of rigid PVC still creates problems. As it was
during the early days, this sensitivity continues to be a major drawback when processing the polymer at high production outputs. Hence,
polymer stabilization and overall additives modification are of enormous
importance to reduce this sensitivity. Aside from the issue of polymer
thermal stability, rigid PVC melt rheology is critical to polymer processability. A knowledge of the rheology of rigid PVC is the key to understanding flow behavior in screw extruders, dies, injection molds, etc.,
and for design purposes. Furthermore, acquaintance with rheological
properties helps one to understand the influence some of these properties have on the profile forming operation and on the appearance of surface defects. In summary, to increase the utilization of the desirable
properties of rigid PVC, processors need to understand, modify, and
manipulate rigid PVC thermal/shear sensitivity and its rheological properties.

v

In keeping with the centrality of the previous issues, the first part of this book sequentially blends discussions of resin, additive modification, and rheology with those of polymer preparation, melt processing, and forming techniques. Because of their paramount importance, melt temperatures, the existence of radial temperature gradients and their effect on melt quality, and temperature measurement techniques are topics thoroughly covered throughout this work. To illustrate how additives and their concentrations are effectively used in the field, the formulations of some rigid PVC compounds are discussed throughout the book.

The second half of the book treats of the major conversion operations used to process rigid PVC and the application of these processes in manufacturing the major lines of engineering products. This section has been authored by a group of leading international experts, outstanding in the subjects they have covered.

Since it deals with the fundamentals of polymer processing and of conversion operations used to manufacture various engineering products, *Engineering with Rigid PVC: Processability and Applications* should prove of great value to polymer chemists, process engineers, and technicians of the rigid PVC industry in their daily polymer and process work. With all the background vital to understand the processing of rigid PVC clearly presented and readily accessible, this book should be an immensely useful resource for process engineers and technicians in developing countries around the globe. Because it covers both principle and practice, *Engineering with Rigid PVC: Processability and Applications* can serve as a helpful reference for all students of polymer chemistry and polymer engineering.

Many individuals have helped to make this book more informative and accurate. The author wishes especially to thank George A. Thacker, Jr., who not only reviewed my contribution but also supplied me with a wealth of information on conical twin-screw extrusion; Victor A. Matonis and J. L. Grover, who provided much help in reviewing my contribution; the contributors who undertook the difficult task of writing Chapters 5 through 9, and Josephine A. Giorgi for editing and Catherine A. Kowalczyk for typing my contribution. Although Monsanto Company is out of the PVC business, permission, support, and encouragement provided by this company and the immediate management to publish my contribution to the book are also gratefully appreciated and acknowledged.

I. Luis Gomez

Contents

FOREWORD *by George A. Thacker, Jr.*   iii

PREFACE   v

CONTRIBUTORS   xi

1 INTRODUCTION   1
   I. Luis Gomez

   1.1   Background   1
   1.2   Rigid PVC Conversion Operations   5
   1.3   Polymer Preparation   6
   1.4   Rigid PVC Rheology   6
   1.5   Rigid PVC Extrusion   7
   1.6   Extrusion Blow-Molding of Rigid PVC Containers   8
   1.7   Injection Molding of Rigid PVC   9
   1.8   Injection Blow-Molding of Rigid PVC Containers   11
   1.9   Calendered Rigid PVC Products   12
   1.10 Thermoforming of Rigid PVC Sheet   13

2 RIGID PVC: POLYMER PREPARATION   15
   I. Luis Gomez

   2.1   Introduction   16
   2.2   PVC Resins   17
   2.3   Additives Enhancing Polymer Processability   22
   2.4   Additives Enhancing Mechanical Properties: Impact
          Modifiers   38
   2.5   Compounding: Equipment Classification and Techniques   42
   2.6   Summary   95
   References   96

3 RIGID PVC RHEOLOGY   99
   I. Luis Gomez

   3.1   Introduction   99
   3.2   Nonnewtonian Flow Behavior of Rigid PVC Melts   102
   3.3   Ideal or Newtonian Fluid Concept of Viscosity   102
   3.4   Rigid PVC Melt Rheology   104
   3.5   Power Law   113
   3.6   Factors Affecting Melt Viscosity, Postextrusion
         Swelling, and Melt Quality   118
   3.7   Melt Stability   135
   3.8   Color Propagation Curves   138
   3.9   Glass Transition Temperature T_g 138
   3.10 Melting Point T_m 140
   3.11 Temperature Dependence of Viscosity   141
   3.12 Pressure Dependence of Viscosity   145
   References   149

4 RIGID PVC EXTRUSION   151
   I. Luis Gomez

   4.1   Introduction   153
   4.2   Extrusion of Rigid PVC   153
   4.3   Modern Single-Screw Machines   155
   4.4   Polymer Residence Time in Modern Single-Screw
         Machines   156
   4.5   Instrumentation-Application to Monitor Operating
         Variables   157
   4.6   Extrudate Temperature T_* 157
   4.7   Melt Pressure   159
   4.8   Flow Function ϕ 160
   4.9   Mechanical Power p   161
   4.10 Analysis of Screw Performance   161
   4.11 Energy Factor   162
   4.12 Mechanical Energy Dissipated as Heat in the Polymer 164
   4.13 Scale-up Rules for Single Screws: Analysis and
         Discussions   166
   4.14 Rules for Semigeometric Scaling   167
   4.15 Flaws of the Semigeometric Scaling Rules   170
   4.16 Drag Capacity of the Metering Section of the Screw   171
   4.17 Analysis of Experimentally Determined Performance
         Data   172
   4.18 Analysis of Two-Stage Single-Screw Extrusion in
         Vented Barrel Extruders   174
   4.19 Characteristic Curves   178
   4.20 Processing Rigid PVC Dry Blends in Two-Stage Single-
         Screw Vented-Barrel Extruders   182

4.21   Two-Stage Single-Screw with a Barrier or Double-
       Channel Section   187
4.22   Vent-Zone Operation   191
4.23   Pumping-Zone Operation   192
4.24   Screw Nose Design   192
4.25   Screw Temperature Control   193
4.26   Extruder Temperature Control   194
4.27   Combined Screw and Die Performance   196
4.28   Flow-Restricting Devices   200
4.29   Plate-out   201
4.30   Twin-Screw Extruder Process   203
4.31   Analysis of Twin-Screw Extrusion Process   208
4.32   Matching the Formulation to the Extrusion Process and
       to the Finished Products   208
4.33   Extrusion of Pipes and Conduits: Single Versus Conical
       Twin-Screw Dry Blend Extrusion Process   212
4.34   Rigid PVC Extrusion of Siding and Accessories   230
4.35   Rigid PVC Extrusion of Window and Door Profiles   236
References   242
Bibliography   244

5 EXTRUSION BLOW-MOLDING OF RIGID PVC CONTAINERS   245
   W. J. Fudakowski

5.1   Introduction   246
5.2   Twin-Screw and Single-Screw Extruders: PVC Dry
       Blend Powder Versus Granules   246
5.3   Two-Stage Extrusion System   251
5.4   Single-Screw Extrusion Systems   257
5.5   Extruder Dies: Parison Control   260
5.6   Bottle-Blowing Machines   264
5.7   Platen-Type Machines   264
5.8   Rotary Machines   271
5.9   Biaxially Oriented PVC Bottle Equipment   278
5.10   PVC Compounds for Bottle Blowing   289
References   300

6 RIGID PVC INJECTION MOLDING   301
   John A. Baclawski and Jerry L. Murrey

6.1   Background and History   303
6.2   Market Trends and Opportunities   306
6.3   Economics   316
6.4   Properties and Characteristics of Current Products   320
6.5   Formulations and Their Effect on Material Properties   324
6.6   Product Design and Mold Construction   344
6.7   Processing Recommendations   349
6.8   Processing Techniques   356
References   360

7 INJECTION BLOW-MOLDING OF RIGID PVC CONTAINERS   363
   Robert DeLong and I. Luis Gomez

   7.1   Introduction   363
   7.2   The Injection Blow-Molding Process and Its Variables   364
   7.3   The Parison Injection-Molding Process   366
   7.4   Cavity Filling in Multicavity Molds   366
   7.5   Parison and Blow Core Temperature Control   367
   7.6   Parison Blowing   368
   7.7   Injection Blow-Molding of Rigid PVC Containers   369
   7.8   Alternate Processes   380
   Bibliography   383

8 CALENDERED RIGID PVC PRODUCTS   385
   A. William M. Coaker

   8.1   Introduction   386
   8.2   Market Factors   393
   8.3   Manufacturing Operations   406
   References   436
   Additional Reading   438

9 THERMOFORMING OF RIGID PVC SHEET   439
   Robert A. McCarthy

   9.1   Introduction   439
   9.2   The Thermoforming Process   440
   9.3   Thermoforming Techniques   443
   9.4   Rigid PVC Thermoforming   449
   9.5   Markets for Thermoformed Rigid PVC   451
   References   453

INDEX   455

Contributors

JOHN A. BACLAWSKI BF Goodrich Chemical Group, Cleveland, Ohio

A. WILLIAM M. COAKER* Plastics Consultant, Morristown, New Jersey

ROBERT DELONG Captive Plastics, Inc., San Bernardino, California

W. J. FUDAKOWSKI Metal Box p.l.c., Wantage, England

I. LUIS GOMEZ Monsanto Company, Springfield, Massachusetts

ROBERT A. McCARTHY Springborn Laboratories, Inc., Enfield, Connecticut

JERRY L. MURREY[†] BF Goodrich Chemical Group, Avon Lake, Ohio

Current affiliations
*BF Goodrich Chemical Group, Avon Lake, Ohio
†Mobay Corporation, Pittsburgh, Pennsylvania

ENGINEERING WITH RIGID PVC

1

Introduction

I. LUIS GOMEZ /Monsanto Company, Springfield, Massachusetts

1.1   Background 1
1.2   Rigid PVC Conversion Operations 5
1.3   Polymer Preparation 6
1.4   Rigid PVC Rheology 6
1.5   Rigid PVC Extrusion 7
1.6   Extrusion Blow-Molding of Rigid PVC Containers 8
1.7   Injection Molding of Rigid PVC 9
1.8   Injection Blow-Molding of Rigid PVC Containers 11
1.9   Calendered Rigid PVC Products 12
1.10 Thermoforming of Rigid PVC Sheet 13

1.1 BACKGROUND

Nonflammability, weatherability, chemical resistance, low gas permeability, rather wide processability range, and low cost are six of the outstanding advantages of rigid PVC. Also, because 56.8% of its weight is chlorine, a nonpetroleum product, the effect of petroleum price increases is not as drastic as it has been on 100% petroleum-based polymers. Consequently, PVC also shows an advantage in cost stability. In summary, it is not pretentious to state that there is no other commodity polymer in the field with such a balance of processability, properties, cost, and cost stability. This explains the vital role it plays in the economy of the developing and underdeveloped countries, since it is the polymer that also offers the broadest spectrum of end uses.

Engineering with Rigid PVC covers the series of conversion techniques carried out on this polymer to increase the utilization of its desirable properties. The technical aspects of the PVC polymer conversion industry normally include machine and equipment design; plant design and operations, with emphasis on streamlined operations, from the in-

1

coming resin and other formulation ingredients to the finished products; product development, including auxiliary product line to complement principal products; and the analysis of PVC conversion operations. In this book, emphasis is placed on polymer processing rather than on the machine and equipment aspects of the process. Polymer processing consists of the polymer preparation, melt processing, and forming techniques.

The machinery and equipment used in compounding and processing rigid PVC have been the subject of several books and many technical articles. Extruder screws, for example, have been described in hundreds of patents. No attempt will be made to review or interpret the extensive literature. Frequent references to processing equipment and the polymer behavior in this equipment will be made throughout this book.

Because deformation and flow are involved in all the polymer conversion operations, rheological data are frequently used in this book. Also, polymer modifications via selective additives will be covered. Reference to key components and to formulations that, because of their use, are becoming classic in the industry, will also be used.

Tailoring the resin to the process and to the finished product is, perhaps, the first step in engineering with PVC. For example, if the resin is to be used in a dry blend extrusion process, a resin with a large amount of porosity and high bulk density is needed. If the resin is for calendering of rigid PVC film and sheet, a lower molecular weight (K value 55 to 60) is normally used. Also, for food contact applications, the level of residual vinyl chloride monomer in the resin is more critical than that for, e.g., siding, gutter, or furniture.

In the last 25 years, a great number of significant developments have brought the PVC industry to a very advanced stage. The days of frequent in-house tooling improvisation, due to the lack of reliable commercially available equipment, are almost gone, and complete machine systems are now readily available. A visit to any of the international fairs, like the Interplas in Birmingham, England, and Kuntststoff in Düsseldorf, West Germany, held every four years, will give the reader a good idea of the great profusion of processing equipment now available. To bring together machine manufacturers and PVC processors was not an easy task and it took a lot of time and frustration. It was not until the personnel involved realized that the many and varied problems associated with PVC conversion operations could be solved only by systematic research and a team effort approach by resin and additives suppliers, equipment manufacturers, and PVC processors that the development of processing machinery kept pace with the expansion and progress achieved in the resin and compound fields.

As one of the first thermoplastics commercially available, unplasticized PVC resin was used to make various articles as early as 1935. The first rigid PVC applications were developed almost exclusively by trial and error, to fulfill explicit needs. Looking back to those years, when minimum design technology was applied, it is fair to say that the users and producers were quite fortunate with polymer performance. As this industry has grown, prediction of the performance of rigid PVC products has received greater attention, and a designer or product engineer now has information available not only to upgrade performance of existing products but also to define the optimum process, formulation, and dimensions for new products.

Rigid PVC residential siding provides a good illustration of the above statement. Before a systematic outdoor weathering program was completed, producers were selling and installing siding with an average 20 years' service life warranty. This writer recently inspected an installation about 19 years old and, despite the fact that the titanium dioxide pigment content of this formulation is significantly lower than the levels presently used, the siding was found to be in very good condition. Because of the dual ornamental as well as functional purposes of this product, vinyl siding offers a unique example of the approach used by the various manufacturers before a systematic approach to design technology was implemented and some correlation between short-term accelerated testing and outdoor performance was established. Steps followed in those early days were as follows.

Step 1: Conceive a part as well as its shape and select the extrusion process.

Step 2: Screen the available rigid PVC, mostly white compounds, on the basis of processability, short-term weatherability, and those engineering properties that relate to performance.

Step 3: Use lab tests to predict field performance of engineering properties, such as retention of color and impact resistance, surface distortion, and thermal expansion.

Step 4: Test under actual part use, such as a model home.

Step 5: Redesign, retest, and tune up compound formulation and the extrusion process to improve engineering properties and to reduce surface gloss and surface distortions, for example.

If part requirements are simple, as with rigid PVC electrical conduits, these steps also become very simple.

Chapter 30 of the Dekker *Encyclopedia of PVC*, Testing of Rigid PVC Products, Analysis of Test Results, will help familiarize the reader with the mechanical, thermal, and environmental behavior of this polymer. This knowledge will also help the reader to use the engineering properties of this polymer more efficiently.

Vinyl siding and its accessories, replacement window and door frames, and rigid vinyl bottles for liquid, syrup, oil, and salad dressing, were perhaps the last major product lines introduced to the market several years ago. The vinyl chloride monomer (VCM) scare halted, at least in the United States, the production of bottles for food-contact application and also had a very detrimental effect on the development of new products. Obviously, major effort was placed by the industry on VCM reduction. During this period, it seems that the role of designers or product engineers was switched to optimization of existing products, processes, and cost reduction. A good example of a cost reduction program is one that has been in progress in Europe with the 1.5 liter bottle for mineral water. Through bottle design, mainly ribbing in the bottle panel to increase stiffness, bottle weight has come down from about 60 to 42 g. Via conditioning of blow-molded preforms and subsequent stretch blow process, which provides biaxial orientation, the industry is at present achieving 1.5 liter bottle weights as low as 39 g (more on this later).

An illustration of product appearance optimization is offered by vinyl siding. It is known that when siding is extruded in single-screw extruders or at elevated temperatures it tends to have a high level of gloss, which for the most part is not uniform, showing marked die lines and other defects. When the siding is embossed with a wood-grain pattern finish technique, which has been progressively perfected, the surface imperfections are masked, and the resultant product shows a very uniform and attractive appearance. Another good example of product appearance optimization is that of vinyl window and door frames. Originally, the areas where the frame components were welded together were very noticeable. At present, due to improved welding techniques, the vinyl windows and door frames seem to be all of one piece.

It is known that weather-resistant impact-modified PVC could not be made glass-clear because of the haziness of the impact modifiers. E. Röhrl, BASF, was able to demonstrate, with the aid of electron microscope photographs, that MBS and ABS formed particles in the PVC matrix, and that their size was dependent upon the shearing action during fabrication. Chlorinated polyethylene (CPE) and ethylene vinyl acetate (EVA) first act as a glue for the original PVC granules and later change into discrete particles in the PVC melt. Certain elastic polyacrylic esters (PAE), generally polybutylacrylates, form discrete agglomerated spheres, 50 to 200 μm in diameter, which is close to the resin particle size, and are rather stable against dispersion and shearing action. Using this rubber as the basis, BASF recently was able to produce a rigid PVC compound, Vinuran KR 3820, which when polyblended in quantities of 5 to 20% with unplasticized PVC, produces a compound that could be fabricated into high-impact, weather-resistant, transparent glass-clear sheets for such uses as cupolas, greenhouses, and window panes.

Before pursuing this discussion, it is pertinent to state that, although by definition PVC is classified as rigid when the elastic modulus E is $\geq 100,000$ psi (689 MPa) at 23°C and 50% relative humidity (ASTM Method D 883), the rigid PVC products, which will be the subject of this work, are (except for some calendered rigid films and some compounds for injection blow-molding applications) manufactured from compounds with $E \geq 350,000$ psi. In fact, for most of the engineering applications that are discussed throughout this work, the higher the elastic modulus, the better the engineering properties, such as heat distortion temperature, impact resistance, and surface distortion.

1.2 RIGID PVC CONVERSION OPERATIONS

Plastics engineers who process rigid PVC products can improve their skills with ease if more attention is given to

1. Resins and raw materials properties
2. Polymer preparation
3. Resin and compound rheology
4. Melt processing and forming techniques
5. Properties of the finished products

When the level of knowledge in these areas is appropriate to their work needs, learning to troubleshoot better, for example, offers plastics engineers the interest, challenge, and satisfaction of any opportunity to use reason and imagination in solving problems. But limited overall know-how or knowledge in an isolated area may confuse and discourage engineers, thus reducing their troubleshooting ability.

Plastics engineers in highly developed countries can improve their skill almost by themselves. There are available many resources to do this, such as technical magazines, professional societies, technical meetings, seminars, and consultants. But others, such as the engineers working in developing and underdeveloped countries, where rigid PVC products are profusely used, mainly for irrigation purposes, are less fortunate. They not only have to wear all sorts of hats, but for the most part do not have access to the needed resources. This situation tends to divert efforts, and emphasis is placed mainly on forming techniques and overall equipment maintenance and not on resin properties, additive modification, or screw design optimization, for example. In countries with problems such as only one resin, limited production of additives, and high import tariffs, the lack of research on resin and additives is justifiable. It is painful to know of a better quality resin when one has to use whatever is produced locally.

The major goals of this work are to synthesize existing knowledge while providing continuity throughout, and to aim all the discussions toward the finished product. Thus, by sequentially blending the discussions on resin and additive properties and rheology with those of polymer preparation, melt processing, and forming techniques, plastics engineers will be helped in their overall knowledge of the process and their troubleshooting ability. To avoid unnecessary labor, we will take advantage of what is known and highly publicized and reduce discussions in these areas.

Extrusion, injection molding, and calendering are the three basic forming techniques of rigid PVC polymer processing covered in this work. These operations, however, are subclassified according to their largest usage, as follows:

Extrusion of pipe and conduit; siding and trim; window profiles,
and so on
Extrusion blow-molding of bottles
Injection molding of fittings, electrical junction boxes, and so on
Injection blow-molding of bottles
Calendering of rigid PVC film and sheeting
Thermoforming of rigid PVC film and sheeting

1.3 POLYMER PREPARATION

Chapter 2 is dedicated to polymer preparation, which is paramount in
rigid PVC polymer processing. Melt compounding, pelletizing, and some
color-compounding techniques are also covered. Since almost every
desirable property of rigid PVC products can be modified to a certain
extent with additives, the practice of blending modifiers with resin is
included. This practice offers great flexibility to processors. The
various commercial PVC resin polymerization processes, as well as the
effect of these processes on some key resin properties, are also discussed.

1.4 RIGID PVC RHEOLOGY

In all the forming operations mentioned above, PVC must flow in order
to be die or mold shaped, and then must solidify while retaining its
shape. Fluidity is achieved by heat (conduction from an external
source) and/or mechanical shear. Chapter 3 deals with certain aspects
of polymer rheology and gives a general outline of how the rheological
properties of rigid PVC compounds help us to understand flow behavior
in e.g., screw extruders, die adaptors, dies, and injection molds.
Data on rheological properties, determined at the proper shear rate
range, are useful not only for designing dies and injection molds but
also for analyzing the performance of existing ones.
  In the single-screw extrusion of viscous polymers like rigid PVC,
traversing thermocouples in the metering zone commonly show melt
temperatures increasing significantly as one penetrates inward from
the extruder barrel. The film of polymer wetting the screw surface is
the hottest, and the film near the barrel wall the coldest. The existence of a radial temperature gradient in the melt can be modeled as a
composite of multiple concentric layers traveling at variable speed,
which decreases from screw to barrel. This also suggests the existence
of a velocity gradient. This velocity gradient is the shearing rate of
the fluid, and these gradients are produced by the motion of the screw,
which in turn is produced by the extruder drive. The internal resistance of the PVC melt to the force applied by the screw is the shearing
stress. In rigid PVC melts, this resistance is mainly caused by the attraction of the pendant chlorine atoms, reinforced by other molecular

molecular bonds and forces. Obviously, to achieve flow of polymer through an extrusion die or into a mold, this resistance to flow has to be neutralized and overcome. In nonnewtonian liquids like rigid PVC melts, the shear stress is not directly proportional to the shear rate, and a unit increase in the shear rate may significantly increase the shear stress. This happens over the entire range of shear rates, including very low or very high values. This obviously creates a significant departure from newtonian fluid behavior.

Another piece of information of great value when processing rigid PVC is the postextrusion swelling or percentage memory. The lower the memory at a given shear rate, the more easily that material is formed and better control on wall thickness is obtained. A high degree of memory, however, is desirable for oriented processes, as in the production of rigid PVC oriented bottles. Also, knowing the shear rates at which melt or surface fracture occur for a specific compound and the computed shear rate factors for a production die system, the maximum acceptable commercial quality output for production can be determined. Furthermore, since the shear rate at which melt fracture occurs can be changed by varying the length to the opening ratio of the die land (higher ratios allow the use of high shear rates before fracture is encountered), this information is of great value for die design purposes.

1.5 RIGID PVC EXTRUSION

Chapter 4 deals with the extrusion of rigid PVC products. Rigid PVC formulations may differ widely from each other in their composition (e.g., type of resin, resin molecular weight, type of additive, and additive concentration) and degree of compounding (e.g., dry blend or pellets). For example, the unmodified, rigid, lead-stabilized formulations for electrical conduits and pipes differ so widely from the clear, impact-modified, food-grade formulations for bottles that, quite frequently, the same machine and extruder screw combination that performs very satisfactorily with one of these compounds could not produce acceptable parts with the other one. Furthermore, the same extruder-screw-die combination that handles a pelletized compound nicely could produce unacceptable parts when the same compound is used as a dry blend. In a few words, there is no universal extruder and screw-design combination that can handle every one of the rigid PVC compounds available. This statement is more true in the case of single-screw machines than with the multiscrew extruders.

For every product, however, there exist all sorts of combinations of extruders, screw designs, and modes of operation that meet minimum performance standards. This explains why, from plant-to-plant, there is such a variety of extruders and screws producing the same profile. If the performance of these extruders is closely analyzed for extrusion output (pounds per hour, lb/hr), power economy (pounds per horse-power-hour, lb/hp hr), and surge and production yield, it is more than likely that very significant performance differences among these machines will be detected.

Extrusion operations are used to manufacture pipes and conduits; residential siding and its accessories (downspouts, gutters, and trims); window and door frame components (the so-called lineals); building panels (corrugated and flat); louvers, rolling shutters, and blinds; film and sheeting for packaging applications; and foamed profiles. Single- as well as twin-screw extruders are used for these operations. There are some trends, however. In very competitive applications, such as the extrusion of pipes and conduits, dry blend extrusion in twin-screw extruders is the preferred route. Significant savings in compounding cost appears to be the reason for this preference. Dry blend extrusion in twin-screw extruders is also the choice for large diameter, heavy-walled PVC pipes. In highly modified formulations with a high level of titanium dioxide (10.0 to 18.0 parts per hundred of resin, phr) like those used to manufacture window and door frames and residential siding and accessories, the manufacturers are divided between single- and twin-screw extruders. The smaller producers, less than 10 Mlb, seem to favor single-screw extruders and pelletized compounds, while the oldest manufacturers continue using single-screw extruders in a dry blend mode via barrier-type screws. In the case of building panels, both systems have been used successfully.

1.6 EXTRUSION BLOW-MOLDING OF RIGID PVC CONTAINERS

All the extrusion blow-molding processes have in common the extrusion of a tubular parison, which is intermittently fed, while hot, to a blow-molding device. In the larger units, molds are mounted on horizontal or vertical moving wheels. The parison extrusion and the blow-molding steps are normally carried out in a continuous operation.

This process is dominated by single-screw extruders. Even in the two-stage extrusion systems, the metering of the melt to the parison die is mostly done by single-screw extruders. Some of the reasons for this preference are related to melt quality: First, the extrudate entering the die must be very well fused; poor fusion of the extrudate may result in rupture of the parison during blowing. Second, the extrudate surface must be free from surface blemishes, such as melt fracture or screw tip marks. Third, for glass-clear applications, i.e., mineral water, oil, or salad dressings, the extrudate should show the highest clarity and surface gloss, usually attainable using high melt temperatures. With the recent introduction of polyethylene terephthalate (PET) bottles for the same applications, these quality requirements are becoming more critical. Fourth, the extrudate should be free of foreign particles. These foreign particles, for the most part, are eliminated with the use of a breaker-plate and screen combination.

Other reasons for the use of single-screw extruders are related to the geometry of the parison dies, which for the most part are small in size and of small-diameter entry. These small dimensions favor the circularity of the single-screw extruder bore over those of the conical of parallel twin-screw extruders, which are not circularly symmetrical.

One feature of this process is the provision for varying the parison geometry along its length (parison programming) to permit efficient utilization of material in the container. This, as expected, results in a cyclic downstream flow resistance, causing the head pressure to vary within some period of time. This period, which could be of the same magnitude as the time for a single revolution of the extrusion screw, imposes a rather severe limitation on the output stability. Another factor, also very critical, is the temperature of the melt. In general, high-temperature, low-viscosity melts usually result in containers with improved glossy surfaces, barrier properties, and strength. The desirability of using a breaker-plate and screen combination is another reason the single-screw machines are preferred for this application.

Minimum or no orientation is obtained in this continuous process. In an alternate process, the parison is sequentially extruded, shaped into preforms, stretched, and blown into bottles. This is done in one parison extrusion head as well as in double heads. This process claims a lot of advantages. Among them are containers exhibiting high impact resistance, improved clarity, and barrier properties due to the orientation introduced in the stretch and blow process. The largest single application of the extrusion blow process is the 1.5 liter bottles for mineral water. These containers are profusely used in Europe.

Progress in the extrusion blow-molding of rigid PVC bottles has been so intimately influenced by that of machine development that the major commercial processes are included throughout Chapter 5. An application of rigid PVC never deemphasized in Europe, extrusion blow-molding, has been heavily influenced by European technology.

1.7 INJECTION MOLDING OF RIGID PVC

Utilization of polyvinyl chloride formulations to produce commercial parts by injection molding has been slow in developing. This was largely due to the shear sensitivity of the early PVC polymers. The early attempts to injection mold the highly viscous, high-molecular-weight (K value 66 to 70) PVC extrusion formulations on plunger-type injection machines were rarely successful. Applications were limited to small parts with relatively heavy wall thickness. At the time, pipe and conduit fittings were produced by machining and forming or fabricating sections of extruded pipe. It was not until the early 1950s, with the introduction of the in-line reciprocating screw machine, that commercial injection-molded rigid PVC pipe fittings started to appear on the

market. Rigid PVC injection molding compounds required a reduced
shear sensitivity and improved thermal stability. Continued PVC polymerization developments led to intermediate-molecular-weight resins
(K value 55 to 60), which were much more suited for use in injection-
molded pipe-fitting formulations. However, even these "improved"
compounds were found to have little utility in the more demanding speciality injection-molding markets.

From these initial days, dramatic processability improvements have
been made through more recent advances in polyvinyl chloride polymerization technology. It became apparent that, although compounds
based on PVC copolymers (vinyl acetate and propylene) were processable at lower melt temperatures, many of the outstanding properties
of homopolymer PVC were sacrificed when copolymers were substituted.
A new low-molecular-weight (K value 50) PVC homopolymer resin,
which combines the inherent properties of PVC with much improved
processability, has evolved. Due to the recent dramatic cost escalation
in petrochemical feedstocks based on crude oil, the timing of this new
resin development has been very significant. The high inorganic content of PVC (56.8% chlorine) has resulted in both current and future
attractive cost positions in speciality injection-molding markets, especially applications in which flame retardancy is required.

Chapter 6 provides a general background and history of rigid PVC
injection molding, bringing one up to the current state of the art.
Great progress has been made since the mid 1970s in developing not
only the specialized technical capabilities of injection molding rigid
PVC but also vast new market potentials. Although injection-molded
pipe fittings are still perhaps the most universal application, a number
of new rigid PVC injection molding markets are the focal point of this
chapter. General material characteristics and physical properties important to each market are also reviewed. Properties of the new breed
of more thermally stable, high melt flow compounds are compared to
the earlier generations of rigid PVC. Recent past and current economic trends that relate to molding rigid PVC are reviewed along with
anticipated cost factors, which are expected to have a favorable future
influence. On the technical side, the general effects of various formulation and additive changes on final compound processing and property characteristics are discussed. Both part and mold design considerations for rigid PVC are reviewed along with machine and processing recommendations. Rigid PVC is known for its past processability
problems in injection molding, but there is strong evidence that it has
turned the corner. The rooted stigma is gradually subsiding as more
and more designers and molders become comfortable in handling rigid
PVC in demanding injection-molding applications on a day-to-day basis.

1.8 INJECTION BLOW-MOLDING OF RIGID PVC CONTAINERS

With some variations, most injection blow-molding of rigid PVC today is done on integral three and four stations of the revolving-turret or shuttle-type (Gussoni-type) machine. This integral machine approach offers the advantage of separate stations for injection, blow, and ejection molding. The modular four-station design, as expected, offers additional flexibility, mainly a conditioning-preblow step, which occurs in Station 2 immediately after the premolded preform is transferred by the code rod from Station 1. In Station 3, conditioned preforms are blown into their final bottle shape and cooled. Bottle blowing is produced by compressed air introduced into the preform through the core. Finished bottles are then ejected in Station 4. This four-station process is also known as the condition injection blow-molding (CIBM) process.

In Station 1, rigid PVC melts are injection molded in a preform cavity where the neck finish is completely formed and the exact amount of polymer required is injected to fill the space between the core pin and the cavity wall. Hot liquid from a temperature control unit is circulated through core and cavity wall passages to cool the molten plastic and hold a predetermined temperature.

The making of preforms by injection molding offers several attractive advantages over the extrusion blow-molding system: primarily, injection molded preforms produce excellent precision control of neck design in the finished bottles. Other overall advantages of injection-blown ware are well known, such as, no scrap or trim (once onstream), higher impact resistance, thinner walls, lighter weight, and high output.

With the availability of special new grades of PVC resins, such as homopolymers and ethylene-vinyl chloride copolymers, which offer improved clarity, heat stability, processability, and strength, the injection-blow process for PVC containers is getting a lot of attention in the large-volume packaging markets. In cosmetics and beauty aids alone, there is the prospect of hundreds of millions of containers in 4 oz and smaller sizes. For containers up to 12 oz capacity, the potential market for PVC in the injection blow process is even bigger.

In Chapter 7, a review of materials as well as equipment modifications tailored to process the new PVC resins is presented. The critical aspects of the process as well as the effect of machine and process variables on the product quality and process yield are also discussed. Discussion of injection-molding stretch and blow processes and their potential advantage of improved container properties via biaxial orientation are also included.

1.9 CALENDERED RIGID PVC PRODUCTS

Calendered unplasticized PVC film (thickness <0.010 in.) and sheet (thickness >0.010 in.), due to their unique combination of properties, i.e., low cost, chemical resistance, barrier properties, strength, low level of toxicity (when properly formulated and compounded), and high gloss and transparency, are ideal candidates for the packaging industry. In this area, significant growth is expected not only because PVC is the most versatile of all the thermoplastics but the cost gap between PVC and other thermoplastics will continue to widen (PVC is only about 44% derived from petrochemicals). The food- and drug-packaging markets presently available to unplasticized calendered PVC are: (1) processed meats (bologna, sliced ham, and luncheon meats; (2) jams, jellies, and cheeses, usually in individual-portion packs; (3) candy, cookie, and fruit trays. The pharmaceutical or drug-packaging market is dominated by single-dose blister packages. The non-food-packaging applications of unplasticized PVC include the following markets:

1. Medical supplies, such as boxes for syringes and thermometer kits
2. Cosmetics and toiletries, such as packaging for lipsticks, eye shadow, eyebrow pencils, and mascara
3. Pens and lighters, razors and blades, and air freshener containers
4. Tools and hardware

Among the nonpackaging usages of unplasticized PVC film and sheet, the following fast growing markets exist:

1. Stationery supplies, such as notebook covers, wallet envelopes, and box lids.
2. Lighting and signs: both applications use the same gage, mostly translucent, highly glossy sheeting.
3. Mobile home skirting, 30 mils thick, opaque sheeting with a matte finish.
4. Floppy disks: a floppy disk serves as the magnetic memory storage component of the small desk-top computers. The disk consists of a magnetic layer sandwiched between two layers of rigid PVC. The expected growth rate is 60% per year.
5. Cooling tower sheet, for cascade and trickle trays in water-cooling towers replacing worn-out asbestos trays, and in new installations.
6. Pipe jacketing: thermoformed rigid PVC jacketing for insulated pipe is another growing market for rigid PVC calendered sheet.
7. Miscellaneous markets, such as credit cards, thermoformed sheet and ceiling tiles, and wall coverings as woodlike finish.

Some discussions on polymer preparation and melt processing for the calendering process can be found in the section dealing with melt compounding (Chap. 2). Either ribbon blenders, Banbury fluxing equipment, mill rolls, and strainer lines, or compounding extruders e.g., (Buss-Kneaders, Kombiplast, or planetary screw extruders) are used to feed the rigid PVC melt to the calenders. In Chap. 8, emphasis is placed on the true calender operation, specific compound formulations, troubleshooting, quality control, and so on.

The use of calenderettes (smaller sized and narrower calenders), which are normally fed by compounding extruders, and which are extremely popular in Europe and South America, is also discussed in Chap. 8.

1.10 THERMOFORMING OF RIGID PVC SHEET

The process of thermoforming is responsible for the past and projected future market growth of calendered and extruded unplasticized PVC sheets and films. This process involves the heating of these PVC materials to a point where their flat shape can be altered to the shape of the mold with the help of vacuum, air pressure, or direct mechanical force. The part is then cooled to about or below the T_g of the unplasticized PVC compound, where it will remain dimensionally stable. At this point, it is removed from the mold and trimmed to its final shape.

One of the major drawbacks of the process is created by the amount of trim generated in the thermoforming process. If the thermoforming operation is near the sheet- and/or film-making process, the trim can be reground and returned, as reworked material, to the calender or extrusion equipment. If the thermoforming process is conducted at customers' plants or the trim cannot be used, the thermoformed part becomes more expensive, because its cost has to include the handling, transportation, and disposal of the trim.

Another drawback inherent to the process is that it is not always possible to maintain a constant thickness throughout the part, especially in the presence of deep draws and sharp corners. Also, the presence of mold as well as stretch markings is another problem, sometimes difficult to eliminate.

Radiant heaters (e.g., clarods, quartz, and ceramics) are the most popular method of heating today. The temperature supplied by these heaters is controlled by thermocouple feedback to a temperature controller. After the material is heated, it is ready for forming by vacuum, air pressure, mechanical force, or combinations of these forming elements.

Vacuum is normally used to evacuate the space between the sheet, preventing air entrapment, and also to assist in the forming operation. It could be used as the only forming method in the so-called straight vacuum-forming process. When higher pressures are used and the female cavity is sealed so that the air pressure will not escape, this is referred to as pressure-forming. This method could be used to assist vacuum- as well as air-forming.

The cooling cycle requires more time than any other stage, since the heat has to be removed to assure that shrinkage of the part will be consistent and the shape of the part will be retained. This is achieved when the temperature of the part is $\leqslant T_g$ of the compound. Aluminum tooling allows running at a much faster rate. Fast cooling, however, can introduce thermal stresses in the part.

Part removal from the mold, as well as the subsequent trimming operation, should be conducted on a programmed sequential basis so that the part shrinkage can be kept under control.

Chapter 9 presents an in-depth review of the process of thermoforming of rigid PVC products and reviews the largest areas of application of these products, i.e., blister or skin packaging.

2

Rigid PVC: Polymer Preparation

I. LUIS GOMEZ /Monsanto Company, Springfield, Massachusetts

2.1 Introduction 16

2.2 PVC Resins 17
    2.2.1 Suspension Resins 18
    2.2.2 Bulk or Mass Polymerization 18
    2.2.3 Emulsion Polymerization 20
    2.2.4 Molecular Weight 20

2.3 Additives Enhancing Polymer Processability 22
    2.3.1 Heat Stabilizers 22
    2.3.2 Processing Aids 33
    2.3.3 Lubricants 35

2.4 Additives Enhancing Mechanical Properties: Impact Modifiers 38

2.5 Compounding: Equipment Classification and Techniques 42
    2.5.1 Low-Shear Horizontal Nonfluxing Mixers: Ribbon Blenders 43
    2.5.2 Color Concentrates 46
    2.5.3 High-Shear Vertical Nonfluxing Mixers 47
    2.5.4 Low-Shear and High-Shear Mixing in Horizontal, Jacketed, Cylindrical Blenders 55
    2.5.5 Hot Melt (Fluxed) Compounding 56
    2.5.6 Single-Screw Extruders 58
    2.5.7 Multiscrew Extruders 62
    2.5.8 Two-State Compounding Extruders 77
    2.5.9 Two-Stage Continuous High-Intensity Fluxing Mixers: Farrel Continuous Mixers 84
    2.5.10 Compounding Lines Based on Batch-Type Internal High-Intensity Fluxing Mixers 85

2.6 Summary 95

References 96

2.1 INTRODUCTION

Polymer preparation consists of a series of processing steps conducted on the incoming polymer to increase its processability range and to satisfy the requirements of the finished product. Since unmodified PVC resins are virtually unprocessable, the need for additive modification is indispensable. Additive modification is also needed to enhance polymer properties and to ensure the service life of the end product, hence increasing its utility. Some additives are used to enhance polymer processability:

Heat stabilizers, both primary and adjuvants or costabilizers
Lubricants (without lubrication, melted PVC sticks to hot metals), external and internal
Processing aids, mainly used to increase "hot melt strength" while reducing melt fracture, allowing processors to extrude or calender at faster rates

Other additives are used to enhance mechanical and electrical properties:

Impact modifiers (Izod impact of unmodified PVC is about 0.6 ft lbf/in., inadequate for most applications)
Heat distortion improvers
Electrical insulation boosters

In addition, additive modification is used to:

Improve outdoor service life (weatherability): pigments such as titanium dioxide, zinc oxide, ultraviolet (UV) light absorbers, and special stabilizers
Improve compound costs—fillers and extenders, such as calcium carbonate and fine-particle hydrated alumina

Some heat stabilizers, like dibasic lead phosphite, provide a unique balance of properties, such as excellent heat stability, improved electrical insulation, and outstanding weather resistance. As seen, dibasic lead phosphite could be classified under every one of the previous categories. Epoxy plasticizers, although used primarily as adjuvant heat stabilizers, also enhance light stability, mainly in rigid PVC formulations containing barium-cadmium heat stabilizers.

Achieving a balance of these additives to obtain adequate processability and heat stability, satisfactory properties and service life, compliance with the various government regulations, and an acceptable level of performance versus cost is mainly the responsibility of the compound formulator. This is the area in which a vast industrial technology has gradually developed over a period of years and is the most publicized. In this area, additive producers and formulators probably have experimented more than in any other segment of the industry. Obviously, the high degree of attention that the rigid PVC additive modification is receiving is shared by every sector of the industry.

If one considers that, in 1979, about 47,000 tons of PVC heat stabilizers were consumed in the United States [1] the attention received by this sector of the industry is justifiable.

The other function of polymer preparation, the actual compound preparation (dry blend and hot melt compound), unfortunately, has not yet received this attention. The importance of producing a consistent, homogeneous dry blend or pelletized compound to obtain quality extrusion or injection molding on a day-to-day basis cannot be emphasized strongly enough. Variability in a compound bulk density, homogeneity, flow characteristics, and degree of fusion (if applicable) translates directly to variability in the subsequent melt processing, requiring constant readjustment of process conditions. For example, an overworked pelletized compound produces an extrudate with surface imperfections called "orange peel." The extruder operator will have to take steps to attenuate this problem, mainly by lowering the screw speed, which will result in lower production outputs.

Unfortunately, the blending and overall compounding operations at most processing plants are two of the least desirable jobs, often staffed with the least experienced personnel or plagued by rapid personnel turnover. But in all candor, who can blame people for disliking a normally dusty atmosphere, the handling of toxic substances like lead stabilizers, or the unpleasant odor of the tin mercaptide stabilizers?

In summary, throughout this chapter, emphasis is placed on:

Dry blend preparation processes, dry blend properties, and the effect of these properties on the subsequent melt processing

Hot melt compounding processes, key process variables, and the effect of these process variables on the subsequent melt processing and forming steps

2.2 PVC RESINS

In 1979, about 3.5 billion pounds of PVC resins were used domestically for rigid PVC applications [2,3]. According to a market study by Predicasts, Inc. [4], U.S. consumption of PVC resins will hit 15.6 billion pounds by 1990 and will become the largest domestic thermoplastic in volume. Building and construction will surpass all other markets during the 1980s and will reach the 9.4 billion pounds mark. Pipe, conduit, and fittings will account for approximately 55% of the PVC used in building and construction applications.

Three types of resins are normally used in the processing of rigid PVC. These resins, which are classified according to the polymerization process, are suspension, bulk, and emulsion resins. Also, by far the large majority of rigid processing is done with PVC homopolymer from these polymerization methods.

2.2.1 Suspension Resins

Most vinyl chloride (about 80%) is polymerized in aqueous batch suspension. A free-radical initiator is dissolved in the monomer droplets, which are surrounded by a layer of suspending agent. During polymerization, primary particles of 0.5 to 2 µm are formed and, by adhering together, form irregularly shaped grains of 100 to 150 µm. Conventional suspension PVC has a compact, relatively nonporous, closed structure. The PVC grains are rather smooth, with a surface skin, "a pericellular membrane," about 0.5 µm thick, according to Tregan and Bonnemayre [5].

Under the microscope, suspension PVC particles appear smaller than bulk, seem to be harder, and look similar to salt crystals. The extent to which suspension PVC develops higher surface smoothness depends upon (1) agitation during the polymerization process, (2) the type of suspending agent, and (3) the temperature. The smoother the polymerized product, the more difficult it is to accept additives during blending and the more difficult it is to flux. Slow-fluxing resins require higher processing temperatures during dry blend preparation and melt processing. In any event, bulk PVC fluxes (fuses) faster.

Berens and Folt [6] and, most recently, Summers et al. [7] have demonstrated that the morphology of rigid PVC, when extruded at low processing temperatures, is related to the primary particles in the PVC resin. This product, extruded at low temperatures and made up mostly of 0.5-2 µm flow units, is brittle and has poor toughness retention upon weathering. Low processing temperatures tend to produce poor polymer gelation (product or profile will disintegrate in dry acetone) and, consequently, poor mechanical properties. At higher melt temperatures (185 to 190°C), agglomeration of the primary particles occurs, giving a rougher surface but a tougher product. When the temperature is raised even higher (above 200°C) a continuous melt occurs, leading to smooth surface and a tougher product with better toughness retention on weathering. This product swells in dry acetone but with no disintegration, which indicates excellent compound gelation. These comments apply to the most common PVC homopolymers of K value 65 to 70. Lower-molecular-weight resins achieve comparable properties at lower temperatures, as do vinyl acetate copolymers of equal molecular weight.

Suspension polymerization is frequently described as water-cooled bulk polymerization. Because the polymerization takes place inside the vinyl chloride droplets suspended in water, water serves as a heat transfer medium and as the medium in which the polymer particles are formed.

2.2.2 Bulk or Mass Polymerization

Particles of vinyl chloride polymerized in bulk, unlike suspension PVC, are not covered by a skin of suspending agent, making additive and plasticizer absorption more uniform and faster, thus shortening mixing cycle time. Furthermore, conventional bulk resin exhibits a greater effective porosity, fusing faster into a homogeneous melt. Bulk PVC particles are snowball-like, having very porous structures with a large surface. Generally, bulk-polymerized PVC exhibits a higher packing bulk density.

Bulk PVC plasticizes sooner than an equivalent K value suspension PVC by about two to three screw flights. It is pure, with no emulsifier or suspending agent and with the lowest water absorbency of all PVC types. Products with excellent clarity and sparkle can be manufactured from bulk PVC resins.

For calendering of rigid PVC films and sheet, or extrusion-blown bottles, bulk PVC exhibits somewhat better transparency and slightly lower impact resistance than other PVC types. Bulk PVC also has a tendency to develop higher melt temperatures in extrusion and injection-molding processes. Proper formulation, such as the proper choice and balance of lubricants, can attenuate these problems.

In general, bulk PVC has a uniform bulk density that, in turn, yields very uniform compounds. Particles of bulk PVC resins show poor mechanical stability. Rough handling in transportation, for example, can break up large particles, grains of 100 to 150 μm, to give primary particles.

Theoretically, bulk PVC should be lower in cost than suspension PVC since no evaporation of water and no suspending agents are needed. There still exist, however, process and quality problems that at the moment do not permit a lower price. The quality problems, mainly batch-to-batch contamination carry-over and presence of an almost permanent brownish cast in the resin have been, in some cases, so severe that some of the older manufacturers have been abandoning this process.

Rhône-Poulenc [8] (France) has announced that a new process for the manufacture of mass (bulk) polymerization in a vertical reactor has been perfected. This process gives increased productivity, lower investment and production costs, better mechanical operation of reactor, sufficiently low monomer residues to meet the strictest standards, no pollution, and about 50% savings in energy.

In processing bulk PVC, optimum packing densities develop because of the type and form of the particles involved. An earlier temperature rise is apparent because the rougher particles do not readily slip over one another, resulting in a frictional heat buildup that could lead to better feeding and better fluxing.

It can be demonstrated that bulk PVC, loaded in a box and heated in an oven, packs and cakes earlier than suspension or emulsion PVC. Since it fuses quicker, higher levels of external lubricants are required to delay this melting, particularly in fusion-sensitive twin-screw extrusion.

2.2.3 Emulsion Polymerization

Emulsion polymerization is performed in the presence of a water-soluble, free-radical initiator and is characterized by high polymerization rates, leading to high-molecular-weight resins. Emulsion PVC has the advantage of not producing flow lines or fish-eyes. It gels less and has antistatic properties. In Europe, where it is mostly produced by a continuous process (it is not necessary to open the polymerization kettle between each batch), it is less expensive than other forms of PVC. This is not true in the United States, where mostly batch processing is used. Due to the mixture of emulsifiers used in polymerization (mostly anionic, in which the monomer-soluble portion has a negative charge and the water-soluble has a positive charge), this PVC resin shows lower transparency. This is also due to the rather high concentration of emulsifiers used.

Emulsion PVC requires the highest amounts of lubricants. Low levels do not produce a glossy surface, and levels of 2 to 3.0% and higher are quite common. In addition, the porosity is higher and proper blending and mixing must be observed if a uniform compound is to be obtained. There exists a specific range of mixing temperatures that yield optimum bulk density, bulk density being time-dependent at that temperature. In the dry-blend properation, lubricants should be added at higher temperatures and later in the mixing cycle. If added at the beginning, lubricants interfere with the absorption and dispersion of other additives.

Rarely are rigid PVC products manufactured from 100% emulsion resins. These resins, however, are sometimes blended with suspension or bulk resins to

Reduce extrudate swelling
Improve surface appearance of building panels and residential siding
    by imparting a matte finish
Improve impact resistance of unmodified rigid PVC products slightly
Reduce plateout tendency
Reduce static charges on resin

2.2.4 Molecular Weight

The physical properties of the finished product as well as the inherent compound heat stability and processing features are all directly influenced by the resin molecular weight (MW), one of the most important variables to consider when conducting preliminary resin selection.

A PVC grade of 50,000 to 70,000 MW (weight average molecular weight), which corresponds to K values between 54 and 63, is more suitable for injection molding, injection blow-molding, extrusion blow-molding, or calendering. (The lower molecular weight contributes to better product color, lower yellow cast, and overall better processability). A PVC grade of 75,000 to 120,000 MW, K values between 63 and

75, is preferred for extrusion into pipe, siding, conduits, and sheet where higher modulus of elasticity, heat distortion temperatures, and impact resistance are desirable.

Regardless of the polymerization process, the molecular weight of PVC resin is expressed in terms of dilute-solution viscosity either in cyclohexanone (0.2 or 1.0% by weight of resin), nitrobenzene (0.4% by weight of resin), or ethylene dichloride. Some of the expressions encountered in the literature are

Relative viscosity $\eta_{rel} = t/t_0$, also known as the viscosity ratio (IUPAC), where

t = efflux time of the solution and,

t_0 = efflux time of the pure solvent

Specific viscosity $\eta_{sp} = \eta_{rel}-1$

Intrinsic viscosity $[\eta]$, also known as the limiting viscosity number (IUPAC),

$$[\eta] = \frac{\ln \eta_{rel}}{C} \quad C \to o$$

where

$\ln \eta_{rel}$ = natural logarithm of relative viscosity

C = concentration, grams polymer per 100 ml solution

Another popular measure of molecular weight is inherent viscosity (IV) or η_{inh}. The relationship of inherent viscosity with its number average molecular weight, $\overline{Mn}$, is illustrated in Fig. 2.1 [9]. Over 80% of all homopolymer resins have an IV of 0.90 or greater, $\overline{Mn} \geqslant$ 44,000. A typical PVC resin with an IV of 1.00 has an $\overline{MW}$ of 90,000 and an $\overline{Mn}$ of 47,000 [10]. The expression

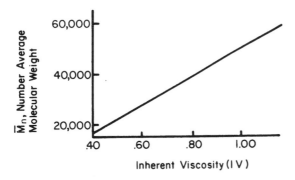

Figure 2.1 Number average molecular weight/inherent viscosity relationship.

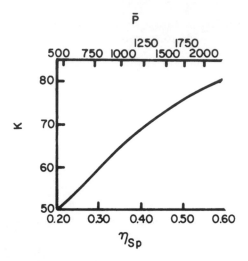

Figure 2.2 Correlation among specific viscosity (η_{Sp}), K value (K), and polymerization degree ($\bar{P}$).

for inherent viscosity η_{inh}, also known as logarithmic viscosity number (IUPAC), is

$$\eta_{inh} = \frac{\ln \eta_{rel}}{C}$$

The ASTM D 1243 method describes the determination of relative, specific, intrinsic, and inherent viscosities of a 0.2% solution of vinyl chloride polymer in cyclohexanone.

The K (Fikentscher) value is calculated from the measurement of the inherent viscosity at 25°C in 1.0% solution of vinyl chloride polymer in cyclohexanone (DIN 53726). This K value is used throughout this book, since it is widely used and accepted throughout a large part of the world.

$\bar{P}$, the average degree of polymerization, is commonly used in Japan and the Far East to measure molecular weight. This $\bar{P}$ value indicates the degree of polymerization (average number of monomer molecules in a chain) and is determined in a solution of nitrobenzene (JIS K 6721). Figure 2.2, presented by Struber [11], shows the relationship among specific (η_{Sp}) viscosity, K values (K), and $\bar{P}$, polymerization degree. Table 2.1 presented by the same author shows the same relationship numerically.

2.3 ADDITIVES ENHANCING POLYMER PROCESSABILITY

2.3.1 Heat Stabilizers

Heat stabilization of PVC is an area of extreme importance for process engineers, since processing unplasticized PVC without thermal degrada-

Table 2.1 Relationship Among Molecular Weight Values in Terms of η_{sp}, K, and $\bar{P}$[a]

Comments	η_{sp}	K	$\bar{P}$
	0.20	50	450
	0.22	51	510
	0.24	53	570
Range for blow-	0.26	55	640
molding, injection	0.28	57.5	710
molding, injection	0.30	59.5	780
blow-molding, cal-	0.32	61.5	850
endering	0.34	63	940
Range for extrusion	0.36	65	1020
of pipe, siding,	0.38	66.5	1100
profile, sheet	0.40	68.5	1200
	0.42	70	1290
	0.44	71.5	1390
	0.46	72.5	1500
	0.48	74	1600
	0.50	75	1720
	0.52	76.5	1840
	0.54	77.5	1960
	0.56	78.5	2090
	0.58	79.5	2230
	0.60	—	2370

[a]η_{sp} = specific viscosity (ASTM D 1243). K = K value, 1% cyclohexanone. $\bar{P}$ = polymerization degree (JIS K 6721).
Source: From Ref. 11.

tion has become possible mainly through the use of effective heat stabilizers. Originally, PVC resins could be fabricated into flexible products only at low processing temperatures and in the presence of plasticizers. In spite of the enormous technical progress achieved in this area, thermal processing of PVC still offers many open problems and still receives a lot of attention.

Regardless of the stabilizing system used, all of them have a common function: arresting or retarding PVC degradation. To effectively perform this function, the following reactions normally take place:

Neutralization of hydrogen chloride atoms with other groups of
    greater stability, and thus elimination of initiation sites
Exchange or displacement of labile chlorine atoms on the polymer
    chain with other groups of greater stability, and thus elimination
    of initiation sites

Reaction with unsaturated sites on the polymer chain
Inactivation of stabilizer degradation products (e.g., heavy metal
    chlorides), which could promote further resin degradation

It is not within the scope of this work to discuss the mechanism of
PVC stabilization or degradation. This author recommends Chaps. 8
and 9 of the Dekker Encyclopedia of PVC and Ref. 12 as excellent
sources on these subjects. Instead, emphasis will be placed on the
following:

1. The various types of key heat stabilizers available to stabilize
rigid PVC compounds
2. The new, improved products that represent the industry answer
to requests for lower toxicity levels and dusting, and for systems
showing cost and performance advantages
3. Typical rigid PVC formulations used in the field and that are presented throughout this work

Distinction has to be made, however, between heat and UV light
stabilizers. Although there is some HCl released during UV light
aging, the rate of the HCl split off is very slow. Something that is
pertinent to state at this point, however, is that products that have
been made from properly stabilized and properly processed compounds
(e.g., no thermal abuses and normal concentration of rework) do show
better UV light stability than products made from similar compounds
containing lower concentration of heat stabilizers or compounds processed with a high concentration of rework or subjected to thermal
abuses. UV light stabilizers act as physicochemical absorbers, through
dissipation of UV energy. They are more effective in clear and translucent applications than in opaque. Formulations containing high loads
of TiO_2 (rutile type) do not need UV absorbers.

Heat stabilizers may be classified in two main categories: Those organic and inorganic compounds containing metals and those, mostly
organic compounds, containing active functions other than metals. The
first category comprises the primary heat stabilizers; the second category includes mostly all the costabilizers, or adjuvants like epoxidized
polyunsaturated vegetable oils, alkyl aryl phosphites, epoxy aliphatic
resins, and some specialized organic compounds like polyols, β-aminocrotonate esters, α-phenylindole, and β-diketone.

The stabilizers containing metals may be classified in the following
groups in accordance with the metal portion in their composition:

Solid barium-cadmium stabilizers.
Lead salts (sulfate, phosphite, and stearate).
Organotin stabilizers (the alkyltin mercaptoesters and the alkyltin
    carboxylates). Included in this group are the main families of
    butyltin, methyltin, octyltin, and the new estertin mercaptoesters
    and carboxylates
Calcium-zinc systems
Antimony stabilizers

Solid barium-cadmium stabilizers

Barium-cadmium solid stabilizers are used primarily in European markets in rigid applications for improving weatherability, as in window and door frames, house siding, and translucent building panels. They are available with a high barium or high cadmium content plus zinc to improve sulfide stain resistance and early color stability. The zinc chloride formed during the HCl and zinc soap reaction, however, has a very strong destabilizing effect, being a strong Lewis acid. Use of an alkyl aryl phosphite can help minimize the adverse effects of zinc. A higher cadmium to barium ratio produces improved clarity and weatherability, which can make these stabilizers more suitable for outdoor applications where good early color and clarity is required. Higher levels of cadmium, however, have a detrimental effect on long-term heat stability. Inclusion of normal lead stearate, up to 0.75 phr, can improve compound heat stability greatly without spoiling product clarity.

These stabilizers are available as salts of saturated fatty acids, like stearic, palmitic, myristic, and lauric. These fatty acids are included to reduce water absorption and, consequently, to reduce hydrolysis tendency during outdoor exposure. They also help to lubricate the compounds. Originally offered only in powder form, these stabilizers are now being sold in nondusting forms, such as pellets, granules, beads, or flakes, which can be shipped, stored, and compounded with minimum health problems. Low dusting properties are achievable by polyblending with PVC resin, spraying with epoxidized vegetable oils or extrusion compounding.

For outdoor applications and to improve hydrolysis resistance, manufacturers of these products are offering polyl-free versions of standard products.

Very seldom are these Ba-Cd soaps used by themselves, since optimum results are obtained in the presence of synergistically effective costabilizers. It is for this reason that medium-chain-length soaps of barium and cadmium are always used in conjunction with organic phosphites or epoxidized vegetable oils, for example. For applications other than outdoors, this system is also modified with such polyols as pentaerythritol or sorbitol. This system, for the most part, is self-lubricated. The Ba-Cd soaps, which are used in high concentrations, have sufficient lubricating action on their own. The same thing can be said of the epoxidized vegetable oils. For building panels, window lineals, and siding applications where surface appearance is of prime importance, however, some extra lubricants are normally added, like partially oxidized low-molecular-weight polyethylenes, montan wax esters, partially saponified montan-wax acid esters, and paraffin waxes. A careful balance of these lubricating materials is essential to achieve optimum output rates and uniform melt viscosities matched to the capabilities of the single- or twin-screw extruders employed. Careful attention should be given to controlling potential plateout problems with Ba-Cd systems.

Solid lead salts

Lead compounds are still a widely used family of heat stabilizers for applications involving some opaque rigid PVC products outside the jurisdiction of the Food and Drug Administration (FDA) or the National Sanitation Foundation (NSF), or similar organizations. The use of lead stabilizers is predominant in European and Far Eastern markets. They function as HCl acceptors, forming water-insoluble, nonionizable, and for the most part, unextractable chlorides. They can provide efficient, low-cost stabilization while minimizing screw wear in multiple-screw extruders for rigid PVC applications. The most commonly used members of this family of stabilizers are tribasic lead sulfate, dibasic lead phosphite, dibasic lead stearate, and normal lead stearate. Because of the excellent heat stability these lead salts impart to rigid PVC compounds, they are used in injection molding of big parts such as appliances and television cabinets [13], and also in some fittings for drain, waste, vent (DWV) pipes and electrical and telephone conduits. Because of the potential health problems, nondusting grades are now readily available. Most current commercial dust-free lead stabilizers are based on proprietary surface treatment to render the powder non-dusting. Nondusting lead stabilizer granules with excellent dispersibility have been prepared by blending, in a high-intensity mixer, tribasic lead sulfate with 1-pph barium stearate, and by blending dibasic lead stearate and tribasic lead sulfate with calcium stearate and isostearic acid. Treatment with 1 to 2% PTFE is also used to create a nondusting "lattice" within the lead stabilizer particles, which is rapidly broken down during high-intensity blending.

As mentioned earlier, lead stabilizer systems dominate the European and Japanese pipe markets, and a very popular form is the so-called one-pack, or coprecipitated prill, which contains all stabilizing and lubricating additives, and even pigments and fillers. These systems are nondusting, and provide the formulator with a very simple two-component formulation: resin and one-pack.

Although lead has historically been a minor factor in the U.S. pipe market, very recent developments and product introductions (such as Halstab's P-1 and P-2) are creating much interest and could in time result in lead becoming a more significant factor in the U.S. pipe market [14,15]. Apart from toxicity concerns, one reason for poor acceptance of the classic lead systems has been their poor early color in the predominantly white U.S. pipe market, even though they provide excellent long-term processing stability. (European and Japanese pipes are primarily gray). The newly developed lead systems (i.e., Halstab's P-1 and P-2) are creating much interest and could in time also provide very good long-term stability compared to tin systems, with significant advantages in basic cost and pound per volume costs.

For most applications, the combination of normal lead stearate and dibasic lead stearate should provide sufficient internal and external lubricity. As reported throughout this book, lead-stabilized PVC compounds containing these two lubricants contribute to a low level of screw wear in twin-screw extruders. This performance characteristic seems to be attributed to the dibasic lead stearate which, due to its high melting point of over 250°C, remains as an internal lubricant-filler protecting the screw surface. If additional lubricity is needed for injection-molding applications, calcium stearate or fatty acid partial glycerides, like glyceryl monostearate (GMS), which act as internal lubricants, are frequently used in conjunction with partially oxidized low-molecular-weight polyethylenes (PE) as an auxiliary lubricant system. For outdoor applications, however, one should refrain from using glycerides, since the presence of free glycerine in technical-grade glycerides may react with the lead stabilizers, causing discoloration. As shown in this book, rather high concentration of these glycerides are used in some injection-molding formulations.

As it has been recently presented by Worschech and Lindner [16], long-chain partial esters of pentaerythritol are not only excellent lubricants for lead-stabilized rigid PVC compounds but do not have the glycerine discoloration problem mentioned above. According to these authors, weather resistance of PVC lead-stabilized products manufactured with penta esters is not negatively influenced. If the partial esters of pentaerythritol are mixed together with the lead salts, the dust level of these stabilizers is greatly reduced.

Lead-stabilized rigid PVC compounds, especially those containing dibasic lead phosphite, have been extensively used overseas for outdoor applications. In fact, it is a common practice abroad to incorporate some dibasic lead phosphite in lead-stabilized pipe formulations, which will be exposed to some outdoor weathering. Low levels of rutile-type TiO_2 (1 to 3.0 parts per hundred of resin, phr) are also included. Despite this ample prior experience, only a few U.S. firms reportedly are testing lead-stabilized formulations for siding, window, and door frames. With the increased availability of weathering data [17-19] comparing leads and other stabilizers in actual long-term exposure, this situation, however, will probably change.

As expected, being solid and relatively high in molecular weight, these stabilizers do not affect the heat-distortion temperature of the compounds and have no effect on the surface distortion of the siding. If anything, and due to their opacity, they help UV screening and heat reflection.

Overall extrusion performance is another factor to be considered in the selection of lead stabilizers for outdoor weathering. Contrary to popular belief, it has been reported [18], and is the author's experience, that a properly compounded rigid PVC formulation containing lead stabilizers should not be susceptible to sulfide staining. The author, while working overseas, had the opportunity to inspect lead-stabilized PVC pipes installed in a viscose rayon plant where levels of H_2S, CS_2, SO_2, e.g., were very high and, to his surprise, did not detect any trace of sulfide staining.

Organotin mercaptide stabilizers

Organotin mercaptide stabilizers ordinarily do not benefit in performance with the use of costabilizers. They do not respond to most of the organic adjuvants or secondary stabilizers frequently used with other metallic compounds. Tertiary phosphites, epoxides, and polyols, for example, are for the most part ineffective when added to compounds containing organotin stabilizers. Furthermore, the thioorganotins have their own built-in antioxidant functionality. All the above contribute to greater simplification in formulating and compounding with organotin mercaptide stabilizers. A very interesting study of the structure-performance relationships in organotin mercaptide stabilizers is presented in Ref. 20.

Some internal lubricants, like liquid unsaturated and solid saturated fatty acid partial glycerides, however, act synergistically with the sulfur-containing tin stabilizers, improving long-term stability. Combinations of sulfur-containing tin stabilizers with these lubricants are widely used in processing rigid PVC film and sheeting via calendering or extrusion.

For glass-clear application, liquid unsaturated fatty acid partial glycerides, like the glyceryl monooleates (GMO), are preferred over the solid glyceryl monostearates because the GMO, being liquid and more soluble in the system, only marginally impair the degree of transparency of the finished product.

Most alkyltin mercaptide stabilizers impart a degree of transparency, perhaps unexcelled by any other type of metallic stabilizer system. This is mainly due to the great solubility that these stabilizers and their chloride decomposition products have in PVC. A severe degree of compounding, although producing some yellowing, does not affect the initial clarity. Almost all tin mercaptide stabilizers are now combinations of mono and dialkyl; the ratios vary depending on the need for early color or long-term stability. Monoalkyltins enhance early color at the expense of long-term color.

Several reasons appear to be behind the great acceptance of the following thiotin stabilizers: the di-n-butyltin and di-n-octyltin-S,S'-bis(isooctyl mercaptoacetates) and more recently, for economic reasons mainly, the methyltin and estertin derivatives of the same thioester. Some of these reasons follow:

1. The overall excellent heat stability, compatibility, and processability, contribute to great efficacy and simplicity in formulating and
compounding, especially for high-output extrusion applications.

2. The rather low order of toxicity of the di-n-octyltin-S,S'-bis
(isooctyl mercaptoacetate showing the lowest level of toxicity to humans), has a sanction from the U.S. Food and Drug Administration.
In Europe, the BGA (West Germany) has sanctioned octyl-, methyl-,
and estertin mercaptoesters.

3. They are not known to cause or to contribute to plateout. The
same is not true for some alkyltin maleates, which can show some plate-
out tendency.

The tin thioesters, however, can have serious performance limitations. Inherently nonlubricating and with a sticking tendency, unplasticized PVC dry blends compounded with these stabilizers when
processed in conical or cylindrical, counterrotating twin-screw extruders, can contribute to excessive screw and barrel wear, significantly higher than the wear experienced with more lubricating lead-
stabilized formulations. In a recent visit to Mexico and South America
(Argentina, Brazil, Colombia, and Venezuela), the writer had the opportunity to talk to the largest PVC pipe producers in those countries,
Monofort in Argentina, Brasilit in Brazil, Tubos Flexibles in Mexico,
and Tubenplast in Venezuela, who corroborated these findings. For
this reason, it is essential to incorporate *higher* levels of internal and
external lubricant additives and to adjust the ratio of internal to external to match the needs of the extruder. For example, twin-screw
pipe extrusion requires *higher* (1.0 to 1.5 phr) external wax levels
and lower (0.4 to 0.8) internal calcium stearate levels, but single-screw
pipe extrusion requires higher internal (1.0 to 1.5) and lower external
(0.7 to 0.9) lubricant balance. Additional use of 0.1 to 0.2 phr low-
moelcular weight polyethylene is also essential as a late-working lubricant in the head and die, *particularly* in large-diameter pipe.

Being liquid, thioorganotins tend to lower the deflection temperature
under flexural load (DTUFL). This effect is related to molecular weight
(lower having a greater effect); generally, 1.0 phr of the commonly
used liquid tin thioesters reduce DTUFL by 2 to 3°F.

They have objectionable odor characteristics during processing.

They have overall poor outdoor weathering. This is more noticeable
in clear and translucent formulations. In opaque formulations containing high loads of rutile-type TiO_2 ($TiO_2 \geq 10.0$ phr), this detrimental
effect, however, is not noticeable.

A major apparent drawback for thioorganotins in comparison with
other stabilizers is their seemingly high cost per pound, although the
newer, more efficient stabilizers can be used at lower levels. The most
meaningful comparisons of stabilizer performance should be based on
equal use-cost; i.e., how much protection does one's money buy?

The reasons for the excessive screw wear are not clearly known.
Some of these stabilizers, however, being polyfunctional, tend to form

cross-linked structures between polymer molecules during processing, thus inducing high melt viscosities and high PVC-steel friction. This, in turn, promotes sticking to the screw surface. Non-sulfur-containing dibutyltin maleates show even greater sticking tendencies. They also tend to react with epoxidized soybean oils, if these are included in the compounds.

The two best known non-sulfur-containing alkyltin stabilizers are the polymeric di-n-octyltin and the dibutyltin maleates. The di-n-octyltin maleate has been sanctioned in some countries to be used for food-contact applications, mainly bottles and films; the dibutyltin maleate is used mainly for outdoor weathering applications. Due to their somewhat poor processability, these materials are rarely used alone; more frequently they have been used in conjunction with mercaptotin products. These sulfur-free products do not show the same compatibility as their sulfur-containing counterparts. For example, lubricants like the glyceryl monoesters (e.g., GMS or GMO), which are very effective with the thiotins, are detrimental to the thermostability of the sulfur-free compounds.

The soaring price of tin, the inability of the International Tin Agreement to control the price, and the strong tendency in the industry to replace heavy-metal stabilizers with low-toxicity compounds and to reduce the level of these heavy metals, have forced producers of organotin stabilizers to make their products more efficiently. Newly introduced estertin stabilizers with a cost-performance advantage of up to 15% seem to fit that efficiency pattern [21].

These estertins represent a new range of stabilizers produced by a new and unique direct process for the production of organotin intermediates, the mono and di organotin chlorides. While retaining the properties of conventional alkyltins, the estertins are claimed to provide

Lower processing odor.

Lower ratings of skin and eye irritation.

Higher LD_{50} values than methyl- and butyltins (lower toxicity levels), and somewhat lower LD_{50} values than the octyltin materials, but much less expensive.

Lower extraction levels than those of the food-approved octyltins.

Significantly lower volatility at higher temperatures than the methyl-, butyl- and even octyltins. This is due to the higher molecular weight and increased compatibility in PVC of the estertin thioester molecule.

Interstab offers Stanclere T-250 SD, T-250 SDX, and T-250 and SDX-1 liquid sulfur-containing organotin stabilizers, manufactured according to the new estertin process, which have NSF approval for use in potable water pipe at levels of up to 0.8 phr. This company also offers Stanclere T-222P and T-233P (other sulfur-containing estertins), specifically for siding applications. Some of these estertins are permitted abroad at levels up to 1.75 phr for BGA-sanctioned applications.

Many of the newer thiotin stabilizers also show increased efficacy in both color and stability and lubricity due to recent developments in the mercaptoester portion of the molecule, including sulfur enrichment and use of "reverse ester" technology based on β-mercaptoethanol (2-mercaptoethanol).

Increased interest in this line of products by most organotin producers suggests that most of the claims on these products have been verified. As of this writing, thioorganotins dominate the U.S. rigid PVC stabilizer markets: pipe and fittings, siding, profile, and clear sheet.

Calcium-Zinc Systems

Ca-Zn stabilizers for rigid PVC applications are mainly available in powder and paste form. They play an important role in food-contact and medical market applications. Their prime use is in the production of rigid PVC products for food packaging. In Belgium and France, drinking-water pipes are basically produced using these stabilizers only. In France, the well-established rigid PVC bottle market (mineral water, fruit juices, vegetable oil, vinegar, and table wine) has been using rigid PVC compounds produced without any addition of heavy-metal stabilizers. Stabilization is exclusively done by means of calcium stearate, containing small quantities of zinc soaps, like zinc octoate, and organic costabilizers.

With very few exceptions, Ca-Zn stabilizers are used in conjunction with other food-grade stabilizers or as complex blends of many different components. In the majority of cases, apart from Ca-Zn soaps, considerable amounts of epoxidized soybean oil or linseed oil are used. Of great importance also is the use of FDA-sanctioned organophosphites, which act as stabilizer adjuvants, improving initial color and color retention through retardation of the "zinc-burning" attribute of Ca-Zn stabilizer systems.

Nontoxic calcium-zinc stabilizers, when compared with barium-cadmium or other stabilizers, show low stabilizing efficiency. The need to synergize calcium-zinc systems or to reinforce them is obvious. Stabilizer producers are well aware of this limitation and have been offering all sorts of versions to improve Ca-Zn stabilizing efficiency. For example, in Europe, alphaphenyl indol, β-aminocrotonate esters, and more recently the stearoyl benzoyl methane, are commonly added to Ca-Zn stabilizers.

Calendering and blown rigid films for food-contact applications are typical examples in which there is a great need for high stabilizing power, improved clarity, color retention, and low odor. Of great importance also (when assessing the suitability of the costabilizers) is the effect that all these additives have on the organoleptic (taste and odor) properties of the packaged materials.

In general, calcium and zinc salts of saturated fatty acids, such as myristic and palmitic, are preferred for food-contact applications. These

salts seem to be more compatible with PVC and have less tendency to
plate out. Bottles produced from these systems show fewer die marks.
As with most of the Ba-Cd stabilizers, polyols like sorbitol and pentaerythritol are frequently added to Ca-Zn salts to improve compatibility.
These polyols, however, quite often can result in impurities called
"nibs" or "bits" in the product and can cause mechanical problems in
the equipment.

Ca-Zn-stabilized PVC compounds are not required by the FDA to
undergo extraction testing.

Commercial pipes extruded from NSF-sanctioned compounds containing Ca-Zn stabilizer systems can exhibit, when compared to other
commercial systems, superior weather resistance. This means that
Ca-Zn pipes can show greater ability to resist discoloration upon outdoor storage. However, their marginal effectiveness as processing
stabilizers exposed to high-output pipe extrusion and poor rework
capabilities (especially in white pipe) has severely restricted their
commercial use in the United States.

In view of the intricacy of government regulations, the complexity
of formulating with Ca-Zn soaps and their costabilizers, the periodical
extraction testing of compounds manufactured with dioctyltin stabilizers, and other factors, some producers of food-grade PVC products
have shown no special interest in doing their own formulation and compounding work. This is also the trend in South America, where most
of the food-grade PVC compounds are imported, although most of these
countries have manufacturing capability. This, however, has not been
the case with rigid PVC pipes and profiles, for which most of the producers prepare their own compounds. Being inherently very competitive and involving higher volume markets, the cost-conscious pipe
and profile producers must invest in in-house compounding capabilities.

As expected, the use of precompacted materials, however, has some
drawbacks; for example, when a situation arises in which improved
heat stability and/or processability is needed, such as extrusion-
blown rigid PVC bottles in a multiple-head die, the process engineers
do not have the option of introducing formulation changes.

Antimony Stabilizers

The alkyl mercaptides and mercaptoacid esters of antimony, like the antimony-S,S',S''-tris(isooctyl mercaptoacetate), have been approved by
the NSF at levels up to 0.4 phr. This concentration is adequate for extrusion in multiscrew extruders. For single-screw extrusion, however, this level is not sufficient, and they have to be used in conjunction with tin stabilizers. In some plants where tin and antimony stabilizers have been used together, cross-staining has been observed.
This phenomenon, however, does not take place with all commercially
available antimony stabilizers. Processors who have been using these

stabilizers, however, report some drawbacks on long-term extrusion runs and in the excessive amount of attention needed in a pipe plant where occasional catastrophic failures and higher scrap rates have been experienced [22].

Although they impart good initial color stability and low melt viscosities, almost equivalent to those of organotins, they tend to form orange antimony sulfide. Current "first-generation" antimony mercaptides are not suitable for weatherable PVC compounds for siding or other exterior building applications, due to lack of UV light resistance. Upon UV exposure, vinyl products containing these mercaptides do show greater color discoloration than the products containing their tin counterparts.

A cost advantage seemed to be the reason behind the initial interest in these products, specifically as low-priced alternatives to tin stabilizers. However, the most recent tin stabilizers introduced to the pipe market are more than cost competitive to antimony.

Until recently, there was only one supplier of these products in the United States, Synthetic Products, which holds patent rights (U.S. patents 3,887,508 and 4,029,618) for some antimony stabilizers. This company has granted licenses to Argus, Ferro, Interstab, and M&T. These companies, however, continue their developmental work on new lower-cost versions of current tins, which they are achieving mainly by reducing tin content. Synthetic Products expects that major gains in the pipe market will continue to be made by antimony stabilizers throughout this decade, and these stabilizers will hold a significant portion of the pipe market in the very near future. However, this does not represent a consensus of stabilizer producers. As stated before, the soaring price of tin and the inability of the International Tin Agreement to control the price is forcing producers of organotin stabilizers to continue their interest in lower-cost versions and alternate systems.

To illustrate the use of the stabilizer systems described above, typical rigid PVC formulations used in the field are presented throughout this book.

2.3.2 Processing Aids

Unlike the impact modifiers, compatibility of processing aids with PVC is very high. This is achieved, however, without any detectable plasticizer action, at least at the low levels at which they are normally used (1.0 to 5.0 phr). As reported by Lutz [23], these polymeric additives perform the following desirable multiple functions:

Promote fusion rate, which is achieved without affecting either the apparent melt viscosity of the compound, the heat distortion temperature, or the impact resistance of the end products. Plasticizers, which were the first processing aids used, lower apparent

> melt viscosity, reduce heat distortion temperature, and even at the
> lowest concentration (2.0-3.0 phr), have a detrimental effect on
> impact resistance.
>
> Increase compound melt strength and homogeneity, which in turn
> reduces melt fracture and allows the processor to extrude or
> calender at faster rates.
>
> Reduce plateout and impart greater ductility to the finished products.
> The latter property can be measured by the increase in tensile
> impact strength and elongation at break.
>
> Reduce surface imperfections such as flow lines.

Faster fusion rates contribute to better heat stability and to higher
usage of rework. Faster fusion rates are desirable in single-screw extrusion and injection-molding processes because they tend to contribute
to lower screw torques and some gains in power economy (pounds per
horsepower hour, lb/hp hr). Twin-screw processors, however, may
disagree with one another, since faster (earlier) fusion can also contribute to higher amperage loads (and torque) with the stiffer melts of
370 to 390°F stock temperatures encountered in twin-screw machines.

Various polymeric processing aids are available in the market: the
well-known all-acrylic polymers, the styrene acrylonitrile (SAN) copolymers based on high styrene (ST) to acrylonitrile (AN) (% AN ~
25), the styrene/acrylate copolymers, and poly-α-methylstyrene (PAMS).
The most universally known and perhaps the oldest ones, however, are
the all-acrylic polymers containing high levels of methylmethacrylate,
specifically Acryloid K120N, known overseas as the Paraloid K120N,
with more than 25 years of field experience. They are very efficient
in all kinds of rigid PVC melt-processing operations. Their presence,
however, is indispensable in the extrusion of flat profiles like PVC
panel and siding, where edge integrity is of vital importance. At present, there are two newer versions of this product: The K125, a
higher molecular weight version of K120N, and the lubricating K175,
and some processors have begun to use combinations of the lubricating
K175 with the most popular K120N. The major drawback of these products is their high cost, which for most of the foreign processors becomes excessive.

Linear, low-molecular-weight homopolymers of α-methylstyrene are
also used as processing aids in rigid PVC products. They seem to be
less compatible in PVC than the methylmethacrylate copolymers. They
offer a cost advantage per pound, but must be used at higher levels
to equal the effects of acrylics. Due to this slight incompatibility, the
α-methylstyrenes act also as lubricants. Because the fusion time and
apparent melt viscosity of rigid vinyl compounds are dramatically reduced by these processing aids, higher filler loads (calcium carbonate
surface coated with calcium stearate) can be used without sacrificing
easy processability and extrusion rates. Pipe compounds for noncritical applications like communications conduit, underground duct, DWV,
and sewer pipe applications can take advantage of the potential increase in filler levels, which could be achieved with these processing
aids. It has been reported [24] that vinyl siding formulations containing these processing aids (at 3.0 phr level) and an acrylic or a chlor-

inated PE (CPE) impact modifier (at 5.0 phr), which have been exposed outdoors for 2 years, show no adverse results when compared to well-known commercial siding compounds.

2.3.3 Lubricants

As documented throughout this chapter, lubricants, although used at relatively low concentrations , are, like the heat stabilizers, indispensable in formulating unplasticized PVC compounds. In fact, many would argue convincingly that proper lubrication balance is even *more* important than stabilizer selection to ensure successful processing of rigid PVC.

As is also shown, most of the unplasticized PVC compounds contain at least two lubricants, and quite frequently, three or more lubricants are added to a formulation containing heat stabilizers with some inherent lubricity. The translucent panel and siding formulations are examples of these cases. The statements made above suggest that the lubrication process is rather complex.

Regardless of the formulation and the lubricant system, lubricants in unplasticized PVC are called to perform several fundamental functions:

Reduce the compound apparent melt viscosity, thus promoting melt flow

Control frictional heat buildup by reducing friction between resin molecules after fusion

Promote release of the melt from the hot metal surfaces of the processing equipment, like screw surface, mill or calender rolls, or injection molding nozzles, thus reducing localized overheating

All these functions have the same overall effect, which is that of lowering the melt processing temperatures. This is, perhaps, one of the most important criteria used to assess effectiveness of lubricants and thus to help in the selection of lubricants, which continues to be more an art than a science.* The term "mechanical stabilizer" has been used to describe the function of a lubricant system.

Lubricants have been ranked according to how much they lower the T_g of PVC. The more internal (compatible) ones produce a greater T_g depression (as much as 25°C at 5 phr concentration); the external have no effect on polymer T_g [25].

In general, the following phenomenon has been experienced: The purest PVC resin produced by the bulk polymerization process produces higher melt processing temperatures than equivalent K-value suspension resins. It seems that the residual components from the suspension polymerization process (including some monomer and even water) are so evenly dispersed on the resin particles that they contribute to some lubricity and thus to lower processing melt temperatures.

---

*Very revealing ways of assessing the effectiveness of lubricants in any extrusion process are to measure ΔT across the melt at the delivery tube and axial temperature profiles at the die. Smaller ΔT values and smoother axial die temperature profiles are respectively expected from the most effective internal and external lubricants.

This overall behavior of unplasticized PVC melts has not been experienced by the process engineers who process other commodity polymers, like low- and high-density polyethylene, nylon, polypropylene, and saturated polyesters, which, in essence, are self-lubricated polymers and thus do not need the presence of commonly known lubricants. Also, polymerization residuals (if any) in these other polymers do not appear to improve processability significantly.

As it has been reported by Illmann [26] and acknowledged by White [27] there exist three general types of lubricants: external, internal, and those exhibiting both internal and external lubricating qualities. In fact, there exists a whole spectrum over which lubricants can be classified as more internal or external, relative to each other.

The so-called internal lubricants have fairly short chains and are polar to some degree. They exhibit the highest compatibility and solubility in unplasticized PVC melts, which, in turn, produce greater dispersability and internal lubricity. To this group belong the sorbitan and glyceryl monoesters of fatty acids, the fatty alcohols, and fatty acids, which, due to their high compatibility, do not impair the transparency of the finished products when used at normal levels (≤ 3.0 phr). These internal lubricants are used in glass-clear applications, like rigid PVC bottles and calendered or extruded rigid PVC films. Among these lubricants the glyceryl and sorbitan esters of unsaturated acids, like glyceryl monooleate, glyceryl monoricinooleate, and sorbitan monooleate, being liquids, are even more compatible than the glyceryl or sorbitan monostearates, which are available as beads.

Fatty acid alcohols, like the mixture of cetyl-stearyl alcohols or dicarbonic acid ester of fatty acid alcohol mixtures, show the highest compatibility and solubility in unplasticized PVC melts. Their use in sulfur-containing tin-stabilized unplasticized clear PVC systems has not grown, however, because these fatty acids alcohol derivatives do not contribute to any improved processability as do the glyceryl and sorbitan esters. Calcium stearate and the N,N-ethylene bisstearamide although less compatible in unplasticized PVC melts than the members of the previous group, are also classified as internal lubricants, although the bisamide wax does possess some external attributes. Because they impart some haze or cloudiness to PVC, they find more use in translucent and opaque applications.

Calcium stearate is perhaps the most popular and widely used additive in formulating unplasticized PVC compounds. Main reasons for the broad use are that it

Imparts heat stability to unplasticized PVC melts
Lubricates internally, as far as promoting fusion goes, but does not reduce melt viscosity as much as do the other esters
Is compatible with every other stabilizer system, specifically, mercaptotin esters, antimony mercaptides, barium-cadmium soaps, lead salts, calcium-zinc stabilizers, and so on
Is fully sanctioned or approved by the FDA

As mentioned before, pipe formulations based on calcium soaps do show excellent weatherability.

All the above justify why calcium stearate is used as part of the lubricant package universally used to formulate rigid PVC mercaptotin-
stabilized pipe compounds. This package normally also contains 165°F
paraffin wax and partially oxidized low-molecular-weight polyethylene,
which act as external lubricants. These are classified as external
lubricants because of their low compatibility, which is due mainly to
their long hydrocarbon chains, with little or no polarity. Consequently,
their concentration should be watched carefully to avoid overlubrication.
The PE waxes provide excellent release properties. Although they impart an objectionable haze in translucent formulations when used at concentrations ≥ 0.1 phr levels, they are so effective that they can be
used at concentrations as low as 250 ppm level, still with good lubrication and without significantly affecting the clarity of the finished
product.

Normal lead stearate and dibasic lead stearate, when used together
as the sole lubricants in lead-stabilized compounds, seem to act, respectively, as internal and external lubricants. Dibasic lead stearate,
which remains unmelted in the PVC melt, seems to provide the internal
lubricity protecting the screw surface from wear.

Hoechst Wax E and OP are reported to have internal and external
lubricating properties. Both are derivatives of long-chain montan wax
acids. In examples of translucent panel formulations presented in
this work, Wax E is used to improve extrudate surface and thus in this
formulation is mainly acting as an external lubricant.

Lubricant Selection

Many factors should be considered when formulating with lubricants.
The most important, however, are those related to processing conditions. For glass-clear applications, the lubricant solubility and overall
compatibility in the PVC melt are also key factors in lubricant selection.

Another factor, also of extreme importance, is that of lubricant
concentration, which is even more critical in the case of external lubricants, since compound overlubrication should be avoided. Other
considerations that should not be ignored are the beneficial effect that
some lubricants have on such considerations as impact resistance, heat
stability, and pigment dispersibility.

The blending of lubricants to obtain the best balance of processability and properties is still a very guarded area of knowledge. For
every major application, however, there are some basic lubricant systems commonly used in the industry. Two specific examples are (1)
the mercaptotin-stabilized PVC pipes and conduits, in which the blend
of calcium stearate with 165°F paraffin wax and low-molecular-weight
partially oxidized polyethylene is used (because of the cost-performance advantage), and (2) the lead-stabilized pipes and conduits, in
which the combination of normal lead stearate and dibasic lead stearate
has been and continues to be the key lubricant system.

The one-pack concept (lead stabilizer-lubricant system), common in Europe, has created some recent interest in the United States. A complete tin external-internal lubricant one-pack, designed specifically for twin-screw pipe extrusion, has been available for a few years (ADVASTAB LS-202 manufactured by Carstab), and other tin-lubricant one-packs, although not complete systems, have been introduced periodically in the U.S. market. Combination, or multifunctional, lubricant systems (without stabilizer) have also generated some commercial interest.

Although the list of FDA-sanctioned lubricants is rather long, the combination of glyceryl monooleate and glyceryl monostearate at 0.25 to 0.75 phr levels is one of the most commonly used for crystal-clear rigid PVC applications like rigid PVC sheet and film. Where high concentrations of external lubricants are needed to help roll transfer and release, the GMO is replaced with a combination of Wax OP (calcium-modified montanic acid ester) and N,N'-ethylene bisstearamide (Acrawax C) at 0.25 to 0.75 phr levels. If even more external lubricants are needed, the inclusion of partially oxidized PE wax at low levels, up to 0.1 phr, could improve release properties significantly.

As already discussed, lubricant requirements are dictated by equipment needs, processing conditions, and product properties. Throughout this book, some of the typical lubricant systems used in the field are presented when discussing the various rigid PVC compound formulations.

2.4 ADDITIVES ENHANCING MECHANICAL PROPERTIES: IMPACT MODIFIERS

Rigid PVC products and profiles manufactured from unmodified rigid PVC resins show good weathering resistance, but they are relatively brittle and notch sensitive. These problems become even more pronounced in cold climates and happen even when a high-molecular-weight resin is used. These disadvantages, however, are overcome with the use of impact modifiers that can be incorporated in the formulation by mechanical processing of the mixture; the rubber is dispersed mechanically in a PVC matrix or by grafting the VCM onto a suitable elastomeric backbone.

These impact modifiers increase the utility of rigid PVC resins and provide an excellent tool for the process engineer to penetrate the market of existing products; e.g., recently, impact-modified rigid PVC replaced flame-retardant (FR) polystyrene as the resin for the manufacture of cabinet backs for color television sets [13].

Efficient impact modifiers have the following common requirements:

The modifier must have a certain limited compatibility, must be insoluble in the PVC matrix, and must have good adhesion between the two phases (the continuous glassy phase and the dispersed rubber phase).

It will form a continuous, rubbery network embedding the primary
   particles. If the mixture is abused during processing (too high
   temperature and/or shear), fusing of the primary PVC particles
   will be brought about, the network will be broken down into
   small, irregular rubber particles, and the impact resistance will
   be totally lost [28].
For clear applications like bottles and film, the refractive index
   should have the same value as that of the matrix.
The modifier must not reduce the good properties of the base PVC,
   such as low permeability and low order of taste and odor.
The modifier must retain its elasticity at low temperature, especially
   for outdoor applications.
The modifier must have good thermal as well as light stability.
For economic reasons, they should be effective at low levels, i.e.,
   4 to 10%.

The impact modifiers most commonly used are as follows.

Chlorinated Polyethylene

Suspension CPE is manufactured by chlorination of high-density PE.
The chlorine atoms are randomly distributed along the PE chains. Impact strength is optimum at a 30 to 42% chlorine content range. The
CPE is used either mechanically blended or grafted, (the CPE serves
as a backbone for the PVC chains). Even under the best conditions,
this rubber imparts appreciable haze to PVC resins. End uses of the
resins containing CPE include opaque and translucent profiles, sheets,
and siding. The CPE at concentrations between 4 and 10% imparts
good weatherability to the modified formulations. The weathering
resistance, however, is not as good as that of formulations containing
ethylene vinyl acetate copolymer (EVA) in equal concentration. Hoechst
holds the initial patents on dry blending of PVC with CPE having a
chlorine content of 25 to 42%. This company has at least 20 years of
outdoor weathering experience with the CPE-PVC blends, which they
offer as a pelletized compound, Hostalit Z, or as a dry blend, Hostalit
HM. In the United States, Dow Chemical offers CPE3614, CPE3623,
and CPE3615, used at 4 to 7 phr. These series contain 36% Cl. Based
on over 10 years of outdoor tests, Dow claims that these CPE products
offer better retention of impact strength over time than acrylic or EVA
modifiers.

Ethylene Vinyl Acetate Copolymer Modifier

EVA containing 30 to 50% vinyl acetate makes an excellent impact modifier for PVC. It also provides excellent weatherability and color retention superior to or equal to the acrylics at equal levels, good low-temperature performance, and reduced melt viscosity. The EVA are available as pellets (from Dupont, Mobay, and U.S.I. Chemicals), which have

to be mechanically polyblended in high intensity, internal fluxing
mixers (Banbury type), or as a graft copolymer of vinyl chloride on
EVA, normally containing about 45% vinyl acetate (powder form).
Powdered EVA, suitable for dry blending, has been offered in developmental quantities by some U.S. producers, but initial materials tended
to agglomerate. The U.S. market for EVA impact modifiers could be substantial if this powder form becomes commercially available.

Graft copolymers of vinyl chloride on EVA in powdery form are very
popular in Europe, where several companies are offering them for extrusion of window profiles and other outdoor applications and injection-molded articles. Chemische Werke Hüls manufactures Vestolit
HIS/7587, which is offered in the United States by Mobay. Wacker-
chemie manufactures Vinnol K550/79, a high-impact PVC concentrate
with 50% EVA, and also Vinnol VK505/68 [29]. Chemische Werke
Buna also offers some graft copolymers of vinyl chloride on EVA (containing less than 35% vinyl acetate).

A new series of graft copolymers of vinyl chloride on EVA is now
being offered by Pantasote under the trade name of Pantaplast. The
"L" is the base resin of the series and is available as a granular free-
flowing material. Terselius and Ränby [28] have reported that unpig-
mented grafts containing 7.5% EVA-45 in PVC are less opaque than the
corresponding mechanical blends when processed at 160°C for 15 min.
These authors have noticed the same overall effect with some CPE: the
grafted PVC copolymers tend to be less opaque than the corresponding
mechanical blends. If the grafts are abused during processing, fusion
of PVC particles breaks down part of the EVA network, and impact
resistance is lost.

The EVA offer other advantages over acrylics: They reportedly
avoid the creation of shiny profiles, whereas acrylics may require processors to use a gloss suppressant and allow higher filler loading.

ABS, MBS, and MABS Impact Modifiers

ABS (terpolymer of acrylonitrile-butadiene-styrene), MBS (terpolymer
of methylmethacrylate-butadiene-styrene) and MABS (acrylonitrile-
methylmethacrylate-butadiene-styrene) comprise about one-fourth of
all impact modifiers used with PVC. These modifiers, and particularly
the MBS, are used in applications requiring good color, transparency,
and thermal stability, as in blow-molding or injection-molding of rigid
PVC bottles. In this area, MBS has replaced the ABS initially used,
which tends to be yellow and hazier. Furthermore, MBS-modified
resins are easier to process than ABS. They have a larger particle
size to allow dust reduction, and a better powder flow in blending
and/or loading operations.

MBS is made by grafting methylmethacrylate and styrene onto a
polybutadiene or butadiene-styrene latex. It consists of about 50%
MMA, 25% butadiene, and 25% styrene; ABS impact modifiers are produced by grafting styrene and acrylonitrile monomers onto polybuta-

diene latex. MABS materials are either polyblends of MBS with ABS or graft copolymers of MMA, AN, and styrene on polybutadiene or butadiene-styrene rubber. Amounts blended with PVC range between 3 and 15%.

The two best grades of MBS available in the market for clear, food-grade rigid PVC bottles are (1) the Kane Ace B-22 and B-28, manufactured by Kanegafuchi Chemical Industry Co., Ltd., in Japan and in Belgium (this company is the oldest and largest producer of MBS and MABS), and (2) the BTA III, manufactured by Kureha, in Japan, and sold in this country by Borg-Warner. This MBS seems to be one of the purest on the market, imparting almost no taste and odor to the packed materials. Kureha claims that the excellent transparency is obtained by tailoring the refractive index of MBS composition to PVC.

All these ABS, MBS, and MABS have very poor weathering resistance. Their polyblends with PVC require either surface protection or heavy pigment loading for outdoor use. These heterogeneous impact modifiers contain grafted and cross-linked rubber particles of the order of 0.1 μm, which are composed of butadiene homo- or copolymers and are evenly dispersed in a hard matrix. By mechanical blending of these grafts with PVC, the rubber particles are evenly dispersed in a homogenized matrix of PVC and the hard polymer of the graft modifier. The overall impact strength of these blends is somewhat sensitive to processing conditions and to a degree of grafting. However, it seems to be less sensitive than when homogeneous polymer modifiers (e.g., EVA or CPE) are used with PVC.

All-Acrylic Impact Modifiers

PVC polyblends containing all-acrylic elastomers as impact modifiers exhibit high impact strength, low die swell, and excellent weatherability. They are used in formulations for house siding, shutters, gutters, and window frames, for example. Acrylate elastomers are frequently graft copolymers of methylmethacrylate on polybutylacrylate or poly (2-ethyl-hexyl)acrylate. Rohm & Haas markets all-acrylic impact modifiers under the Acryloid KM trade name. Their best-known products are the KM323B, the KM330, and the most recent KM334 (weatherable PVC siding, window profiles). Because of gains in overall processability as well as weathering resistance, these all-acrylic modifiers have been holding their own and tenaciously competing with the more conventional EVA and CPE impact modifiers. A modified acrylic from M&T, Durastrength 200, is said also to offer process aid functions while providing low-temperature impact.

Elastic polyacrylic esters (PAE), generally polybutyl acrylate, are the basis of a new, high-impact transparent rigid PVC compound recently introduced by BASF for glass-clear sheets for greenhouses, roofing, and window panes, for example. This product was initially intended as a replacement for glass panels in cupolas or roofing for sports arenas, for example.

Inorganic Impact Modifiers

High-purity precipitated calcium carbonate imparts a modest improvement in PVC impact strength. However, because of its low cost, it is
widely used generally with other polymeric modifiers. The calcium carbonate used is of ultrafine particle size and is surface treated with
stearic acid, up to 3.0%, to attenuate its scrubbing action; this filler-
impact modifier reduces die buildup (plateout) tendency. The finer the
particle size, the better the impact strength, and equipment abrasion
is much reduced.

    Three major suppliers of this product are ICI, Pfizer Minerals,
Pigments and Metal Division, and Vermont Marble Omya, Inc., Division
of Pluess-Staufer. ICI produces Winnofil S, a calcium carbonate with
a 0.075 μm particle size. Vermont Marble Omya offers Omyalite 95T
and 90T, with 1.0 and 1.2 μm particle sizes, respectively. Pfizer offers the Super-Pflex 200, average particle size 0.5 μm, and the Ultra-
Pflex filler, with 0.07 μm average particle size. This is reportedly
the only domestically available, precipitated, surface-treated calcium
carbonate with such fine particles. Calcium carbonate has achieved its
widest acceptance in rigid PVC pipe, tubings, and fittings applications,
including DWV, pressure and sewer pipe, and in electrical and telephone conduit. The highest loading of calcium carbonate commonly
used in U.S., rigid PVC profile formulations is said to be around 7 to
8 phr. As expected, some processors are investigating higher loadings,
15 to 20 phr. Some conduit compounds contain as much as 25 to 40 phr.

    To obtain improved impact and weathering resistance without impair-
ing other properties, Matsushita Electric Works has developed and
patented a calcium carbonate and zinc maleate composition. For further
enhancement of impact strength, another Matsushita patent uses dibutyltin maleate in combination with small amounts (0.3 to 1.5%) of poly-
methylmethacrylate.

    The special attention given to the impact modifiers in this book is
because they are the second largest component of any impact-modified
rigid PVC formulation and they play a major role in expanding the utility of rigid PVC products, like residential siding, glass-clear panels,
and window frames. Their growth is responsible for the growth of
rigid PVC as an engineering thermoplastic. The additives normally
used as heat distortion improvers and electrical insulation boosters, however, are not covered in this book.

2.5 COMPOUNDING: EQUIPMENT CLASSIFICATION AND TECHNIQUES

For the engineers designing new mixing and compounding operations for
rigid PVC, the task is greatly helped by the abundance of equipment
and technology available in this field. Once they have information on
the line of products to be manufactured, the type of compound (dry

blend or hot melt compound) to be used, the expected productivity, and
the space available, the choice is plentiful and professional help is
abundant. For example, a thorough review of the continuous melt
compounding equipment available from the various machine suppliers
is compiled in Refs. 30 and 31.

For engineers working with existing equipment, the task is more
difficult. In this area, help is mainly provided by process engineers
and compound formulators who tailor the existing formulations to the
available equipment. Via additives, PVC polymers have such a wide
processability range that there is always a possibility of adjusting formulations to equipment and process. This implies a great plant-to-plant
variability in equipment and processes, which is the actual case. Some
trends and also some apparatuses are common among the various installations. Specifically, high-intensity vertical nonfluxing mixers are
common to modern, efficient dry blend compounding lines. These lines,
however, differ greatly from each other in material handling and in
the type of dry blend coolers used. The following discussion describes
the various compounding equipment and techniques used in the field
to process rigid PVC.

2.5.1 Low-Shear Horizontal Nonfluxing Mixers: Ribbon Blenders

The low-shear horizontal blenders were inherited from other compounding operations. They provide distributive blending only, a
uniform distribution of all particles throughout the mix, with the original particle identity retained for the most part. At present, when
designing new plants, engineeers very probably will substitute these
units with more efficient ones for the premixing operation. Typically,
an outer spiral ribbon conveys material in one direction, countered by
an inner spiral ribbon that moves the material in the opposite direction.
The mixing action that results can be defined as convective and diffusive. These ribbon blenders are usually equipped with drives having
a ratio of 1 hp to 10 to 20 lb product. As the machines increase in size,
the shaft rpm goes down and the percentage of material moved by the
mixing elements decreases. In any event, the ribbon blender has insufficient radial and axial flow to transfer all the material into the zone
of higher intensity, and its overall efficiency is low.

Available as part of existing internal high-intensity fluxing mixers
(Banbury mixers) and two-roll mill compounding lines, or as part of
two-stage twin (four screw) extruder where final mixing must be accomplished in the top set of compounding screws, ribbon blenders have
a place, and a commercial process can be developed around their limitations. For example, to minimize the effect of a variable heat history
on end-product color in large blenders, the process engineers use
either a neutral (no heating or cooling) or a blender, operating at low
tempertures ($T \leqslant 60°C$), since for the large blenders (size $\geqslant 4000$ lb
rigid PVC dry blend), there is always a significant difference in dwell

time between the beginning and end of blend. This happens if the line does not have surge hoppers and the large blender is used to feed directly to the line.

The practice of controlling the blend temperature reduces potential color problems. In extruded profiles, like siding, downspouts, rain gutters, and window frames, which are both functional and ornamental, the presence of a variable yellow cast is quite detrimental to their application.

Borderline blending efficiency for dispersability of high concentrations of titanium dioxide (≥ 10.0 parts) or small concentrations of inorganic pigments like chromium oxide is also expected from ribbon blenders. This situation, however, can be appreciably improved as follows:

1. Specify easy dispersion pigments, which are for the most part available from vendors.
2. Add these pigments through a screen on blender cover (to break lumps and other agglomerates).
3. Mix with the resin or resins only before any other additive is incorporated and while resin is heated.
4. Thereafter, start a gradual addition of all the other dry powders in the following order: heat stabilizers, processing aids, impact modifiers, and last, lubricants.

Another of the well-known limitations of these blenders is their inability to produce a good dispersion of small amounts of liquids, such as epoxidized soybean oil, liquid organotin stabilizers, or organic phosphites. The same thing happens with low-melt-point additives like esters of fatty acids or paraffin waxes. Again, this situation is greatly improved if these liquids are preheated to 100 to 110°C and gradually sprayed on the resin while blending at the beginning of the cycle. If a very low melt-point lubricant, such as diglycol stearate and some glyceride esters, are used in the formulation, they should be melted and mixed with the liquid components and sprayed together. If no other liquid components are used in the formulation, the possibility of replacing the waxy additives by higher melt-point lubricants available in powder form should be considered. Waxy soft additives do not disperse well and stick to the blender walls. Once the liquid ingredients have been sprayed, blending should continue for at least 15 min more.

In opaque formulations containing liquid heat stabilizers, the titanium dioxide and any other inorganic pigments are added at least 10 min after the liquid components and mixed for at least 15 min while the blend is heated. At this point, the processing aids, the impact modifiers, and the lubricants are added.

Ribbon blenders, in general, do not disperse agglomerates of organic substances like the benzophenone and benzotriazole UV absorbers well. In this area, however, help can also be obtained from the vendors.

The inclusion of a small percentage of a synthetic fine particle size silica, like Cab-o-Sil or Syloid 244 helps the dispersability of these products enormously. Furthermore, the presence of silica, acting as surface scrubbers, reduces the plateout tendency.

Obviously, operating the ribbon blenders at low temperatures does not remove moisture or other volatiles. This will have to be done at the internal fluxing mixers during fusion and during the mixing operation on the two-roll mill, or in two-stage twin-screw extruders with the aid of vacuum at the discharge port between the two extruders.

In compounding units consisting of small jacketed ribbon blenders and unvented compounding extruders, most of the moisture and volatiles have to be removed in the blender, which has to be operated at higher temperatures, 100 to 110°C, since residence time for the compound in the hopper area may not be long enough to remove high levels of moisture or any other volatiles. This is sometimes done even if the compounding extruders are vented. The higher temperatures in the blender are also desirable to increase bulk density and uniformity of the dry blend. Residence time of the dry blend in the blender at the higher temperatures should be watched and controlled very closely to minimize any potential color problem. This color problem is mainly manifested by a marked yellowing of the finished product, which for the most part is not objectionable in products of only functional use.

The overall mixing capability of the ribbon blenders for a rigid PVC dry blend preparation could be greatly improved by selecting the mixing ribbons best for the job. Modified interrupted ribbons provide better mixing action than the continuous type. Another way of improving the degree of mixing is by running the mixing shaft at higher rpm. This will need larger-capacity motors. Optimum ribbon design for a particular job could be determined by running dispersion trials at vendor testing laboratory facilities. Some of these testing facilities have ribbon blenders with an interchangeable library of main shafts and ribbons, which allow a quick evaluation and thus an optimization of these mixing elements.

Another use of the jacketed blenders are as coolers for batches processed in high-shear vertical mixers. Their cooling efficiency, however, is low.

These blenders are also used in the preparation of an additive master batch for the high-shear vertical nonfluxing mixers or other lines. If an automatic weighing operation for the additives is not available, quite frequently, the additives in powder form, like some lubricants, processing aids, and small concentrations of fillers, are preblended in a resin matrix and then added as a concentrate to the resin batch in the high-shear mixer. This type of addition greatly reduces batch-to-batch variability.

Best results with ribbon blenders are obtained with freshly mixed powder. In reality, however, this is not easily achieved, since in most

cases ribbon blenders are part of high-capacity lines where the freshly
made blends are transferred to silos and stored. There is one serious
pitfall, however, when doing this. Since the bulk storage of powder
blend in silos requires transfer of material through air lines, and cold,
low-shear blends contain particles of different sizes and bulk densities, some separation may occur in the transfer lines due to different
velocities of transport. Also, material in a silo can, in time "strati-
fy," as heavier particles settle lower and lighter particles work toward the top. Obviously, the end result can be erratic performance
in the extruder. Watch for amperage surging.

Despite the limitations mentioned above, the special attention given
in this section to ribbon blenders is because they have some features
very desirable in production lines, i.e., large capacity, low maintenance, minimum operator attention, and greater availability.

2.5.2 Color Concentrates

Carbon blacks and some organic pigments, like phthalocyanine blue
and green, if added to the ribbon blenders, are very difficult to flush
out of the system and can contribute to serious color contamination
problems. If the Banbury or any other internal fluxing mixer is manually operated or if the installation of this mixer permits, these pigments should be added as dry color concentrates to the internal mixer.
A technique that has been used in batch operation processes is to prepare the color concentrate in a resin base so that the same amount of
color concentrate, 1 part to 100 parts of dry blend, is added to the internal mixer charge regardless of the pigment system. The pigments
are, therefore, present in the concentrate at 100 times their concentration in the final product. These color concentrates are normally prepared in a vertical high-intensity mixer. For example, if the color
formulation calls for 50 ppm of a phthalocyanine green, the color formulation is as follows:

Ingredient	Color concentrate, parts per 100 resin	Color formulation, parts per million resin
Resin	99.0	
Pigment	0.5	50
Dispersing agent (Mark WS)	0.5	

If the color formulation calls for 250 ppm of TiO_2, rutile type, and 400
ppm of phthalocyanine green, the color concentrate will be

Ingredient	Color concentrate, parts per 100 resin	Color formulation, parts per million resin
Resin	93.0	
TiO$_2$, rutile type	2.5	250
Pigment	4.0	400
Dispersing agent (Mark WS)	0.5	

So, regardless of the pigment system, a fixed amount of these color concentrates is added to each internal mixer charge.

2.5.3 High-Shear Vertical Nonfluxing Mixers

These mixers are classified as dispersive blenders because they produce a complete random distribution of all particles throughout the mix, leaving, for the most part, no original particle identity.

All the vertical high-shear nonfluxing mixers have several common features.

1. A bottom-mixing impeller assembly rotating at high speed, about 900 rpm.
2. Two types of flow patterns produced by the motion of the rotor blades (impellers): 'one around the circumference of the bowl (centrifugal action) and the other a lifting cascading action from bottom to top. The mixing comprises convective, diffusive, and shear actions.
3. Drives with a ratio of approximately 1 hp to 4 lb resin, which results in rapid heating of the resin powder. Friction created by the shear produces a significant amount of heat. The average heat input is about 30°F/min. The major portion of the consumed power is used to generate the dispersive mixing needed to shear the powders.
4. A vertically mounted stationary blade that acts as a deflector baffle of the powder and as a shield for a thermocouple.
5. Two-speed motors are normally used to minimize high-starting torque.

The best-known vertical high-shear nonfluxing mixing units are the Henschel, the Diosna, the Papenmeir, and the Littleford-Lödige designs. Henschel blades taper and are perpendicular to each other. Henschel is now offering a dual-drive system for the mixing impellers, which are separately driven. They counterrotate in concentric shafts, contributing to reduced cycles [32]. The distance between the bottom and upper blades can be varied. These streamlined blades, when compared with the Papenmeir blades, are easier to clean, but because of their smaller surface are in contact with the resin particles, the rate

of heating is slower. This limitation, however, has been compensated
by the counterrotating action of the impellers in the newer units. The
amount of energy dissipated in one turn of the rotor blade is directly
proportional to the cross-sectional area of the blades [33]. A simple
experiment using lab-size mixers could corroborate the previous statement. For example, for a fixed amount of resin, the time needed to
reach a certain temperature, e.g., 240°F, could be used to measure
the effectiveness of different blade designs. The shorter the time,
the more efficient the rotor blade design. The Papenmeir vessel, which
is slightly tapered from bottom to top, not only enhances the mixing
action but also seems to help self-cleaning. Any of the high-shear,
nonfluxing vertical mixers available, the Littleford-Lödige, the Dreis,
the Diosna, the Henschel, and the Papenmeir, are more than adequate
to process rigid PVC dry blends. Figure 2.3, courtesy of Plastics
Compounding, presents four different impeller design approaches.

Major functions of these high-shear mixers are (1) densification of
the powder as measured by the increase in bulk density: as seen later
on, bulk density has a significant effect on extrusion output and extrusion stability; (2) homogenization of the powder, which is not only the
statistical distribution of the additives within the powder but also the
uniform coating of the PVC particles with the additives: both functions
are greatly helped by running the powder blends at high temperatures,
$T \geqslant 240°F$; and (3) volatiles removal, mainly moisture: the vortex action, the large surface exposure, and the vertical configuration of the
vessels provide maximum release of volatiles.

Blending Procedure

In essence, operating procedures are the same for all these vertical
high-shear mixing units. For a particular mixer, these procedures
vary according to the formulation and also according to the subsequent
melt processing step: either melt-compounding, dry blend extrusion,
or dry blend injection molding. The following is a mixing procedure
used successfully to prepare an impact-modified, white, opaque rigid
PVC compound for window profiles and siding. But before describing
the blending procedure, it is advisable to determine the maximum resin
charge that can be processed without overloading the mixer. The
higher this charge, the higher the friction and the shorter the cycle
time. The formulation, which was used in the United States during
the early days of the siding industry and which is similar to that used
by some European manufacturers, consists of the following components.

Two resins: A suspension homopolymer, low to medium molecular
weight (K value ~62), about 60.0% by weight, and a copolymer in
which an elastomer, superchlorinated PE (10.0% by weight)
serves as a backbone for PVC chains chemically attached to this
substrate. This copolymer offers a very convenient way to incorporate an elastomer in a dry blend form.

A solid barium-cadmium organic complex, mainly salts of fatty acids,
  in powder form suitable for outdoor weathering, about 2.0 phr.
An alkyl aryl phosphite, about 0.3 phr.
An epoxidized soybean oil as a costabilizer (1.6 phr).
Calcium stearate, 0.25 phr.
Diglycol stearate, 0.50 phr.
Low-molecular-weight PE, 0.50 phr.
An all-acrylic resin as processing aid, 2.0 phr.
A rutile-type, sulfate process titanium dioxide, 8.0 phr.

The optimum blending procedure found for this type of formulation
is as follows:

1. Resins are charged to the mixer either with the swivel cover
open or through the feed hopper in the cover.

2. After closing the main swivel cover and tightening securing
screws, the slow-speed button and, almost immediately, high-speed
buttons are activated. This minimizes high-starting torque.

3. Run at high speed until the temperature, as shown by the indicator, reaches 170 to 180°F; at this point, add the solid heat stabilizer, processing aid, lubricants as an "additive master batch," or if
automatic weighing facilities are available, add, in this order, epoxidized soybean oil, organic phosphite, lubricants, nonlubricating powders, processing aid, and solid heat stabilizers.

> Note 1: The additive master-batch may be prepared in an unheated
> blender from PVC resins, processing aid, heat stabilizers, and
> lubricants. Due to the lubricating properties of the Ba-Cd sta-
> bilizers, there is no risk in preblending these additives with
> other lubricants or processing aids.
> Note 2: Regulate the rate of addition so that the amperage indicator
> does not exceed the safe amperage limit.

4. At about 210°F, add the titanium dioxide in the same manner
through the feed hopper while watching the amperage indicator.

5. Keeping the feed hopper uncovered to help volatile removal,
continue mixing until the temperature reaches 230 to 240°F.

6. Turn on the cooler, cover the feed hopper, and proceed to dump
the blend.

7. If reworked material (regrind)* of the same formulation is available, the cooler, if it is a jacketed ribbon blender, is the best place to
incorporate the regrind.

8. The blend is then cooled to 100 to 105°F. Normally, for formulations in which color variation is not objectionable, the blends are
cooled to 120°F. Again, the appropriate cooling procedure should be

---

*Reworked material is the ASTM D883 terminology for clean regrind
that meets the requirements specified for the virgin material.

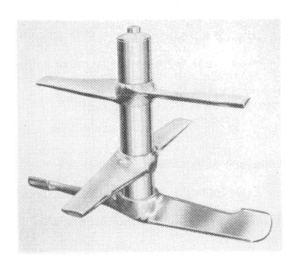

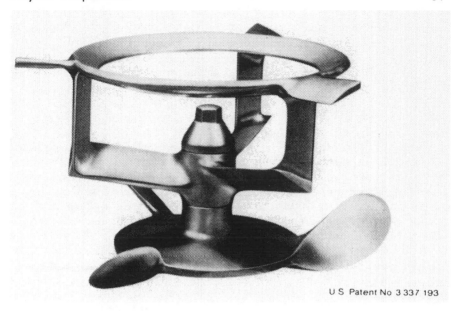

U S Patent No 3 337 193

Figure 2.3 Different manufacturers use different impeller designs with bottom blades to lift material and upper blades to bat the material back down.

used to get the highest bulk density in the shortest time. A more stable extrusion has been observed when the dry blend is conditioned for 24 hr before it is used. Normally in production, however, the powder blend is blown to a silo and thus further cools to the ambient temperature. The reason or reasons for this behavior are not clearly known. It seems that the low-molecular-weight additives, liquids, and lubricants distributed on the surface of the particles have more time to diffuse into the resin particles, thus reducing their lubricating effect in the feed and early part of the transition zone of the extruder. A thin film coating the resin particle is beneficial; a thick one could act as a binder and be detrimental to resin flow. For formulations based on other types of heat stabilizers, such as organotins (liquids and solids) and lead salts, the dry blend procedures are somewhat similar: liquid stabilizers are added at first to the preheated resins, 140 to 150°F, followed by lubricants (stearates and waxes) 165 to 185°F, and nonlubricating powders (process aids, impact modifiers, and fillers) at 185 to 195°F. Nonlubricant powders are added late to minimize abrasion. The pigments are added last at about 210°F. Drop to cooling is done at 240°F. *Note:* If the bulk density of the resin is low, it may be advisable to go to 250 to 270°F. Also, stabilizer addition at 170°F may help if the resin is not as absorptive to liquid stabilizers. Preheating the resin particles seems to enhance its absorptive capabilities, necessary for liquid additives.

As mentioned throughout this work, low-VCM PVC resins, which have been subjected to more severe drying cycles or steady steam stripping, may have different bulk densities, and the surfaces of their particles may be harder, less absorptive to additives, and much drier than ever before, causing occasional static charge problems during handling and blending. These changes require some modifications in blending cycles. For example, if bulk density is high, the processors then have to use lower drop temperatures, or higher drop temperatures if bulk density is low. For harder-surfaced resin particles the contact time between resin and additives has to be prolonged or the stabilizers have to be added at higher temperatures. To minimize static charge accumulation during high-shear mixing, a dilute solution of glycerine or any other antistatic agent added to the resin or resins during blending works very well. Figure 2.4 represents the effect of final blend temperature on compound bulk density. The data, which were generated in a production-size high-intensity mixer [34], show how effective are drop temperatures over 255°F. The main objection to this approach is an increase in yellowness in the final product.

When using the more efficient organotin stabilizers [thioesters like the di-n-butyl or dimethyltin-S,S'-bis(isooctyl merceptoacetate) or maleates like dibutyltin or di-n-octyltin maleate], which, due to their high efficiency and high cost, are used at very low levels, the dispersibility of these additives in the preheated resin is of prime importance. They should be added first and dispersed by themselves without the inter-

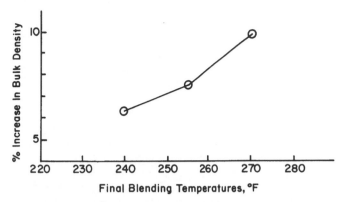

Figure 2.4 Effect of final blend temperature on compound bulk density (percentage increase over PVC resin).

ference of lubricants or processing aids. In formulations based on lead stabilizers in powder form, the additive master-batch approach works very well. To optimize the mixing process, the behavior of the various formulation components should be studied very closely. For example, suspension PVC resins do not behave like the bulk-polymerized PVC. Suspension resins generate higher friction, and higher temperatures are needed to improve the absorption characteristics of the powder. As shown above, this is even more pronounced in the case of low-VCM suspension resins. Bulk resins are formed of porous particles without any skin. These particles are easily coated and impregnated by additives. Bulk PVC resins are the easiest to densify and gel in high-shear mixers.

Mainly to increase blending capacity, there are alternatives to single-batch high-intensity blending procedures. Specifically, we are referring to the so-called double- or triple-batch operations. In these methods, the high-intensity mixer is charged to capacity with resin and blended according to the conventional method described thus far, but with double or triple amounts of additives. This master batch is then dumped into the cooling blender, which already contains one or two additional PVC resin charges at about room temperature. The resin is then blended with the master batch, which produces faster cooling. Blender capacity is thus increased, but at the expense of powder uniformity. One-half or two-thirds of the resin particles will not have achieved optimum bulk density, nor will they contain any absorbed additives. Some of the other potential segregation problems during material transfer and/or storage have already been discussed in this section. A double- or triple-batch approach to compounding is essentially a compromise dictated by economics and production capacity needs.

Dry Blend Properties

Bulk density, molecular weight of the resin, and powder flow properties
of the blend are three of the most important properties to be considered
in any dry blend extrusion process. These properties have such a
decisive influence in the extrusion process that, in twin-screw extruders and for a given screw speed, the output can be predicted closely
if these properties are known. Dry blends with high bulk densities
based on medium to high- molecular-weight resins (K value 65 to 70)
also encourage a more thorough mixing in the extrusion process. The
need to be aware of bulk density variations is most important.

As mentioned before, the very low residual vinyl chloride monomer
(RVCM) resins may be less easily processable than standard resins.
Very low RVCM resins are intended mainly for food-contact applications. Impurities, such as volatiles, or emulsifying agents, are so
evenly dispersed on the resin particles (like a protective colloid
around particles) that, even if they are present in minute amounts,
they contribute to better processability and lower melt temperatures.
This is also true with some other polymers, such as the copolymers of
acrylonitrile (AN) and styrene containing high levels of AN ($\geqslant 70\%$),
which can be processed as dry blends. In general, a low level of
residual acrylonitrile monomer (RANM) contributes to poor resin processability.

If dry blend properties are critical in twin-screw extrusion, they
become even more critical in single-screw extrusion, where the positive gear pump effect of the twin-screw extruders does not exist.
In single-screw extrusion, powder conveying is mostly a function of
the friction forces of the polymer against the screw and against the
barrel. Low friction forces between the powder and the barrel cause
breakdown of conveying, i.e., slippage.

Some instability in the output rate of single-screw extruders has
its origin in the feed section of the screw. Channels partially filled in
the feed section would create instability in the melting process. Likely
sources of the screw feed problem are (1) unsteady powder flow from
the feed hopper to the screw, and (2) unbalanced powder flow properties. A properly designed crammer force-feed in the extruder hopper
and a well-balanced dry blend result in better flow-rate control and
less powder segregation in the hopper. This, in turn, results in
better feeding of the screw and better overall output stability.

In this section, only dry blend properties like bulk density, powder
flow, amount of fines, and lubrication, which have a major effect on
feed problems, will be considered. In the long run, it pays to conduct
a systematic research program to determine the effect of each of these
variables on extrusion output, output variability (surging), presence
of bubbles in the extrudate, and so on.

Dry blend properties can be divided into two categories [35]: the
fundamental properties and the associated bulk properties. The

fundamental properties, which are mostly resin related, can be
listed as follows: grain shape, grain size and size distribution,
nature of grain surface, moisture content, coefficient of friction, and
chemical effects. The associated bulk properties are bulk density,
packing, grain density, angle of repose, permeability, and rate of
discharge.

Free-flowing rigid PVC dry blends seem to have in common:

1. Absence or very low concentration of fines, particle size below
10 μm
2. Uniform-size particles
3. High bulk density
4. No static charges on resin
5. Smooth surface
6. Spherical particles

Flow properties can be altered by surface changes: a thin uniform
liquid film promotes flows by making the particle surface more smooth;
a thick and uneven film can have the effect of binding the particle
surface, thus reducing flow. Process engineers can more easily modify properties 1, 3, and 4 listed above. In the case of the other three
properties, however, they have to work with the resin manufacturers
in order to meet the requirements. The emphasis is given in this work
to high-shear vertical mixers, blending procedures, and dry blend
properties reflects the industry trend. This author recommends Ref.
33 as a commendable presentation of the effect of the high-intensity
mixing process on extrusion output, power economy, and extrudate
quality.

2.5.4 Low-Shear and High-Shear Mixing in Horizontal, Jacketed, Cylindrical Blenders

These mixers, although they resemble the ribbon blenders, are if
properly modified perhaps the most versatile mixers in the field. They
consist of either plow-shaped mixing tools or the so-called bakery plows,
which are like sections of ribbons mounted radially on a horizontal
shaft that rotates at a rather high rate of speed. These plows, although producing low-shear mixing, mechanically fluidize the materials
and achieve homogeneity by intersecting trajectories from adjacent
plows. The high-shear mixing is provided by the choppers or the impactors installed in the blender walls [36]. Each chopper consists of
four blades—two tiers—mounted on the extender motor shaft on the
inside of the mixing drum. These blades resemble the Henschel blades.
The so-called impactors consist of a series of pins mounted peripherally and perpendicularly in a moving plate. This plate rotates inside
a pinned stationary cage. Both high-shear units rotate at 3600 rpm.
These mixing tools, which supplement the mixing action of the plows,
contribute to raising product temperature and to improving additive
dispersibility.

Product heat buildup of the plows is only moderate, usually in the
range 1 to 2°F/min. Heat buildup, however, could be significantly
improved by circulating steam or hot water through the jacket while
turning the choppers on.

As described in the U.S. patent 3,831,290, this type of blender
can be easily modified to increase its utility for any dry-blend preparation process. Two of the most useful modifications, as described in
this patent, are two circular openings at the top end of the vertical
walls of the blender. The inlet, is used for the circulation of hot
air into the blender while an air classifier is attached to the outlet to
remove the fines detrimental to the extrusion process. The circulation
of hot air (temperature 220 to 240°F) during the heating and blending
process helps to remove volatiles, which could include some VCM, also
helps fines removal, and acts as another heating element.

These modified blenders are also very efficient when partially
loaded. Loads as low as 25% of total working capacity can be processed
adequately. The basic blenders, however, are commonly used as
coolers. When used as coolers, the choppers and the hot air system
are eliminated and the jacket wall is constructed from a thinner-gage
steel.

The operating procedure for these coolers, which are normally
located right below the vertical high-shear mixers, is quite simple:
tap or cold water is circulated through the jacket at all times while
the blender shaft is kept running at the low rpm setting. The volume
and temperature of water will be determined according to the particular cooling need.

2.5.5 Hot Melt (Fluxed) Compounding

Mainly for economic reasons, the extrusion of rigid PVC directly from
powder blend is gaining wide acceptance throughout the world. For
example, the production of rigid PVC pipes and conduits in new plants
is made almost solely from dry blends. For some other applications,
such as extrusion of residential colored siding and accessories and extrusion of rigid PVC for food-contact applications, melt-compounded
materials are still widely used. For these applications pelletized rigid
PVC, if properly processed, can consistently produce a profile or product with great homogeneity and good gelation (degree of fusion). These
properties, which are related to melt quality, are very desirable in any
of the previous applications. For example, in the case of residential
siding and accessories or window frames, these properties contribute
to improved impact resistance and weatherability. In the case of containers, these properties contribute to lower oxygen and water permeability, which in turn contributes to better taste and odor of the
contained products. Improved barrier properties contribute to less additive migration, i.e., antioxidant from the impact modifier into the
packaged product.

Some other reasons for the continued use of the melt-compounded materials, in spite of the higher cost, follow.

For some processors with no compounding facilities, limited testing capabilities, and conventional extrusion and/or injection molding machine lines (unvented, short-screw extruders), melt-compacted resin compounds (pelletized) offer the most trouble-free route and also a better utilization of the melt processing equipment.

For other processors with compounding facilities and some old extruders and/or injection-molding machines, the use of melt-compacted materials reduces cleanouts and contaminations in the compounding operation and constitutes a better feed for the old extruders. Pelletized PVC has a relatively high and uniform bulk density and can, therefore, be fed much more easily into extrusion and injection molding machines.

Some processors of rigid PVC products for food-contact applications prefer the pelletized materials and the use of shorter extruders for better control on the properties of the finished product, i.e., color, surface appearance, and odor and taste of the contained product when applicable. Dry blends of the same compounds require, for the most part, vented and longer extruders to remove volatiles and to deliver a melt with improved thermohomogeneity, respectively. The vented section of the screw, which is always designed to be partially filled only, is a potential source of polymer stagnation and degradation.

Regardless of the reasons for the use of prefused pelletized compounds, there will always be a place for these compounds, and processors will be making them in all sorts of melt compounders, mainly compounding extruders, or high-shear, high-intensity melt mixer lines. In describing compounding lines based on extruders, it is necessary to introduce, at this point, some key elements of the extrusion process and the most important features of the extrusion machines. Some of these features, however, are described in more detail in Chap. 4.

Melt compounding of rigid PVC formulations in screw extruders is presently dominated by multiscrew extruders. Multiscrew machines accept difficult to feed rigid PVC materials more readily than do single screws, and they convey such materials to a strand die at lower temperatures and less temperature gradient across the melt. This is achieved with less supervision of processing conditions.

A large portion of the heat in a single screw is contributed by energy from the drive motor. In general, the higher the screw rpm, the higher the output; but also, the higher the shear, the higher the melt temperature and temperature gradient across the melt. In a multiscrew extruder machines, the mechanical energy source contributes 54% and the thermal energy 46% to the melting of the material [37]. These two sources of energy, producing almost equal inputs, contribute to better temperature control. This illustrates why twin-screw extruders have much smaller motors than do equivalent output single-screw machines, which depend mostly on mechanical energy for the melting

process. As will be shown later in this book, the output of single-
screw machines is sensitive to die restrictions and is highly pressure
dependent. The output in multiscrew extruders is less sensitive to
die variations and varies almost linearly with the screw speed.

2.5.6 Single-Screw Extruders

If the compounding needs are clear-cut and do not require the sophistication of continuous high-intensity melt compounders, some proces-
sers would do well to consider the relatively inexpensive option of
single-screw extruders.

Compounding in single screw extruders is dominated by two-stage
screws in conjunction with vented barrels. Two-stage screws improve
the quality of the extrudate by allowing gases and vapors, for example,
to escape from the melt. The most common screw size has an overall
length of 24D. The screw consists of two screws in series on the same
shaft. Longer screws, with lengths of 30 to 34D, are used for a direct
dry blend extrusion. For compounding purposes, the longer extruders
contribute to better melt thermohomogeneity and output; however,
they may produce higher compound yellowness due to the longer residence time at temperatures.

All the screws for vented extruders have in common a vent or decompression zone (see Fig. 2.5), where the screw channel is quite
deep, creating a zone of sudden decompression to help release of volatiles with the aid of a vacuum pump and trap. Special attention should
be given to the vent zone, since it is designed to be partially filled
only, and polymer stagnation or degradation may occur. Again, due to
the characteristics of the screw at the vent, a total flushing of the
stagnated materials is never obtained. For compounding, the vent
zone of the extruder is normally operated at atmospheric pressure. At
the higher screw speed, however, volatile removal at atmospheric pressure is not sufficient to produce a bubble-free extrudate. In most
cases, the use of high levels of vacuum at the vent eliminates this
bubble problem.

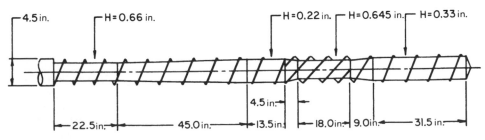

Figure 2.5 Representation of 4.5 in., 32/1 L/D PVC screw.

To analyze screw performance as judged by the compound quality, a piece of information of great value is the nominal shear rate $\dot{\gamma}$ in both metering zones and its effect on melt temperature as the screw speed N increases. This nominal shear rate is obtained using the following equation:

$$\dot{\gamma}_{1m} = \frac{\pi DN}{H_{1m}} = \sec^{-1}$$

where

D   = screw diameter (in.)
N   = screw speed (rev/sec)
H_{1m} = first metering zone channel depth (in.)

Since the second stage metering section H_{2m} is deeper than the first one, $\dot{\gamma}_{2m}$ will be lower. This deepening of the channel is necessary to give the second stage, at zero die pressure, a greater pumping capacity than that of the first stage and thus reduce the risk of melt flow through the vent opening. During pelletizing, die pressure should be monitored closely, since it could increase over a certain value, reducing the output of the second stage below that of the first and resulting in back flow out of the vent. Normally, pelletizing dies operate at lower pressures. A faulty heater in the die, die adaptor, or metering zone, or a sudden restriction in strand die output, because some die openings are plugged with contamination or degradation, could create this situation.

As seen in this simple equation, the $\dot{\gamma}$ value increases when the metering channel depth becomes shallower. These higher shear rate values increase melt temperature and extrudate yellowness. Deeper channels produce lower $\dot{\gamma}$ values and lower melt temperatures, as well as better extrudate color. Deeper channels also produce higher output and output per revolution (W = output/revolution). Torque T will be higher, since $T \propto W/E$; where E = energy economy, lb/hp hr. The channel depth may be limited by the torque limitations of the screw and/or extruder drive.

Another design element to be analyzed is the balance between heat removal by barrel cooling and avoidance of excessive-shear heat generation imparted by the screw. With moderate barrel cooling, the power economy E can be about 8 to 10 lb/hp hr or higher when low flow resistance is offered by the die.

When substantial barrel cooling is used, this economy can drop to about 6 lb/hp hr, and since screw torque is inversely proportional to E, the screw must be proportionately stronger. Small-diameter screws (below 3.5 in.), normally have sufficient strength in the feed zone of the screw without a special high tensile strength screw. In larger diameter screws, such as the 4.5, 6, and 8 in., which are the sizes

normally used to pelletize rigid PVC compounds, the screw strength
as well as the operating procedures deserve more attention. Maximum
barrel cooling and cold starts, for example, should be avoided.

An approximate expression for maximum shear stress τ_{max} in the
feed zone of the screw can be of help in analyzing potential mechanical failures:

$$\tau_{max} \cong 19.3 \times 10^6 \frac{W}{E} D_o^{-3}$$

where

    W = output per revolution (lb/rev)
    E = energy economy (lb/hp hr)
    D_o = root diameter of screw (in.)

This equation is obtained by combining a common expression for torque:

$$T = \frac{(360)(33,000)}{\pi} \frac{W}{E} \quad in.\ lb \tag{2.1}$$

with the usual relationship of maximum shearing stress τ_{max} in the
screw feed zone as a function of torque, T, root diameter of the screw
D_o, and the polar moment of inertia I (in.[4]):

$$\tau_{max} = \frac{TD_o}{2I} \tag{2.2}$$

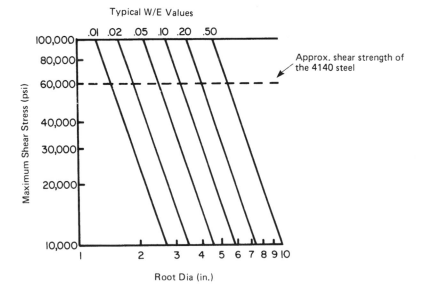

Figure 2.6 Maximum shear stress (psi) in the screw feed zone versus
root diameter (inches) for various values of the parameter (W/E).

and with an approximate equation for I:

$$I \simeq \frac{\pi D_o^4}{32} \tag{2.3}$$

τ_{max} values can be experimentally determined. Using the expression
for τ_{max}, a plot of τ_{max} versus D_o can then be drawn for various typical values of the parameter W/E (0.01, 0.02, 0.05, 0.10, 0.20, and
0.50 (see Fig. 2.6). Also shown in Fig. 2.6 is a horizontal line representing the shear strength of the steel used to manufacture the
screw, i.e., about 60,000 psi. To examine the safety factors for
strength of a given screw, one proceeds as follows. First D_0 is calculated from the screw diameter D and the channel depth of the screw
in the feed zone, while E and W values are calculated from the experimental data. From the values of D_0 and W/E, one may obtain the
τ_{max} of the screw using either Fig. 2.6 or the formula for τ_{max}. This
value is then compared with the shear strength of the steel.

As shown, the maximum shearing stress varies inversely as the
third power of the root diameter. The equation for τ_{max} is intended
to help compounders of rigid PVC, using two-stage single-screw extruders, in the analysis of the safety factors for strength of the
screws in their machines when compounding under various conditions.
The information presented so far should also help to improve this
performance.

Typical extruder drives for compounding rigid PVC formulations
are:

Extruder size, in.	Extruder length, D	Motor drive, hp
$4\frac{1}{2}$	24	100-125
6	24	250-300
8	24	400-450

All these compounding extruders are equipped with strand dies. In
some cases, the strands are cut at the die face by a set of rotating
knives mounted in a head hinged to the die. The pellets thus cut can
be of variable shapes, from a cylinder to thin lentils, depending on
output, extrudate swelling, extrudate melt temperature, formulation,
and rotating cutting knife speed, among other factors.

They are either air cooled or swept along by a stream of water onto
a shaking conveyor. More frequently, the strands are sequentially
immersed in a water bath and the moisture swept away with air knives.
Then the partially cooled strands are cut into pellets. As seen either
process for cutting the strands needs no special attention.

Mainly, for economic reasons, there is a tendency in the field to operate the compounding extruders at their maximum melting capacity.

This situation contributes to the following problems: (1) melt deteriora-
tion, since temperature gradient across the melt increases with the rpm;
(2) nonuniform flow of the melt through strand dies, causing rippling
of strands, with the faster flow; (3) potential screw mechanical failure;
and (4) some potential back flow out of the vent.

2.5.7 Multiscrew Extruders

The use of the multiscrew extruder for processing rigid PVC compounds is believed to have originated in Italy. Some time in the 1930s,
Roberto Colombo of Turin produced a successful two-screw arrangement in which the screws intermeshed and thus formed a positive
pump. The first machine consisted of screws of the same length, with
identical threads and constant pitch, rotating in opposite directions.
This arrangement ensured a positive feed and, by virtue of the mutual
wiping action of the flights, prevented the material from adhering to
the screws. It also became possible to achieve high pressures at the
die and to maintain these pressures consistently. At an early date,
Colombo abandoned the principle of opposite rotating screws and
adopted and patented the corotating screw machines. Serious screw
wear in the early Colombo counterrotating machines appear to be the
main reason for the switch. These machines identified by the trade
name RC are manufactured in Italy by Lavorazione Materie Plastiche
and in England by R. H. Windsor, Ltd. As part of a complete line of
manufactured rigid PVC building panels (using an annular slot die),
these extruders were introduced in the United States early in the 1960s.
    The number of variables on a twin-screw machine is obviously great-
er than that of a single-screw machine. It is not only possible to alter
flight profile, compression ratio, the pitch, and the root diameter, for
example, but also the relative rotation of the screws. Apart from the
Colombo machine, there is now available a comparatively large number
of other important multiscrew extruders, based generally on the counterrotating screw system. Perhaps these different approaches were
mainly use for patent reasons. Some of the most interesting machines
follow.

> The Cincinnati Milacron multiscrew extruder system. This company
> offers the cylindrical twin, the conical twin, and also the two-stage
> twin-screw machine. A similar two-stage twin-screw extruder
> has been intermittently manufactured by Egan Machinery Corp.
> The Krauss Maffei system also offers among their line of twin-screw
> extruders a conical twin-screw extruder. This company recently
> introduced a new twin-conical screw design, the so-called double-
> conical screw system, which consists mainly in changing depth
> of thread over the entire screw length (more on this later).
> The Maplan twin-screw extruders showing improved drive construction
> The Amut, which is available as part of complete lines to manufacture

mainly rigid PVC profiles for building applications, such as lou-
vers, siding, window frame components, and moldings.
The NRM, under Weber License, offers the Pacemaker III twin-
screw extruder.
Reifenhauser-Nabco with their twin-screw under the Bitruder trade
name.
The Kestermann.
American Leistritz twin-screw extruders.

All these machines are used for both compounding and direct extrusion,
mostly from powder. Several makers, concentrating on compounding
only, have used the multiscrew principle in machines designed specifically for this purpose:

The Werner-Pfleiderer, mainly their Komblipast system.
The Welding Engineers, Inc., which offers a twin nonintermeshing
screw machine, the screws of different lengths.
Planetary-Gear Extruders, very popular in Europe where they were
designed mainly for compounding rigid and plasticized PVC.
They are offered by Berstorff, EKK Kleinewefers Kunststoffmas-
chinenen, and Krauss-Maffei. Krauss-Maffei also offered a planetary extruder concept for high-output PVC powder extrusion.
Compounding on some of the best-known machines is descussed
in greater detail in this chapter.

Corotating, Intermeshing, Parallel Twin-Screw Extruders: RC Extruders

As with the single-screw machines, the multiscrews have to perform
various functions around their main function, changing feed material
bulk as it passes from the hopper to the die. In the RC machine,
which in this section is used as the prototype, the compression ratio
is brought about by changing the pitch and diameter of the screws in
three distinct zones. The rear section of the screws nearest the
hopper has a coarse pitch and the largest outer and smallest root
diameter. The pitch in the middle section is smaller, and the outer
and root diameter are somewhat smaller and larger, respectively. The
front section, the metering zone, is the longest with the smallest pitch.
    Other design features of these screws are (1) the threads in all
the three sections are deeply cut, (2) the flanks are tapered, and (3)
the intermeshing of the screw is almost to the full thread depth along
their whole length. The barrel of the machine is cut in three steps
of different bore diameters to fit the three sections of the screw.
These three steps, with the progressive reduction in diameter, contribute to obtaining compression but in a discontinuous manner.
    In a single-screw extrusion, the screw rotating inside the barrel is
not able to push the material forward by itself.

If for some reason the compound filling the channels of the feed zone sticks to the screw, the screw becomes a rotating cylinder and provides no forward action. To be pushed forward, the material should not rotate with the screw or, at least, should rotate at a slower rate than the screw. The only force that can keep the material from turning with the screw and make it advance along the barrel is the friction between the material and the inside surface of the barrel. The more friction and the less rotation of the material with the screw, the more forward motion or transport. To yield sufficient production with a low friction factor, the screw must have a large diameter and turn at high speed. However, a large-diameter screw rotating at high speed develops high shear, as shown in the shear rate equation:

$$\dot{\gamma} = \frac{\pi DN}{H}$$

Furthermore, large-diameter screws rotating at high speed require large motors, which may reduce the energy efficiency of the extruder.

In corotating twin-screw extruders, the screws approximate a positive displacement pump and do not depend on friction against the barrel to move material forward. The L/D ratio is of less importance for the moving of material because the conveying action is not appreciably affected by differences in head pressure.

The shear for these extruders can be calculated with the same formula as that for single-screw extruders, provided the actual diameter D used for single-screws is replaced by the equivalent diameter D_e for twin screws [37]. Because the two intermeshing screws overlap, the periphery of the screws actually working the material approximates the circumference of the equivalent screw diameter. Since H, which is the channel depth, is three to four times greater than in the single-screw machines, and the screw speed N is much lower (15 to 20 rpm), the total shear rate $\dot{\gamma}$ is also significantly lower, as much as 80% lower. These screw characteristics keep material at lower temperatures and the material is consequently more viscous. Also, the temperature gradient across the melt is significantly lower. Extrudates showing lower swelling values and more uniform color are always associated with these extruders.

In the late 1950s, the writer worked overseas with a line of these RC extruders, which were equipped with one set of screws to do both pelletizing and profile extrusion. Unvented, these extruders mainly produced rigid PVC pipes and electrical and telephone conduits. Every machine was also equipped with pelletizing heads, consisting of a strand die and die face cutting knives. Since the plant was almost compound self-sufficient, there was an abundance of pelletizing facilities. Dry blend preparation was conducted in a steam-jacketed ribbon blender where moisture and volatiles removal used to take place. Some of these machines had horizontal feeding screws that worked very well with all kinds of pellet sizes and also with powder.

Figure 2.7 Line of RC twin-screw extruders, late 1950.

Even during those early days, machine reliability was excellent, requiring minimum maintenance. The largest extruder, the RC10, for example, pelletized continuously (see Fig. 2.7). Operating procedures for machine shutdowns and start-ups were very simple, and for the most part, no flushing of the extruders was necessary.

As explained before, one of the reasons for the excellent machine reliability with rigid PVC is that the resin is almost totally moved from one screw to the other and then downstream. Very little resin passes between the screw gap, which in the corotating screw is called the "wiping gap." Also, when head pressure is low, material leakage over screw flights, the so-called overflight, has such a minor effect that, from a practical standpoint, it can be ignored. Since almost all the material is transferred from one screw to the other one at each revolution, there is always a complete renewal of material on the screw channels. The resin path itself is a long one, allowing ample opportunity for heat exchange between the polymer and the barrel walls.

The writer worked with these extruders for four consecutive years and remembers very few occasions when the screws had to be pushed out and cleaned because of polymer degradation. This relative absence of polymer degradation was mostly due to the excellent self-cleaning characteristics of the corotating machines, which do not tend to accumulate material at any point around the screws. In addition to the machine self-cleaning charactistics, other elements were responsible for such a good performance. For example, the initial formulations

used were heavily stabilized until personnel were properly trained.
Then, these formulations were gradually adjusted. An example follows:

Pipe and electrical conduits	Initial formulation
Suspension PVC resin, K value 65	100.00 parts
Tribase (tribasic lead sulfate)	4.00 phr
DS-207 (dibasic lead stearate)	1.00 phr
N-Lead stearate	1.50 phr

Resins used were the well-known Solvic 229 (Solvay), Vipla KMO
(Montecatini), and Vestolit S65 (Chemische Werke Hüls AG).

Pipe and electrical conduits	Production formulation
Suspension PVC resin, K value 65	100.00 parts
Tribase	2.50 phr
DS-207	0.75 phr
N-Lead stearate	1.00 phr

As seen, these formulations contain no impact modifiers: notched
Izod impact ~ 0.6 ft lbf/in. (notch). The formulation containing the
lower level of stabilizers produced the compound with the best properties. A technique adopted from Europe was that of blending the
suspension homopolymer (70.0% by weight) with a higher molecular
weight emulsion resin (30.0% by weight), in order to

Improve impact resistance: about 0.2 ft lbf/in. improvement was
    obtained
Improve product surface and overall processability
Control postextrusion swelling

For pipes, conduits, venetian blinds, and other profiles for outdoor
weathering, part of the Tribase was replaced with Dyphos, and a rutile, nonchalking TiO_2 was incorporated in the formulation as follows:

Ingredient	Formulation
Suspension PVC resin, K value 65	100.00 parts
Dyphos (dibasic lead phosphite)	2.00 phr
Tribase (tribasic lead sulfate)	1.50 phr
DS-207 (dibasic lead stearate)	0.15 phr
N-Lead stearate	1.00 phr
TiO_2 RA-NC	2.0 to 5.0 phr

Surprisingly enough, these formulations, based on lead stabilizers, have remained unchanged through the years. Even the blending of emulsion and suspension resins continues to be a common practice.

Another formulation that was frequently compounded to feed reciprocating injection molding machines was the following:

Ingredient	Formulation
Suspension PVC resin, K value 65	100.00 parts
Acryloid K-120 (all-acrylic process aid)	4.00 phr
Tribase	3.00 phr
DS-207	1.00 phr
N-Lead stearate	0.50 phr
Epoxidized soybean oil	2.00 phr
Glycerides of fatty acid esters	2.00 phr

Using this pelletized compound, these injection molding machines with cold-runner type molds were continuously run to supply all the fittings and junction electrical boxes that complemented the line of electrical conduits. The runners were ground and reworked in either these machines or in the extruders.

The degree of compounding had a very decisive influence on injection-molding cycles and the parts quality of these small injection-molding machines. Underfused compound, as judged by pellets showing some disintegration when immersed in dry acetone, produced appreciably longer cycles and frequently parts with yellow or brown streaks.

The power economy of these extruders during pelletizing was as high as 18 lb/hp hr.

Counterrotating, Intermeshing, Cylindrical Twin-Screw Extruders

The early cylindrical, intermeshing, counterrotating, twin-screw extruders (some of them are still around) suffered, among other problems, from poor output rates. This was mainly due to the following reasons:

1. The overall capacity of the feed zone was low, which limited the amount of dry blend per screw revolution transported in that zone.
2. Screw channel depths could not be increased much further without weakening the screws, which were bored for cooling.

Also, motors were underpowdered, and close attention had to be given to start-ups, temperature profiles, amperage readings during extrusion, and flushing operations. Machine shutdown procedures had to be developed to minimize degradation of the PVC, which always remained in the extruder even after the flushing operation. Last, but not least, was another very serious problem: an excessive amount of screw and barrel wear.

Before proceeding to discuss the compounding operation in these machines, let us describe the mixing mechanism in these extruders. The material forms a bank at the point where the two screws meet. In the bank, a portion of the material called the "calendered fraction," is squeezed through the gap between the screws and, at that point, subjected to high shear. Corotating screws offer a substantially higher maximum gap shear rate than counterrotating screws, but the fraction compounded in the gap, per turn, is much lower. The other portion of the bank, the so-called conveyed fraction, is moving axially downward and conveyed by the screw flights with shear rates similar to those of single-screw machines. Statistically, it is necessary that the material passes through the high shear gap area in a controlled fashion, but the gap itself presents some problems.

On the one hand, a narrow gap results in extremely high, and sometimes uncontrollable, shear conditions for the product passed through the gap. Also, a narrow gap, which is desirable for polymer processability, could create excessive screw wear. This situation increases with the size of the screws. Although most of the early limitations discussed have been minimized, potential screw wear at the higher screw speed especially for the bigger machines, however, remains. This is due to two factors: (1) Large diameter screws experience much greater radial velocity at a given rpm. (2) Counterrotation (material diverging at the top of barrel and converging at the bottom of the barrel) creates an upward-outward 10 o'clock-2 o'clock thrust of screws against the barrel, and this is where wear is most severe. This phenomenon especially happens in the powder compression zone, just before the vent, and also in the melt metering zone.

In spite of the potential screw-wear problems, these machines have been used successfully in the compounding and extrusion of rigid PVC. There are several reasons for this successful application. The screws are operated at very low screw speeds and with relatively low drive

power. The degree of compounding and compound uniformity are reproducible on a day-to-day basis, and a great part of the energy required for melting the product is derived from outside conductive
forces of heat, typically high-intensity heater bands.

In the early 1960s, the writer had the opportunity to work with
one of the first Anger APM counterrotating twin-screw extruders
brought to this country. It was an A2-120 (120 mm screw diameter)
equipped with two sets of screws: the so-called panel screw set to
extrude PVC sheets from dry blend and the compound screw set to
pelletize from dryblend. The two-screw pairs had approximately the same
configuration, the main difference being their pitch. Obviously, the
panel screws had the shorter pitch to provide for a higher compression of the stock. For the same formulation, the shorter pitch screws
(panel set) contributed to lower production output. At $11\frac{1}{2}$ rpm, they
produced 400 lb/hr (of a residential panel formulation); the compound
screws produced about 550 lb/hr.

This A2-120 machine came with a pelletizing head (a strand die and
a die face pelletizer) for compounding and with a sheet die for the extrusion of rigid PVC panels. In this section, however, emphasis is
placed on the use of this extruder as a pelletizer. The pelletizing head
is connected directly to the extruder and worked on the die face cutting
system. The plastic stock emerges through a perforated plate in the
form of round strands and is cut, while still warm, directly at the die
by rotating cutters. The center of the die plate is a solid torpedo
shape, which is immersed in the plastic stock and surrounded at the
base by several rows of circular holes. At compounding outputs higher
than 500 lb/hr, some peeling of the strands was always observed. The
ones with the higher degree of peeling, the inner ones, happended to
be the strands with the higher stock temperature, about 15°F higher
than the outer strands. This temperature gradient was not expected
from a twin-screw extruder and was attributed to the frictional heat
in the die torpedo, which was not heated.

The following process steps, which were used with this machine,
when applied to similar extruders, will help to make their operations
more trouble-free.

Better mixing in the fused stage was obtained by using high temperatures in the three back zones and lower in the front, starting in the
first decompression zone. Very early, it was discovered that the extruder temperature profile had a marked influence not only on the mixing obtained during compounding but also on the stability of the run.

The machine start-up was conducted in a starve-feeding mode.
Initial extruder temperature profile was 5°F higher than standard,
and no cooling water was used in the extruder zones.
It was also found out, as described before, that a conditioned dry-
blend with high bulk density and based on medium- to high-
molecular-weight resins, promoted not only better compound uniformity but also better extrusion stability and optimum resin
bulk density; 580 to 600 g/liter.

During machine shutdown and to avoid PVC degradation in the two decompression zones of the screws and in the die adapter, a purge compound was developed that proved to be very helpful. This purge compound was formulated as follows: 100.00 parts of low-density PE pellets, 0.10 phr of epoxidized octyl tallate, and 20.00 phr of CaCO$_3$ (Omya BSH or similar). The epoxidized octyl tallate (Drapex 4.4 or similar) was added to the PE pellets to wet their surface, thus promoting adhesion of the CaCO$_3$. This blend was prepared as needed. This flushing compound seemed to have a double function: scrubbing action and coating protection of the PVC left in the extruder. The CaCO$_3$ provided the scrubbing action, and the low-density PE coated any PVC inside the extruder, thus minimizing any PVC degradation during the heating cycle of the next start-up.

The practice of using a purging compound during machine shutdowns continues. At present, however, most of the purging compound used is based on PVC resins [34]. A typical "purge" formulation follows:

Ingredient	Formulation
PVC (K-68)	100.0 parts
CaCO$_3$	100.0 phr
Calcium stearate	4.0 phr
165 Paraffin wax	3.5 phr
High-efficiency tin stabilizer	10.0 phr

Many formulations were pelletized in this Anger extruder, for experimental and also for commercial use. Some of the formulations pelletized for commercial applications are shown in Tables 2.2, 2.3, and 2.4. As seen in Table 2.2, this formulation contains several lubricants. In addition to the inherent lubricity of the Mark 99, three other lubricants, normal lead stearate, partially oxidized low-molecular weight PE, and Wax E, are included in the formulation. This apparent extra lubrication was needed because the PVC melt is subjected to very complex flow patterns, resulting in greater frictional heat and shear degradation.

In the translucent formulation, the pigmentation was added as a color master batch, prepared as described in Sec. 2.5.2. A color formulation that was in great demand contained 1.25 ppm of a light stable red pigment (Ciba-Geigy Cromophtal Red GR) and 250 ppm of an optical brightener. This combination produced a very light and bright pink color cast.

In Europe, this formulation, without the rubber modifier (the chlorinated PE) and as dry blend, has been extruded directly into panels in

Table 2.2 Translucent, Impact-Modified Panel Compound[a] with Ba-
Cd-Pb Stabilization

Ingredient	Formula
Suspension PVC homopolymer (K value 65)	60.00 parts
Copolymer of chlorinated PE (~10.0%) and PVC	40.00 parts
All-acrylic process aid (Acryloid K-120N)	2.00 phr
Mark 99 (Ba-Cd heat stabilizer)	2.00 phr
N-Lead stearate	0.75 phr
Organic liquid phosphite	0.30 phr
Epoxidized soybean oil	1.00 phr
Wax E (montanic acid ester)	0.25 phr
Partially oxidized low-molecular-weight PE	0.20 phr
Benzotriazole UV absorber	0.35 phr

[a]Similar formulations, but containing EVA graft copolymer as the impact modifier are being used by European compounders to formulate
a weatherable dark brown color for windows and door frames. For
this application, the pigment system used is based on Ciba-Geigy
Cromophtal Brown 5R.

a similar Anger extruder and Anger panel line system. Formulations
similar to the one in Table 2.4 are widely used by European compounders to formulate weatherable white profiles. Compound undergelation

Table 2.3 Translucent, Impact-Modified Panel Compound with Tin
Maleate Stabilization

Ingredient	Formulation
Suspension PVC homopolymer (K value 65)	60.00 parts
Copolymer of chlorinated PE (~10.0%) and PVC	40.00 parts
All-acrylic process aid (Acryloid K-120N)	2.00 phr

Table 2.3 (continued)

Dibutyltin bis(monoalkyl maleate)[a]	1.5-2.0 phr
Wax OP	0.35 phr
Wax GL-3	0.35 phr
Partially oxidized low-molecular-weight PE	0.35 phr
Tinuvin 327[b]	0.35 phr

[a]This product was diluted with up to 30% stearamide wax.
[b]A substituted hydroxyphenylchlorobenzotriazole that does not form a complex with metals like Sn and Co because the hydroxyl group is blocked by a tert-butyl radical, does, however, impart a light yellow cast to the extruded compound.

(underfused) is a latent problem in these twin-screw extruders. Formulations containing high loads of TiO_2 > 8.0 phr processed at high compounding rates have the tendency to be underfused. These "underfused" compounds, when extruded into siding, produce a very

Table 2.4 White, Opaque Impact-Modified Formulation with Ba-Cd Stabilization

Ingredient	Formulation
Suspension PVC homopolymer (K value 65)	60.00 parts
Copolymer of chlorinated PE (~10.0%) and PVC	40.00 parts
All-acrylic process aid (Acryloid K-120N)	2.00 phr
Mark WS (Ba-Cd solid)	2.00 phr
Epoxidized soybean oil	1.60 phr
Calcium stearate	0.25 phr
Partially oxidized low MW PE	0.25 phr
Di-glycol stearate	0.50 phr
Rutile-type TiO_2 with moderate chalking	8.0 phr[a]

[a]This amount was subsequently increased to 10.0 phr. This formulation was used to extrude both siding and opaque white panels.

dull, uniform surface which, for the most part, is quite tolerable.
Sections of this siding, when immersed in dry acetone, show considerable microroughness in the entire surface. These underworked pellets, when blended with properly fused pellets, help to reduce the
surface gloss of the siding, masking some of the surface imperfections,
like die lines or shiny areas.

Counterrotating, Intermeshing, Conical Twin-Screw Extruders

The conical twin-screw extruder concept appears to first have been
manufactured in Austria by Anger AGM. Mainly for efficient transport,
the large-diameter rear-feed zone has a volume capacity greater than
the compresson and metering zones. Thus, a greater amount of powder
is transported to the feed zone per rpm, and a higher output rate is
achieved with minimal shear and frictional heat accumulation. There
is a reduced shear rate toward the die. In general, they are, however, more difficult to manufacture and cannot take advantage of the
building-block principle, in which screw and barrel consist of numerous elements that can be arranged in any possible sequence. Properly computer-controlled machines, tools and engineering concepts, however, can and do produce conical barrels and screws that are more than
cost-competitive with equal-output cylindrical twins. Also, the building-block concept contributes to overall reduction in strength of
screws, particularly torque capability.

Because these screws are tapered, there is enough space between
the shafts to fit two large helical bevel gears that drive the screws
through a simple two-shaft system rather than a threee-shaft gear
system with tandem bearings. The conical design also permits the
screws to be bored for cooling or heating with minimal loss in strength.
In fact, the greater rigidity of conical screws gives more resistance
against twisting and bending, which can be one cause of screw and barrel wear. Development of back pressure and resistance to screw rotation occur primarily in the metering zone, where the PVC compound
is almost completely fused. Back pressure is directly related to the
cross-sectional area of the metering zone, which means that, for extruders of comparable output rates, the conical design has a smaller
metering zone cross-sectional area and develops a lower back pressure.

The tapered screw design puts more surface at the feeding section
for heat transfer to cold material. As the PVC compound moves toward
the die, the tapered screw offers less surface and less heat to the
material. Each screw has a bored core through which oil is pumped.
The oil transfers any unwanted frictional heat from the metering zone
back to the feeding zone. Some new extruders have screws with
two-circuit temperature controls: one to heat the feed section of the
screw, the other to heat or cool the metering zone.

Although these extruders are mainly used for profile extrusion, like
rigid PVC pipe tubing, siding, and window frames, they are also commonly used for pelletizing rigid PVC compounds. For this application,
minimum attention and space are needed, and one operator can take
care of various machines at the same time. With the Cincinnati Milacron

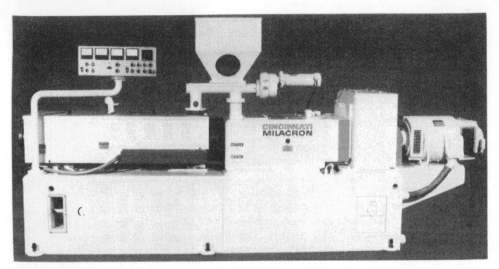

Figure 2.8 CM 80. (Courtesy of Cincinnati Milacron, Cincinnati, Ohio).

Figure 2.9 CM 90 with pipe head. (Courtesy of Cincinnati Milacron, Cincinnati, Ohio.)

CM90 machine (length = 27.7D; 90/177 mm diameter front/back; 125 hp dc drive) the output rate for some rigid PVC pipe is up to 1600 lb/hr* and for pelletizing over 2000 lb/hr. Energy economy as high as 26 lb/hp hr has been reported. See Fig. 2.8, a CM80 machine without the pelletizing head, and (Fig. 2.9), a CM90 with a pipe head.

The most difficult rigid PVC formulations, such as those containing a minimum amount of heat stabilizers, i.e., rigid PVC for food-contact applications, have been successfully compounded in these extruders. This is achieved mainly because of the following basic processing capabilities of these machines.

Excellent temperature control with uniformly short residence time
Minimal shearing and frictional heat accumulation, due to minimal
    clearance between flights
Efficient transport of materials
Streamline discharge
All the above contribute to a minimum color formation (yellowing)
    during compounding.

Tables 2.5 and 2.6 show some typical clear food-grade rigid PVC bottle compounds that have been successfully pelletized in these extruders. In the mid-1970s, the writer used the CM 111† (a precursor of the present extruders) to process a highly viscous (apparent melt viscosity, η_c at 100 sec^{-1} and at 450°F between 23 and 28 KP) methacrylonitrile styrene copolymer (90:10) that contained, except for 10 ppm of a phthalocyanine blue pigment, no other additives. The writer was very impressed with, among other things, the blue color cast of the extrudate, which indicated no or minimum color formation during the melt processing. A green color cast was normally obtained when extruding these materials in some other machines.

Krauss-Maffei [38], another manufacturer of these conical twin-screw extruders, has recently introduced a new screw design approach, the so-called double-conical, which is based on the changing depth of flight over the entire screw length. In conventional conical screws, the depth of the thread remains unchanged. Owing to the considerably greater volume in the feeding and preplasticizing zones of this new design, the dwell time and surface of the melt are increased, thus allowing the adjustable heating to become more effective. The greater reduction of the cross section toward the metering zone ensures a uniform compression of the material. The volume in the venting zone can be increased by as much as 50% so that higher outputs can be coped with.

---

*Output rate is based on a CM formulation with a bulk density of 32 lb/ft^3 running 10 in. SDR 26 pipe on a CM PH-1020 die.
†Length = 22D; 80/130 mm diameter front/back; 50 hp drive.

Table 2.5 Clear, Food-Grade Rigid PVC Formulation with Di-Octyltin Heat Stabilizer

Ingredient	Normal additive usage
PVC suspension homopolymer (K value 60 to 63)	100.00 parts
Acryloid K-120	3.00 phr
MBS (BTA III, Kane Ace B 28)	12.00 phr
Di-n-octyltin maleate	1.00 phr
Di-n-octyltin-S,S'-bis(isooctyl mercaptoacetate)	0.25 phr
Glyceryl monostearate	1.00 phr
Glyceryl monooleate Blue toner	1.00 phr

Table 2.6 Clear, Food-Grade PVC Formulation with Ca-Zn Heat Stabilizer

Ingredient	Normal additive usage
PVC suspension homopolymer (K value 60 to 63)	100.00 parts
Acryloid K-120	3.00 phr
MBS (BTAIII, Kane Ace B 28)	12.00 phr
Complex soaps of Ca-Zn[a]	2.00 phr
Epoxidized soybean oil	3.00 phr
FDA-sanctioned organic phosphites[b]	0.75 phr
Mixtures of unsaturated fatty acid esters	0.75 phr
Glyceryl monostearate	0.50 phr
Blue toner	

[a]These soaps may be used with organic nitrogenous stabilizers such as amino-crotonate esters and α-phenyl indole. Upon request, polyol-free (low plateout) versions are available.
[b]Such as tris-nonylphenyl phosphite.

Rigid PVC pipe has been extruded at 25% higher output over a conventional conical screw extruder with uniform-depth thread.

2.5.8 Two-Stage Compounding Extruders

As shown at the 1979 Kuntstoff Fair [39], a great proportion of the compounding extruders on display were specifically tailored to process PVC. Among these extruders, the so-called two-stage extruders were the items of most interest. All these machines are based on the separation of functions: mainly, the feeding, compression, and melting are conducted in one extruder, which is separated from the extruder metering the PVC melt. These extruders have separate drive motors, and the ability to vary the compounding rpm independently of metering removes some of the constraints on operating conditions mentioned above. The rpm difference between the two stages can be adjusted to the fusion characteristics of the particular compound, maximizing volume transport with minimal shear buildup.

Andouart, a French manufacturer of both single- and twin-screw extruders, has combined a twin-screw extruder as the first stage in an "L" configuration with a single-screw extruder. This unit has been specifically designed either to process bottle-grade rigid PVC compounds for drinking water, wine, vinegar, and other food-contact application, or for the dry-blend extrusion of rigid PVC bottles (the extrusion-blow application of these extruders is also covered in Chap. 5).

Another two-stage system for compounding rigid PVC formulations was recently introduced by Buss-Kneader. This system, built around a Buss-Kneader of 7D length, has a vertical regulating screw between kneader and discharge screw so that the degree of gelation can be adjusted. Vacuum degassing is located at the transition to the discharge screw. (More information about these machines is given later in this section).

EKK-Kleinewefers offers a different two-stage design approach. The EKK two-stage extruder is particularly designed for the preparation of PVC requiring a high degree of process technology, mainly food-grade PVC. The first stage consists of a planetary roller extruder designated as PWE; the second stage, of a single screw. In the first stage, the dry blend is conveyed into the feeding zone either by a dosing screw or by a force-feeding screw. A 3D feeding screw conveys the powder into the planetary roller system of short length, where it is plasticized. Depending on the machine size, the number of planetary screws may vary between 7 and 18 screws. These toothed screws are also driven by a toothed main screw, which intermeshes with them. While driven by the main screw, they float in the plastic material. One revolution of the main screw produces up to 1100 tooth engagements, and many thousands of rolling gaps are therefore available during extrusion for an intensive, continuous processing of the plastic material under optimum thermal conditions. The planetary screws, which also intermesh with a fixed internally toothed barrel, are retained in the planetary roller system by a stop ring on the outlet side. Easily exchangeable stop rings, having different internal diameters (i.d.) are available. For a fixed stop ring i.d., the dwell time and the back pressure can be varied by a manually adjustable main screw extension, which changes the distance of the outlet end. The desired degree of plastification is quickly reached by this dwell-time regulator. The efficiency of the planetary roller, system can be further increased by varying the number of planetary screws.

At the outlet end of the planetary roller system, fixed knives cut the cleft strands into short lumps. Vacuum venting in this transition area ensures that any volatile matter is extracted from the plastics.

The role of the second-stage extruder, which is in tandem but at a lower level than the first one, is to meter the molten polymer through a conventional pelletizing head. The L/D of the second screw, a two-start type, is 5. For all these machines, extrusion output for compounding rigid PVC ranges from 880 to about 6000 lb/hr. Power economy, varies between 7.8 and 10.0 lb/hp hr. Berstorff offers a similar system for pelletizing rigid PVC compounds. They are mounted in an "L" configuration.

Planetary gear extruders are also used as part of calendering lines to preplasticize rigid PVC dryblends. BASF, in their proprietary Luvitherm calendering process to produce unplasticized PVC films, uses one of these compounding extruders as the only source of the melt, which feeds a five-roll calender. An emulsion PVC resin of high molecular weight (K value 78) containing a small concentration of diphenylthiourea (0.2 phr) as thermostabilizer, 2.0 phr montan wax as lubricant, and three parts impact modifier, is the basic formulation for this film. Since unplasticized emulsion resins are not easy to process and the amount and efficiency the diphenylthiourea used as heat stabilizer is borderline, it is assumed that the planetary extruders BASF uses process these formulations under optimum thermal and plasticizing conditions. These compounding extruders, however, are not yet well known in the United States.

Kombiplast Compounding Extruder

The Kombiplast is one of the best known two-stage compounding extruders available. This unit, which was developed for compounding purposes only, consists of a twin-screw extruder unit feeding a single-screw section. Plastification, some venting, and homogenization are performed in the twin-screw extruder unit, featuring corotating, intermeshing, and self-cleaning screws with screw speeds up to 325 rpm. In the fluxing section, the screws consist of individual screw and kneading elements, which can be changed to suit specific compounding requirements of the various rigid PVC formulations. In this

section, there is no pressure buildup. Final plasticizing and pressure
buildup for extrusion are performed in the single-screw section, which
operates at a relatively low screw speed. Like all the other two-stage
compounding extruders, each stage is driven with its own motor and
is equipped with separate temperature-control systems. Some processing characteristics of the Kombiplast with rigid PVC dry blends follow.
The twin-screw compounding extruder possesses

Excellent feeding
Positive conveyance with close control over residence-time distribu-
    tion
Intensive shear in the kneading sections without pressure buildup
    and good control over intensity and location of shear
Pressureless transfer of the fluxed product into the single-screw
    discharge unit
Devolatilizing port at the transfer point

The single-screw discharge unit possesses

Carefully controlled pressure generation with a low degree of
    shear and good longitudinal mixing in the single-screw dis-
    charge unit
Streamlined flow of melt through the die plate
Pelletizing on the basis of hot melt cutting without cooling water

To process rigid PVC dryblend, the three most popular models are
(listed in increasing size) the KP500, the KP800, and the KP1500.
Following are some power economy data for these extruders. These
values, however, do not represent the maximum power economy attainable in these extruders.

Model	Typical power economy, lb/hp hr
KP500	11
KP800	10.5
KP1500	9.0

As shown, the smallest of these units produces the highest power economy.

Baker-Perkins MPC/V Compounding Extruder

Baker-Perkins offers another line of two-stage compounding extruders,
the MPC/V, recently revamped. The twin-screw extruder is corotating

and intermeshing and is coupled with a crosshead extruder provided with a strand die. Among the design changes are: (1) a rectangular clamshell barrel designed for electric-cartridge heating and cored for water cooling; (2) twin-screw shafts connected to the variable-speed drive through a single gear box (a combination of gear reducer and splitter); (3) two barrel valves that can be adjusted to restrict melt flow, thus intensifying mixing: these valves actually throttle the melt along the barrel; when partially or fully closed, the valves act like a barrier or reverse screw flight to restrict flow and thus enhance shearing and mixing [40]; and (4) a provision made for vacuum venting behind the feed section of the extruder where melt from the close-coupled mixer enters the extruder. These units are designed for adiabatic operation (minimum heat history).

Buss-Kneader

Although these machines are basically classified as single-screw compounding extruders, modifications in the screw design and extruder barrel are so extensive that, in this work, they are included in a separate category. Well known in Europe and Japan for at least 20 years, they are not yet well known in the United States. The screws of conventional single-screw machines, for example, have a continuous flight and a simple rotating motion. Most of the mixing is conducted within the screw channels. In the single-screw kneaders, the screw flight is interrupted by three gaps per turn. This provides the kneading tools on the screw. As the screw rotates, it simultaneously reciprocates. Corresponding with the gaps between the sections of screw flight are the kneading pins, or teeth, installed in the kneader barrel. When the plastics material passes through the narrowing wedge-shaped gap between the flight and kneading tooth, it is deformed or sheared. The rotation plus the reciprocation (axial oscillation) of the kneading screw produces a relative motion between kneading teeth and section of flights. This relative motion first carries the flights past the kneading teeth, which shears the stock trapped in the shearing gap. As the axial and rotary motion progresses, the kneading flights pass through the gaps between the teeth, shifting the stock around the kneading teeth axially to produce a local mixing effect. In addition, the laminar layers formed in the shearing gap are broken up intermittently, yielding a mixing action comparable to the lapping effect on mixing rolls.

Under normal operationg conditions, the kneading flights and the kneading teeth create a self-wiping system. The kneading flights wipe the interior surface of the barrel and all the kneading tooth surfaces.

The combined radial and axial mixing effect of the Buss-Kneader and similar units permits adequate processing in barrel lengths that are only one-half or even less than one-half the length required with conventional single-screw machines.

Starting from a standard processing length of 7D, the barrel can
ordinarily be extended modularly by 4D sections up to a total length
of 19D. All single-screw kneader barrels are manufactured using the
clamshell design for the purpose of inspection and cleaning.

Heat-transfer synthetic oil continues to be the preferred heat-
transfer medium, which circulates into individual heat transfer zones
to permit differentiated thermal control. Quite frequently heat is controlled by mixing oils of different temperatures. Both the cooling
and heating oil are thermostatically controlled.

The kneader screws, which are of constant flight depth, are divided
into feeding, compression, and plastication zones. The screw shaft
is cored for heating or cooling possibilities.

From the kneader the PVC melt is delivered to a slow-rotating discharge screw of short L/D and low or no compression ratio. In some of
the older versions, the discharge extruder (pelletizer) was in a cross-
headed position. In the modern versions, the pelletizer extruder is
in tandem with the kneader and a vertical regulating screw provides
the connection between the two units. The vertical screw, which was
developed for heat-sensitive materials like rigid PVC, provides control
over the degree of gelation. This model has degassing capabilities between the regulating and discharge screws. See Fig. 2.10, the WKG
10-14 (100 mm kneader, 140 mm discharge screw) and Fig. 2.11, the
KG20-25 (200 mm kneader, 250 mm discharge screw) machines.

The feed hopper is a conical shape with infinitely variable speed
control of the feed screw for agitating and metering the dry blend into
the kneader.

The writer was involved in the evaluation of an earlier version of
the Buss-Kneader, which was manufactured in Japan. The PR200
model had a G-45F gearing and an AS200 cross-head extruder with die-
face pelletizing head or granulator, which was used to pelletize a translucent rigid PVC compound containing no impact modifier or processing
aid. The pelletizing capacity was 600 to 770 lb/hr. Starting from a
cold system, about 4 hr was needed for heating up before material was
processed through the unit.

These units are said to offer the following processing advantages:

1. Short residence times, mainly with the shortest screws, i.e,, 7
length/diameter with minimum or no pressure generation.
2. Precise temperature control within a 0.5 to 1°C variation, which is
typical of oil-heating media.
3. Overall low-pressure operation. Even in the discharge screw, the
product pressure is very low.
4. Positive displacement of material.
5. Very compact units, high plant-floor utilization.
6. Low labor requirement.
7. High compound efficiency, ~98%.
8. Ease of disassembly. Color change cleanup is quick and easy.

Figure 2.10 WKG 10-14. (Courtesy of Buss Condux, Elk Grove, IL.)

Suspension PVC resins with high porosity, a K factor range of 66
to 69, and a bulk density between 0.45 and 0.55 kg/liter seem to produce a better compound gelation in the single-screw kneaders. Geon
103EP has been extensively used in Japan in these compounding extruders.

Lubricants and lubricating stabilizers are the additives that perhaps
affect processability the most in the kneaders. If dry blend does not
gel under the desirable mild conditions used in kneading, less lubricant
or a less efficient lubricant should be used.

Figure 2.11 Buss-Kneader KG 20-25. (Courtesy of Buss-Condux, Elk Grove, IL.)

Several potential processing problems have been observed during the operation of this earlier unit.

1. Powder leaks at the junction of the barrel halves on the main screw.
2. Output of the discharge extruder surges with every stroke of the main kneader. This results in varying length pellets and small amount of shavings and fines at the low point in the surge cycle.
3. Pellets from center holes ran slightly overgelled.

The power economy claimed by manufacturers with rigid PVC is about 0.027 kWh/lb, which is equivalent to about 27 lb/hp hr, comparable to other twin-screw compounding extruders.

Many innovations and design changes have been introduced recently that rectify the above-mentioned difficulties. With segmented screw elements, greater flexibility exists to optimize the process at relatively high screw speed. Since stationary teeth in the housing act as a second screw, there is a self-wiping effect achieved, as observed in the twin-screw systems. The main difference is that the kneading-mixing action is in an axial direction in the kneader as opposed to radial direction

observed in twin screw systems. Due to higher screw speeds and improved solids seal arrangement, the problems mentioned earlier in the
older units are eliminated or minimized.

By physically separating compounding from pumping through the vertical regulating screw, more control of gelation can be made, as well
as easy change from one type of material formulation to the other without hardware modification. Absolute control over temperature at each
step of the operation is essential in heat- and shear-sensitive materials,
such as rigid PVC, if long-term runs are to be achieved without burning the product.

2.5.9 Two-Stage Continuous High-Intensity Fluxing Mixers: Farrel Continuous Mixers (FCM)

These machines are a further development of the Banbury mixers. In
these machines, the first section of each rotor is the feed screw, which
runs in separate bores. The next section begins at the end of the
feed screw and extends as far as the peak, or "apex," of each rotor.
This section, helical in shape, is called the "forward helix" because it
pushes material forward toward the discharge end of the chamber. The
next rotor section starts immediately after the apex. It is also helical
in shape, but in the opposite direction, so it is called the "reverse
helix." This part of the rotor forces material back against the new
material coming in, which helps mix it more thoroughly. The material
is moved toward the discharge end by the combination of the reverse
helix, which forces softer material backward in the chamber, and the
forward helix, which presses stiffer material forward. The net result
is a forward motion of the mixture. The discharge zones, also called
the pump section, are shaped as smooth cylindrical continuations of the
kneading section. The two shafts are counterrotating and are run with
variable and different speeds. One design change recently introduced
in the FCM is a single-drive gear reducer that eliminates the connecting
gears on the rotors [41] reducing bearing and shaft loading as well
as the effects of mixing loads on the connecting gears. Another mixer
improvement is the adoption of a clamshell barrel, which permits steam
and water (standard temperature-control media) to be conveyed to the
barrel by rigid pipe instead of flexible hoses.

In the new two-stage versions, the back pressure and degree of
mixing are now controlled by the extruder-screw speed [42]. This
may result in a rather pressure-sensitive operation. Depending on the
degree of mixing required, the output rate can fluctuate over a considerable range. Furthermore, special attention has to be given to
start-up procedures and the buildup of the proper working volume for
each formulation, since these machines do not work well on the starve-
feed mode. The shearing forces generated by the rotors and the mixing chamber walls permit adiabatic operation of the FCM. To avoid
product overheating, the rotor shafts are hollow so that water cooling
can be carried out.

The continuous mixer and cross-head extruder have been close coupled by means of a short-heated transition piece. The transition piece eliminates the need for the adjustable discharge orifice previously used to control back pressure and mixing. It also eliminates the problems associated with exposing polymer to the air in the conveyor belt used to carry the melt from the mixer to the extruder. The present system has the mixer at right angles to the extruder.

In their special way, these machines guarantee good distribution and compounding of the various components of rigid PVC formulations. This is true for simple rigid PVC formulations. But for heavily modified compounds containing high levels of impact modifiers and pigments, the residence time in these internal mixers, at the higher output, is not always long enough to obtain the necessary degree of mixing and melting. Some modified rigid PVC formulations may require the lower shear and longer residence time possible only in a batch mixer. In the newer FCM, designated the FMX lines, the PVC melt is protected from air exposure, and potential oxidation, contamination, and heat loss. These units are more suitable to compound and pelletize some critical PVC formulations, such as some clear food-grade compounds.

Stewart Bolling & Co. and Japan Steel Works offer a similar line of two-stage, continuous high-intensity fluxing mixers. The Stewart Bolling unit has as its special feature the cross-head extruder's transverse screw, enveloping the discharge end of the rotors. This eliminates the need for a discharge orifice, which tends to give resistance to the melt flow. As in the FMX unit, pressure in the mixing chamber is controlled to some extent by the speed of the extruder screw. Rotors in this unit are driven from the discharge and near the mixing section where the most torque is required. The two-stage continuous fluxing mixers from Bolling are designated as the Mixtrumat series and are offered for high-output compounding only. The smallest unit in the series has an estimated output of 2500 lb/hr of rigid PVC. Energy economy fluctuates between 0.02 and 0.05 kWh/lb, or 33 to 14 lb/hp hr.

The CIM-S unit is the Japan Steel Works' new version of two-stage continuous high-intensity fluxing mixers. They are quite similar to the Farrel and Bolling units already described. The plasticated melt is conveyed to the melt extruder directly below the mixer by means of a discharge chute. The chute can be vented.

2.5.10 Compounding Lines Based on Batch-Type Internal High-Intensity Fluxing Mixers

In batch-type, hot-melt compounding processes centered on Banbury internal high-intensity fluxing mixers and two-roll mill mixers, great flexibility of operation exists. Also, compounding lines built around these mixers, which were inherited from the rubber industry, show great variability in the number of components. Not only the actual arrangement of these mixers varies but also the type and number of auxiliary equipment. The writer has seen compounding plants with six of these lines and no two were alike. Some of the common equipment arrangements are

  Ribbon blender-Banbury internal mixer-two-roll mill-dicer at the same level, on the same floor. This is common in smaller lines.

  The ribbon blender and the Banbury in the same floor above the two-roll mill and the dicer.

  The ribbon blenders in one floor, feeding by gravity a Banbury below, which in turn discharges onto the two-roll mill at the ground floor, where another set of rolls or a short extruder (strainer) with a strand die is also located. This arrangement is typical of the lines based on automatic Banburys.

Despite the increasing popularity of continuous compounding equipment, batch-mixing machines still have their place [43].

Regardless of these line variations, a good-quality rigid PVC compound can be manufactured in any of these lines. They are less sensitive than any other compounding lines to resin and dry blend property variations. Thus, resins with a wide range of particle sizes, moisture content, and bulk density, as well as dry blends, cold or warm, showing also a wide range of properties, can be successfully processed in these compounding lines.

Key elements to the success of these compounding lines include

  Availability of fast and reliable quality control tests, which could be conducted by the line.

  Good correlation between the tests predicting degree of compounding (fusion) and the extrudate quality obtained in the subsequent melt-processing step.

  Reliable and accurate pyrometers to check Banbury drop temperatures, and roll mill surface temperatures. Stock temperatures in the Banbury as well as during the mill operation should be accurately controlled within the prescribed limits.

Two simple quality control tests for pellets, the so-called acetone immersion test and the squash test, can be of great help in adjusting Banbury operating conditions. In the acetone immersion test, the pellets, e.g., 0.5 g pelleted compound, to the nearest pellet, are immersed in a mixture of three volumes of *dry* acetone and one volume of *dry* methyl ethyl ketone (MEK). The corked tube is turned end over end through 360° once every 5 sec (a total of 60 revolutions) for 5 min. This operation can be done by hand, but more reproducible results are obtained if it is done on a motor-driven wheel turning at 12 rpm. Immediately after the last turn, pour off the supernatant liquid and flocculated compound (snow) into a clean, empty tube, leaving behind the pellets. The amount of snow generated is inversely proportional to the degree of compounding. For semiquantitative assessment, immediately after the last turn, pour off the liquid and snow into a clean,

empty 13 X 100 mm test tube. Stopper the tube and compare the concentration of suspended snow with snow-concentration in a set of previously prepared sealed test tube standards. These standards are
made up from a series of pellet samples spanning the range from under
compounded to over compounded, as observed from their performance
in the extruder.

The conditions of the test may be varied to suit specific individual
requirements. It is more suitable, however, for diced pellets than for
cylindrical or lentil-shaped pellets. The edges of the diced pellets
provide an accessible area for solvent attack. The composition of the
test solvent, the degree of agitation, and the ratio of compound to solvent are conditions that can be varied and then standardized when a
satisfactory arrangement is reached. The test is run at room temperature. Figure 2.12 shows a set of tube standards, representing, from
left to right, extreme undercompounding to extreme overcompounding.
Except for the loss of solvent, the standards have not shown a tendency to change over many months of use.

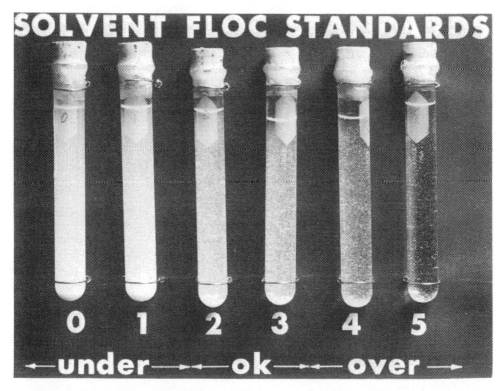

Figure 2.12 Solvent floc standards.

In Europe, however, many prefer to use methylene choride in place of anhydrous acetone, feeling it is a more severe test. Any moisture present in acetone severly hinders its solvating power, resulting in a poorly fused compound, appearing to pass the test.

Shrinkage Index of Rigid PVC Pelletized Compound

This is another test that provides a semiquantitative method for the degree of compounding of rigid PVC pelletized materials. This, in turn, is indicative of the extrusion performance and, specifically, of the extrudate quality. Visual examination and measurement of shrinkage of a thin slab, made from rigid vinyl pellets under controlled conditions of temperatures, pressure, time, and physical dimensions, makes it possible to determine whether the rigid PVC has been over- or underworked during processing.

As expected, there are many variations of this test in the field. The test procedure described here is a simple one, which could be run by the compounding line while the line is in operation. A hydraulic press, with 8 X 8 in. platens, fully cored for both high pressure steam (140 psi) and water and a mold 8 X 8 X 3/8 in., consisting of steel plates hinged at the rear, are the only two pieces of equipment needed for this test. The inside surface of the upper mold plate is flat; the inside of the lower plate has a 5 in. diameter depression in the center, approximately 30 mils deep. This is surrounded by a deep well to take the excess compound as flushing.

To run this test, the press is always kept at 175 ± 1°C and the mold is always preheated for at least 2 min for a series of tests. With 25 of pellets charged into the center of the mold cavity, the mold is closed and returned to the press, where pressure immediately is built up to 40,000 lb (~10 sec). At 40,000 lb, pressure is held for 40 sec. The pressure is then released, the mold is removed and opened, the molding flash is cut off, and the molded disk is placed for a minimum of 30 sec between polished plates.

For reading the shrinkage index, three guide marks with wax pencil, approximately 90° apart around the edge of the specimen, are made on the specimen. The two opposite guide marks should be on straight line through the center of the specimen. The 5 in. mark of the rule is placed on the edge of the opposite guide marks (mark 1) with the 0 to 5 in. section of the rule on the disk and the zero end of the rule at the other opposite guide mark (mark 3). Note the reading at this point to the nearest 1/32 in. and report the shrinkage index as the number of 1/16 in. units from the zero mark on the rule to the edge of the disk. Repeat the shrinkage index measurement, starting with the point midway between marks 1 and 2, also at mark 2, and at the point midway between marks 2 and 3. Average the four values thus obtained.

Shrinkage index disks are also compared to standard sets and graded from 1 to 5 for uniformity. Grade assignments are very subjective. Contrast of dull and shiny areas, super rough areas, and off-color pellets (yellower) are used for the grade rating. Obviously, overworked rigid PVC compounds show higher shrinkage but also the highest variability on disk surface uniformity.

A typical squash test specification for a rigid PVC compound with an adequate degree of mixing area follows:

1. Pellet uniformity, maximum 3.0
2. Shrinkage index, minimum 3.0; maximum 6.5

Ribbon Extrusion Test

Another test frequently used to assess the degree of compounding is the ribbon extrusion test. In this test, about 5 lb of each compound is processed in a 1 in. 20/1 L/D extruder, provided with a 1/4 X 1/2 in. ribbon die or similar ribbons. Once the ribbon is cooled, it is compared with the standard ribbons. The degree of surface lumpiness and gloss are normally used to assess the degree of compounding. This test, however, is more reliable to detect overworked than underworked compound. Another disadvantage of this test is that it must be run continuously to keep all the extrusion conditions constant, which contributes to a significant use of material at the end of a compound run.

These valuable quality control tests, although discussed in this section, could be adapted for any of the melt-mixing compounding techniques described so far. The reasons they are included at this time are that they are indispensable in a batch operation to adjust conditions during start-ups, and also, since the Banbury compounding lines are, for the most part, in different locations than the extruders, the use of these tests tend to minimize off-grade compound and scrap generation.

Banbury Operation

The most critical factor in Banbury operation is control of the total heat and shear history of the compound up to the point where it is dropped to the rolls below. These factors can be kept under control by using the true stock temperature at drop as a guide and by using any of the quality control tests mentioned above on the pelleted compound as primary indicator of the integrated heat-shear history. For example, if the acetone-MEK test is used, too much snow (pellet flocculation) calls for longer mixing time and higher stock temperature at drop. Very low or no flocculation requires a downward adjustment of stock temperature at drop by using a shorter mixing cycle. In order that stock temperature may be used in this manner as a guide, it is necessary that the temperature and rate of flow of the Banbury cooling water, the steam or hot water temperature through the jacket, and the ram head pressure be kept constant at all times. The Banbury charge, which is a function of the bulk density of the dry blend, should be adjusted accordingly. To keep the degree of mixing in the Banbury under control, it is necessary to maintain the Banbury batch size within a narrow range. For example, for a 3A Banbury and for a dry blend requiring an optimum Banbury batch size of 160 lb, it is recommended to keep the load within plus or minus 5 lb. Banbury batches larger than 170 lb produce a somewhat dry drop, with a profusion of small chunks. Banbury batches smaller than 150 lb might deliver somewhat overworked Banbury drops.

It is also recommended to use Banbury drop temperature as another indicator of the degree of mixing. It should be adjusted, however, according to the Banbury drop consistency. These adjustments on Banbury operating conditions are conducted at the beginning of each run, and it is wise to do so while flushing the line with the same dry blend to be compounded.

Product quality is dependent upon doing a good, intensive mixing job in the Banbury without either excessive heat buildup or excessive shear. These conditions are extremely important in compounding lines with only one set of rolls downstream, since the Banbury is called upon to carry out the major share of the mixing job. In these compounding lines, the Banbury drop load must be delivered to the rolls with the longest possible mixing time in the fused state, while still avoiding overworking of the stock.

Mill-Roll Operation

Compounds containing high loads of TiO_2 or other fillers and a modest concentration of lubricants are very sensitive to mill-roll operation. In these formulations, like the ones used for residential siding and its accessories, the mill-roll operation becomes the most critical phase of the whole compound process. Great care should be taken to minimize residence time of the stock on the rolls. Mill-roll surface temperature must be accurately controlled within the stated limits. High mill-roll heat is the cause of "nerve" or loss of plasticity in the compound, resulting in extruded products with excess orange peel in the surface and requiring a drop in extrusion rate for correction. A compound that has suffered too high an exposure to heat on the mill rolls is difficult to extrude; also, it is more difficult to set the proper gage and dimension in the extrusion process.

In mill-roll compounding, as in Banbury mixing, one must select conditons and techniques of processing that provide a maximum of mixing or homogenization with the least possible heat history and, particularly, with the lowest "peak" temperature. It is a matter of experience that stock that reaches a peak of T°F just prior to being stripped from the rolls and cooled will have built into it strains that require the compound,

on later melt processing (extrusion or injection molding), to reach at
least T°F in the extrusion process to release these strains; otherwise,
these strains show up as an irregular surface in the extruded product.

The mill-roll clearance must also be accurately controlled within the
stated limits; otherwise, we have no control over the amount of frictional
heat. If two sets of mill rolls are used, the clearance of the mill roll
directly under the Banbury is narrower at the side where banding occurs to help fusion and maintain more uniform feeding of the ribbon to
the other set of rolls. To reduce ribbon breaks between the two sets
of rolls and between the rolls next to the dicer and the dicer, the mill
roll operator should work the stock adjacent to the ribbon by cutting
and rolling up small sections of the banded stock and feeding it end-
wise to the roll nip in the area of the ribbon.

First Set of Mill Rolls

This is the set directly under the Banbury mixer. Normally, rear roll
surface temperature, as checked by an accurate surface pyrometer,
is kept lower than the front roll. For example, if the front roll surface
temperature is 290 to 300°F, the rear roll is maintained between 260
and 270°F. Stock temperature depends on the above temperature setting, mill-roll clearance, mill-roll rpm, and the formulation. A typical
mill-roll clearance on the side where the ribbon is cut is 0.160 to
0.180 in., and the other side is 0.170 to 0.190 in.

Second Set of Mill Rolls

This is the set of rolls near the dicer. Again, the rear roll surface
temperature is kept lower than the front. For example, if the front
roll surface temperature is 280 to 290°F, then the rear roll surface
temperature is 250 to 260°F. As soon as a blanket is formed, the mill-
roll spacing in the nip is adjusted to a tighter clearance on the ribbon
side, about 0.010 in. less than the other side. The normal nip gap
on the ribbon side is between 0.145 and 0.165 in. About a 5 to 7 in.
width ribbon is threaded to the water bath before reaching the dicer.
The section of the ribbon that must be immersed in the water bath is
determined by the surface temperature reaching the dicer as well as
the kind and size of the dicer.

Roll inventories must be kept low, particularly in the stripping
area. Cold stock in this area causes ribbon breaks. Large pieces
should not be fed to these rolls. In emergencies, small pieces may be
transferred to the ribbon area to keep the line operating. Ribbon
widths may be adjusted accordingly.

Cooling and Dicing

Some water cooling of the rigig PVC ribbon before dicing is necessary to
facilitate dicer operation. A cold and thick ribbon (thickness ≥ 180 in.)

will jam the dicer, and too hot a ribbon causes a very high pack-out
temperature. The following highlights some of the more pertinent processing points of cooling and dicing operations.

1. For the start-up the water bath should be at about 100°F.
2. The ribbon should pass through the water bath, through the air-
knife wipers, and into the dicer.
3. After starting, the ribbon heats the bath, and it will be necessary
to add cooling water to maintain the desirable water temperature (105
± 5°F is commonly used).
4. On breaks, the ribbon must be kept tight in the water at all times.
A loose ribbon quickly develops kinks, making it difficult to pass
through the rollers and feed to the dicers.
5. Air jets should be used at all times, in order that the ribbon passing
   to the dicer should be dry to the touch.
6. Any tears or holes in the sheet should be watched at the dicer.
Loose ends tend to curl, causing jamming of the dicer.

Dicing is commonly produced by feeding the ribbon by powered
pull rolls over a saw-toothed bed plate set at a 45° angle to the line
of flow. A number of meshing fly knives rotate against the bed plate
cutting the ribbon between them. Since the edges of the cuts are
parallel and at right angles to the edge of the belt, there is almost no
waste of material. Rigid PVC compounds containing high levels of TiO_2
and fillers such as calcium carbonate tend to dull the dicer blades, producing tailing and stringers, which cause pelletized compound to bridge
in hoppers. Dicer blades must be kept sharp and properly aligned to
produce clean cubes. Normally, there is some formation of a small
percentage of fines and oversize, which must be screened out of the
compound. Normal specification for pellet size is 98% passing U.S.
Standard No. 4.

Pack-out

A pack-out temperature over 140°F should be avoided. Compounds for
building products application show appreciable change in color when
pack-out temperatues over 140°F are used.
    If one considers the

1. Power, electric, and steam needed to operate these compounding
lines
2. Labor involved
3. Testing and frequency of testing
4. Floor space taken by the complete line

the overall line efficiency and power economy (pound/horsepower
hour) of these batch-type compounding lines are rather low. It has
been reported [44] that continuous mixers use only 60 to 70% of the energy required in high-intensity batch mixing.
    When compared with continous melt-compounding equipment, the high-
intensity batch mixers described above show significantly longer resi-

dence time. It has been estimated that, from hopper to exit, continuous mixers of the twin-rotor type may have a cycle of 1 to 2 min, compared with 3 to 4 min for an equivalent batch operation. The shorter residence time is advantageous for some glass-clear rigid PVC formulations in which color formation could be objectionable. In opaque rigid PVC formulations, containing high loads of TiO_2 and impact modifiers, like residential siding, lower shear and longer residence time, however, may be more advantageous.

A more consistent product quality is also characteristic of continuous mixers, since they show only slight fluctuations in amperage usage. Batch types, and mainly the Banbury mixer, start low and swing high when fluxing occurs. Furthermore, the stock on the rolls is always exposed to temperature gradients from side to side, an uneven degree of mixing, and nonuniform residence time.

This group of batch-mixer units offers perhaps the maximum versatility for compounding rigid PVC formulations, as it is capable of handling the most varied mixing chores, like combining resins of different particle sizes, bulk density, and molecular weight; rubber materials, either in crumbs or as graft copolymers; small amount of plasticizers or waxy additives; variable loads of inorganic pigments and fillers; and variable amounts of rework. The key to this flexibility is the Banbury type of batch mixer, in which the fluted rotors constantly reorient the mass. As an intensive mixer, it is probably the only mixing tool that approaches what can be considered an ideal mixer. This explains why the Banbury rotors design approach was incorporated into the continuous, intensive, high-shear mixer.

The writer worked with several of these batch-type mixer lines, compounding a series of commercial rigid PVC formulations. Some of these compounds follow:

1. A translucent, intermediate impact, ultraviolet-stabilized rigid vinyl compound based on a Ba-Cd-Pb heat stabilizer system and a graft copolymer of chlorinated polyethylene as the impact modifier.
2. An opaque, intermediate impact, white rigid vinyl compound based on a Ba-Cd heat stabilizer and a graft copolymer of chlorinated polyethylene as the impact modifier. This formulation is for the profile extrusion market.
3. A Type II high-impact rigid PVC formulation similar to compound 2 but containing higher levels of the chlorinated PE graft copolymer. This compound was available for the profile extrusion market.
4. A Type I white opaque rigid PVC formulation for outdoor applications based on a lead system as the heat stabilizer and containing an EVA graft copolymer as the impact modifier.
5. A glass-clear formulation based on mercaptotin stabilizers, containing no impact modifier for indoor application.
6. A glass-clear food-grade PVC compound containing di-n-octyltin maleate and di-n-octyltin mercaptoacetate as heat stabilizers and MBS (terpolymer of methylmethacrylate-butadiene-styrene) as the impact

modifier. This compound is for bottle applications.

7. A glass-clear food-grade PVC compound similar to compound 6, but containing Ca-Zn heat stabilizer instead of the di-n-octyltin maleate.

8. An opaque, pastel-colored rigid PVC compound, containing high loads of TiO_2, tin maleates as the heat stabilizers, and a chlorinated PE graft as the impact modifier. These compounds are for outdoor profile applications.

In a few words, changes from translucent to opaque, from unfilled to highly filled compounds, from formulations containing no rubber to rubber-containing compounds, different type of rubbers, either grafted or mechanical polyblended, and so on, can be handled with relative ease.

Rigid PVC Compound Forms

The pellets manufactured from the compounding equipment described above are available in the following distinct shapes.

> Face-cut pellets: round shape and lentil shape
> Cylindrical pellets
> Diced pellets

Face-cut Pellets The various types of face-cut pellets are produced by extruding through a multiple hole strand die and slicing the strands with rotating knives that wipe the surface of the pellets, as described elsewhere in this chapter. By adjusting extrusion rate and blade speed, one can vary the length of the pellets. At the same extrusion rate and knife speed, however, pellet shape may vary with the compound formulation, which determines the degree of postextrusion swelling. Also, in this kind of process there is an appreciable size variability between the pellets from the outer and inner strands. Cutting is carried out under an air spray as the cooling medium to prevent the pellets from sticking together after they are formed. This method of pelletizing is very well suited to the rigid vinyl field and is the one offered by most of the manufacturers of twin-screw extruders. Any type of rigid PVC formulation can be readily handled, including formulations containing no impact modifiers. Also, lentil-shaped pellets are preferred by some of the European compound manufacturers. Smaller-size pellets require less screw torque and produce faster fluxing in processing equipment.

Cylindrical Pellets These are also manufactured by extruding round strands through a multiple hole die. The strands are successively air or water cooled, pulled by pinch rolls over a stationary bed knife, and sliced by rotating fly knives. Rigid PVC compounds, even the unmodified ones, can be easily pelletized using this process. Banbury compounding lines, using extruders with strand dies (strainer) instead of dicers, produce this kind of pellet. Also, compounding extruders,

using strand dies with holes only in the center of the die, thus producing fewer strands, use this pelletizing system.

Diced Pellets Cubical or diced pellets are produced, as explained before, dicing the continuous ribbon, which is fed from a set of two-roll
mills or from a continuous flat belt extruded in a strainer with a flat
die. They can be also produced from compounding extruders equipped
with a flat die. For reducing screw torque and for faster fluxing in
processing equipment, the smaller-size cubes are preferred.

Pellet Post Lubrication Although not a remedy for all maladies, a
technique that has been successfully used to attentuate the problems
associated with the extrusion of somewhat overworked pellets is that
of postlubrication of pellets. Common extrusion problems, such as
presence of flow lines in extrudate, frequent breaker plate changes,
presence of gray streaking in opaque, white profiles, and high amperage readings, can be reduced, using this simple technique. Furthermore, platcout due to the presence of certain lubricants can be
reduced by lowering lubricant level and using postlubrication. Wax
E, a montan acid ester, is one of the most effective lubricants for this
applicaton. A typical formulation for post lubrication is octyl epoxy
tallate 0.025%, (Drapex 4.4 or similar), and Wax E, 0.045-0.070%.
The blending procedure is as follows. Drapex 4.4 is added and blended
with the pellets for 10 min. At this point, the Wax E is sprinkled on
the surface of the wet pellets and blended for 20 minutes. Any of the
eqipment used for blending pellets with pigments in powder form can
be used for this application.

2.6 SUMMARY

So far, we have described a series of machines and processes, continuous or batch, that are used to compound rigid PVC formulations.
The list could go on and on. Rigid PVC, if properly modified, has
such a wide processability range that almost any melt-mixing apparatus
in the field may be adapted to compound these materials. This is seen
in the field, where the compounding equipment varies greatly from
plant to plant.

   As would be expected from such a product, there is also great
variability in processing equipment: from the short-barreled, low-
compression screw extruders, which can do little more than remelt
the compound and meter to the die, to the long-barreled, higher-com-
pression screws, which can process from partially fused compacted
pellets to dry-blended formulations. All the above leads us to the conclusions that there is always available a suitable processing equipment
to process any rigid PVC compound, regardless of its shape and degree of fluxing.

Despite their high initial cost, the present trend in the industry, however, is for the resins suppliers and custom compounders, as well as some larger processors of finished products, to develop their own compounding process built around specialized continuous mixers.
Such a trend results in the development of the most economic production methods. These lines offer very high power economy, and the potential of trimming down raw materials and conversion costs.

REFERENCES

1. Plastics Compounding *January/February*, 1980.
2. Society of the Plastics Industry (SPI), Monthly & Annual Statistical Reports, 1979, 1980.
3. Modern Plastics *January*:78, 1980.
4. Predicasts, Inc., Plastics World *March*, 1980.
5. R. Tregan and A. Bonnemayre, Plastiques Modernes et Elastomeres *23*:220, 1971.
6. A. R. Berens and V. L. Folt, Polymer Eng. Sci. *January*:1968.
7. J. W. Summers, et al., Polymer Eng. Sci. *January*:1980.
8. *Information Chimie*, 1979/09.
9. B. D. Bauman, *ACS/CSJ Conference*, Honolulu, April 1979.
10. W. D. Davis, *In Encyclopedia of PVC*, Vol. 1, Chap. 5, p. 189. Marcel Dekker, Inc., New York, 1976.
11. V. R. Struber, *Theory and Practice of Vinyl Compounding*, The Argus Chemical Corporation, New York, 1968.
12. D. Braun, Pure Appl. Chem. *53*:549-566, 1981.
13. Plastics Eng. *November*:6, 1980.
14. Modern Plastics *59*(9):62, 1982.
15. Plastics Compounding *November/December*, 1981.
16. K. Worschech, and R. A. Lindner, *SPE ANTEC 39th*, May 4-7, 1981.
17. J. L. Dunn and M. H. Heffner, Outdoor weatherability of rigid PVC, *SPE ANTEC 31st* May 7-10, 1973.
18. G. L. Levy, M. H. Heffner, and R. C. Gross, Outdoor weatherability of rigid PVC. Part II, *SPE ANTEC 33rd*, 1975.
19. Lead stabilizers push into siding market, Plastics Technol. *February*:1976.
20. L. R. Brecker, Pure Appl. Chem., *53*:577-582, 1981.
21. Interstab; Plastics Technol. *July*, *June*:14, 1979, and *July*:68, 197ȿ
22. Plastics Technol. *July*:73, 1981.
23. J. T. Lutz, Jr., Plastics Compounding *January/February*:34, 1981.
24. Amoco Chemicals Corporation Resin 18, Bulletin R-25a, Chicago.
25. L. F. King, and F. Noël, Polymer Eng. Sci. (2):112, 1972.
26. G. Illman, Waxes as lubricants in plastics processing, SPEJ, 121:71-76, 1967.
27. E. L. White, Lubricants, in *Encyclopedia of PVC*, Vol. 2. Marcel Dekker, Inc., New York, p. 650.

28. B. Terselius, and B. Ränby, Pure Appl. Chem. *53*:421-448, 1981.

29. Wacker Chemie G.M.b.H., Modern Plastics Int. *October,* 1979.

30. A. T. Murray, Plastics Technol. *November*:83, 1978.

31. A. T. Murray, Plastics Technol. *December*:65, 1978.

32. Plastics Compounding, *January/February* :20, 1981.

33. C. Guimon, Improvement in the extrusion of rigid PVC, SPE ANTEC *25th*, May 15-18, 1967.

34. G. A. Thacker, Jr., *Rigid PVC Extrusion*, Cincinnati Milacron Chemicals, 1976.

35. G. M. Gale, RAPRA Members Journal *November*:273, 1974.

36. Littleford Bros., Inc., Bulletin 210, Florence, Kentucky.

37. F. Martelli, Plastics Compounding *March/April*:69, 1980.

38. Krauss-Maffei Austria AG, *Plastics Processing Int.* 1980.

39. Plastics Compounding, *November/December*, 1979.

40. Plastics Compounding, *November/December*, p. 35, 1982.

41. Plastics Technol. *September*:132, 1978.

42. Plastics Technol. *December:*68, 1978.

43. Plastics Compounding *January/February:*20, 1981.

44. Plastics Technol. *November:*78, 1978.

3

Rigid PVC Rheology

I. LUIS GOMEZ /Monsanto Company, Springfield, Massachusetts

3.1   Introduction   99

3.2   Nonnewtonian Flow Behavior of Rigid PVC Melts   102

3.3   Ideal or Newtonian Fluid Concept of Viscosity   102

3.4   Rigid PVC Melt Rehology   104
      3.4.1 Degree of Departure from Newtonian Fluid Behavior   104
      3.4.2 Rheometers   105

3.5   Power Law   113

3.6   Factors Affecting Melt Viscosity, Postextrusion Swelling, and
      Melt Quality   118
      3.6.1 Effect of Crystallinity   120
      3.6.2 Complex Modulus Behavior of PVC   128
      3.6.3 Effect of Fillers   129
      3.6.4 Melt Fracture   131

3.7   Melt Stability   135

3.8   Color Propagation Curves   138

3.9   Glass Transition Temperature T_g 138

3.10 Melting Point T_m 140

3.11 Temperature Dependence of Viscosity   141

3.12 Pressure Dependence of Viscosity   145

References   149

3.1 INTRODUCTION

In Chap. 2, the dry blend preparation processes, the properties of
these dry blends, and the various melt-compounding techniques used
in the field to process rigid PVC formulations were discussed in some
detail. The quality of the finished product, however, depends not
only on the polymer preparation aspect but also on the efficiency of
the melt processing and shaping tools, such as extrusion dies, injection molds, and blow molds. Especially critical is the need to understand and be able to relate PVC melt flow pressure and temperature
and deformation properties to optimum screw, die, and mold design.

This chapter deals with certain aspects of polymer rheology and
gives a general outline of how the rheological properties of rigid PVC
compounds help one to understand flow behavior in screw extruders,
die adapters, dies, injection molds, and so on. The melt properties of
rigid PVC under shear at processing temperatures and pressures, for
example, is useful information for die designers. Mainly, this information helps one define an appropriate taper flow path in the body of
the die that could achieve an acceptable pressure drop while minimizing potential flow defects (melt fracture). The judicious use of tapered flow channels, in turn, contributes to: (1) considerable increase in the output rate Q of the extruded products, (2) extrudates
free from surface rupture (melt fracture), (3) longer extrusion runs
because of fewer undesirable deadspots, and so on. The choice of taper angle, however, also depends on the tensile properties of the
melt as well as on its shear properties, because tensile stress in the
flow direction reaches a maximum value at the narrow end of the taper, where shear rate also is maximum. If this tensile stress exceeds
the tensile strength of the melt, rupture is very likely to cause melt
fracture.

In the extrusion of viscous polymer melts, like rigid PVC, traversing thermocouples within one screw flight commonly show melt temperatures increasing significantly as one penetrates inward from the
inside wall of the extruder barrel. This means that polymer next to
the screw is hotter than it is next to the barrel wall, and these temperature differences increase with increasing screw speed (rpm).

Such large temperature gradients have several effects. First, the
exposure to high temperatures of the polymer passing through an extruder increases, resulting in eventual adverse effects for thermally
sensitive materials like rigid PVC. Second, the melt delivered by the
screw will not be at uniform temperature, although some equilibration
of melt temperature occurs in the extruder head. Such temperature
inhomogeneity can create quality problems in subsequent molding and
forming operations. Finally, the calculation of screw performance
characteristics, if one assumes a uniform radial temperature, can be
significantly in error. The above-stated comments, however, relate
more to extrusion in single-screw machines than for multiscrew extrusion.

The existence of a radial temperature gradient in the melt can be
modeled as a composite of concentric polymer layers at various tem-

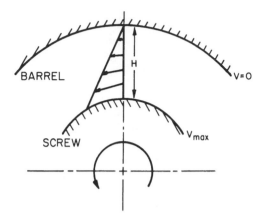

Fig. 3.1 Velocity gradient across the melt at the metering zone.

peratures, the temperatures decreasing from the screw root to the
stationary barrel. One can also conclude that these layers travel at
variable speeds, which also decrease from the screw root to the barrel. Consequently, each incremental layer differs from adjacent layers in its rate of flow. The hotter layers move faster than the colder
ones. This sets up a radial velocity gradient across the melt, which
also corresponds to the temperature gradient. At this point, we can
assume that the rigid PVC melt in the metering section of the screw
has the following overall characteristics (see Fig. 3.1).

Laminar flow.

The PVC resin particles surrounded by the dispersed additives and
modifiers move in parallel planes.

The velocity of any of these planes is proportional to its relative
position from a fixed parallel plane, e.g., the extruder barrel.

Before continuing this discussion, let us consider how melted rigid
PVC polymer flows in any extrusion process. At the metering zone,
where the melt temperature is normally above 182°C (360°F), the polymer is way above its glass transition temperature but below its crystalline melting point (>210°C). The force applied by the motion of the
screw, which in turn is produced by the extruder drive, moves the
molten polymer away from the screw toward the die area to relieve the
stress. As the polymer is forced to move, it tends to untangle and
orient itself parallel to the direction of the flow; thus, the polymer
chains tend to separate from each other. This motion reduces the van
der Waals forces, which, due to the high polarity of PVC, are rather
high. For the resin particles to move, they must slide over each
other. As the screw rpm continues to increase, higher orientation of
the polymer molecules also occurs, while the van der Waals electro-
static forces continue their decrease. There is also a breakdown of
the original resin particles. A point can be reached where an increase of shear rate will produce a greater increase in the shear
stress by several units.

3.2 NONNEWTONIAN FLOW BEHAVIOR OF RIGID PVC MELTS

In the previous section, from a measurable parameter, which is the temperature gradient across the melt, we visualized the melt flow behavior of rigid PVC in a metering zone of an extruder at open discharge as well as an implicit concept of its melt viscosity. Almost without exception commercial rigid PVC compounds, in the melted, fused state, exhibit nonnewtonian flow behavior over the entire range of shear rates experienced in the various commercial melt-processing operations. The use of the idealized newtonian fluid model, however, represents a convenient starting point in characterizing the flow behavior of rigid PVC polymers, since melt studies on PVC resins at either elevated temperature (amorphous flow region) or prolonged time periods are difficult. Invariably, stabilizers and lubricants must be incorporated into the resin, and the resultant flow measurements and die swelling properties reflect the behavior of a stabilized resin or compound. The nonideal behavior of rigid PVC melts is best defined and visualized in terms of deviation from these ideal conditions. Furthermore, many of the concepts developed for the idealized materials prove to be of value in interpreting the behavior of true polymers.

Due to the presence of chlorine in the monomer units, PVC resins exhibit rather high polarity. This polarity, which is desirable because it contributes to very attractive properties, such as low permeability to water and gasses and high chemical resistance, is however, one of the sources of some of the difficulties encountered when investigating this polymer. Thus it has been demonstrated by Utracki [1] that the polar forces are also responsible for the particular behavior of PVC above the glass transition temperature, $T_g \sim 80°C$.

In melt rheology the word "stress" is not used in the sense of a force acting on a melt. Rather, stress is the internal resistance of a melt to an applied force. In rigid PVC melts, this resistance is mainly caused by the attraction of the pendant chlorine atoms, which is reinforced by other molecular bonds and forces. Thus, to achieve a faster flow of polymer through an extrusion head, the resistance to flow of the multiple layers of molecules must be neutralized and overcome. Shearing stress is the measure of this resistance to flow. In ideal fluids (newtonian), it is assumed that the shearing stress τ producing the distortion of any volume element of the fluid (shearing strain) is directly proportional to the rate of shear $\dot{\gamma}$.

3.3 IDEAL OR NEWTONIAN FLUID CONCEPT OF VISCOSITY

The simple Fig. 3.2 illustrates the behavior of an ideal fluid under the influence of a shearing stress F/A. In this experiment it is assumed that the fluid fills the space between the two parallel plates, the movable plate of area A and the fixed plate. When the shearing

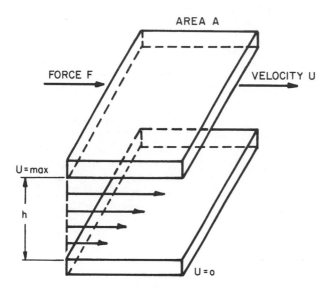

Fig. 3.2 Concept of newtonian viscosity.

force is imposed on the upper plate bounding the fluid, it moves one
plate relative to the other at a constant velocity u, as long as the force
remains constant. This force acts perpendicularly to one side of the
upper plate. Once the shearing force is removed, however, the fluid
exhibits no tendency to return to its initial shape.

   It is not hard to assume that the layer of liquid next to the stationary plate has zero velocity, while the one in immediate contact with
the upper plate has velocity u in the direction of the force. Since the
shearing force is transmitted uniformly through the liquid to the lower plate, each layer of height h moves relative to the next with a uniform change in velocity with distance. The velocity of any layer is
then proportional to its distance from a fixed parallel plane, say, u =
kh. The rate of change of velocity across the liquid is the slope of
the line connecting the velocity vectors, du/dh, which is the velocity
gradient or shearing rate of the fluid. For a given force, the velocity
of the moving plate is directly proportional to the applied force and
inversely proportional to the area of the moving plate, since for a
given force, the larger the area the slower the velocity. This can be
written as follows:

$$F = fA \frac{du}{dh}$$

The proportionality constant f denotes the coefficient of viscosity of
the fluid and is designated μ for a newtonian fluid and η for a nonnewtonian fluid, such as a rigid PVC melt. The previous equation

may also be written in another manner as follows:

$$f = \frac{F/A}{du/dh} = \frac{\tau}{\dot{\gamma}}$$

The numerator is the shear stress τ, and the denominator is the shear rate $\dot{\gamma}$. For newtonian liquids, this can be written as $\tau = \mu \dot{\gamma}$. In a newtonian liquid, the shear stress is always directly proportional to the shear rate. In thermoplastic materials like rigid PVC melts, this is not the case, and a unit increase in the shear rate may cause a nonlinear increase in the shear stress (force). An illustration of the above is presented in Fig. 3.3 using logarithmic coordinates that show that PVC resins exhibit nonnewtonian flow behavior over the entire range of shear rates. PVC melts are not like most thermoplastic melts, whose slope of the shear stress and shear rate curve approach unity at low or very high shear rates.

3.4 RIGID PVC MELT RHEOLOGY

3.4.1 Degree of Departure from Newtonian Fluid Behavior

From the previous discussions, it is logical to infer that melt viscosity has a dominant effect on the processability of rigid PVC resins. Consequently, determination of the viscosity is of extreme importance. Capillary flow measurement provide a convenient means of examining the parameters that influence the viscosity and thus the overall processability of rigid PVC resins.

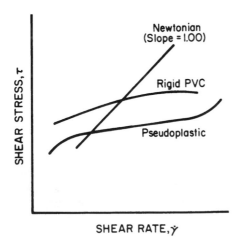

Fig. 3.3 Flow curves, logarithmic coordinates.

The degree of departure from newtonian fluid behavior may be obtained quantitatively from the logarithmic shear-stress shear-rate plot. Taking logarithms of the equation describing newtonian behavior, which for convenience is written as below:

$$\tau = \mu \frac{du}{dh}$$

one obtains

$$\log \tau = \log \left(\mu \frac{du}{dh} \right)$$

or

$$\log \tau = \log \frac{du}{dh} + \log \mu$$

when compared to the equation of a straight line,

$$y = nx + b$$

one sees that it is identical, if one defines

$$y = \log \tau$$

$$n = 1.00$$

$$x = \log \frac{du}{dh}$$

$$b = \log \mu$$

Thus, the shear-stress shear-rate relationship of a newtonian fluid becomes a straight line with a slope n of unity. This straight line is the flow curve for a newtonian fluid. The slope of a logarithmic flow curves of thermoplastic materials, such as rigid PVC melts has been termed the flow-behavior index, or power law flow index. The degree or extent of nonnewtonian behavior is the amount by which this flow-behavior index differs from unity. For rigid PVC melts, as well as all nonnewtonian fluids, the flow-behavior index must be determined over the actual range of shear rates in which it is to be used.

3.4.1 Rheometers

Any capillary extrusion rheometer in which molten rigid PVC resins can be forced from a reservoir through a capillary die and in which temperature, pressure, output, extrusion time, and die dimensions can be measured may be satisfactory to measure maximum shear stress and maximum shear rate at the wall of the capillary and from these values obtain the apparent melt viscosity, which, as shown before, is represented by the Greek letter η.

Another desirable feature of a suitable rheometer for measuring the flow characteristics of rigid PVC melts is the capability of measuring over a wide range of shear rates (10^1 to 10^4 sec^{-1}).

A good assessment of the flow behavior of rigid PVC resins and compounds is obtained by determining the melt viscosity η over a range of shear rates and at various temperatures: ideally, both below the T_m (the crystalline melting point), and above the T_m.

The percentage memory (also known as postextrusion swelling, percentage puff-up, and so on) as a function of shear rate, temperature, and concentration of inorganic filler, for example, is another valuable piece of information. This can be obtained with a rheometer and helps in defining lip gaps for extrusion dies, optimum haul-off speed for a given profile, degree of orientability of extrudates, and other parameters.

If one knows the shear rates at which melt or surface fracture occurs for a specific compound and the computed shear rate factors for a production die system, one can determine the maximum acceptable commercial quality output for production. Since for a given product dimension, e.g., pipe or profile wall thickness, the die exit gap is fixed, other means of safely increasing the shear rate values for that particular die should be known and explored. This knowledge, if properly applied, allows higher extrusion speeds and, consequently, higher outputs (Q). This can be achieved, for example, if the exit land length L is increased. If the extrudate yellowness is already too high, this approach should be avoided. Changes in taper angle, such as increasing the length of the taper or decreasing the die opening at the entrance of the taper section, should be then considered (see also Chap. 4).

When processing shear and thermosensitive materials like rigid PVC, the so-called stability factor, which is a measure of the material's ability to maintain a constant rate of flow or melt viscosity under a given set of conditions, is a very discriminatory piece of information that could be of great value in selecting the best compound for the job. Rigid PVC materials often tend to degrade or cross-link at the temperature of processing. The stability is measured from the plot of the apparent melt viscosity or rheometer extrusion time versus various residence times, e.g., 5, 10, 15, and 20 min. The slope of this line can be used as the measurement of the stability of the rigid PVC compound evaluated. A material with slope 0 is ideal for production, since the melt viscosity remains unchanged and long production runs can be made with minor processing changes.

Flow Mechanism in Capillary Rheometers

To understand the mechanics of flow in simple tubes such as capillary rheometers, let us visualize the velocity profile at the entrance to a tube, when the fluid flows from a larger tube or reservoir. At this

point, the profile is essentially flat. The velocity then rapidly
changes to a value of zero at the wall, and as the flow proceeds down
the tube, a velocity profile is built up that becomes parabolic for newtonian fluids, flow-behavior index, n = 1, and becomes flatter than the
parabolic profile for pseudoplastic, n < 1.

At the tube wall, the velocity of the fluid is zero, and the shearing
stress is maximum. Since shear rate du/dh (velocity gradient) is
given by the slope of the velocity profile at any point, the shear rate
also reaches its maximum value at the wall. At the center of the profile, the velocity u is maximum, and the shear rate and, hence, the
shear stress are zero. These conditions prevail when the flow of the
fluid is steady and laminar, and the fluid is isothermal. Based on the
previous discussion, one concludes that the viscosity calculation
should be conducted at the capillary walls where shear rate and stress
values are maximum. Beyond the entrance region, the velocity profile change shape in isothermal flow through a capillary.

In capillary extrusion rheometers, the melt viscosity η is determined
from the volume of flow in unit time Q and the pressure drop ΔP
through the capillary.

The derivation of the flow equation through a capillary is greatly
simplified by assuming a one-dimensional, steady, laminar flow of an
incompressible fluid in a horizontal, circular tube of constant diameter. If one considers a force balance made on an element of fluid of
radius r and length L in the center of a circular pipe, as shown in
Fig. 3.4, and assumes that the hydrostatic forces (F = PA) acting on
this element of fluid, given by the expression $\Delta P \pi r^2$, are countered
by the shearing (drag) forces over the surface of the element, given
by the expression $\tau 2 \pi r L$, the force balance can be reduced to the expression below, since under steady flow conditions (no acceleration)
in a horizontal tube, the two forces must be equal.

$$\Delta P \pi r^2 = \tau 2 \pi r L$$

In this equation ΔP represents the frictional pressure drop over a
length of tube L or

$$\tau = \frac{\Delta P}{2L} \, r \tag{3.1}$$

At the wall, this equation becomes

$$\tau_w = \frac{\Delta P}{4L} \, D \tag{3.2}$$

where τ_w is the shear stress at the wall. These latter two expressions
show that the shearing forces on a fluid vary from zero at the center
line, r = 0, to a maximum value at the wall.

The expression (3.1) is also used to calculate the pressure drop in
die openings of constant gap opening h and land length L as follows:

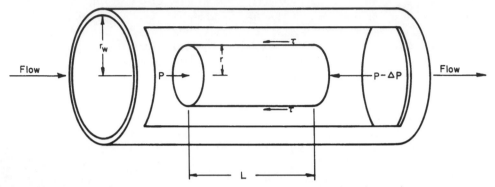

Fig. 3.4 Forces acting on an element of fluid flowing in a pipe.

$$\Delta P = \frac{2L}{h} \tau \qquad (3.3)$$

This is the basic formulation used to determine the pressure drop in the land area of a slot die.

The velocity distribution u in laminar flow is found by substituting the equation

$$\tau = \mu \left(- \frac{du}{dr} \right)$$

into equation (3.1) and then integrating [du/dr is modified with a minus sign because du/dr is negative (u decreases as r increases)]. Using this approach one obtains

$$\frac{\Delta P}{2\mu L} \int_{0}^{r} r \, dr = - \int_{u_{max}}^{u} du \qquad (3.4)$$

and the integrated result is

$$\frac{\Delta P}{4\mu L} r^2 = u_{max} - u$$

or

$$u_{max} = u + \frac{\Delta P}{4\mu L} r^2 \qquad (3.5)$$

Since u = 0 at $r = r_w$ from the assumption of no slip at the wall, it follows that

$$u_{max} = \frac{\Delta P}{4\mu L} r_w^2 \qquad (3.6)$$

or

$$u = u_{max} \left[1 - \left(\frac{r}{r_w} \right)^2 \right]$$

Maximum velocity u_{max} is at the center of the capillary.

Since equation (3.6) is the equation of a parabola, one can see that the velocity profile is parabolic for a newtonian fluid. The magnitude of the velocity varies over the cross section as the coordinates of a paraboloid of radius r_w and height u_{max}. The volume of a paraboloid of revolution is

$$V = \frac{1}{2} \pi r_w^2 h$$

since $h = u_{max}$, this expression becomes

$$V = \frac{\pi r_w^4 \, \Delta P}{8 \mu L} \tag{3.7}$$

The volume of flow in unit time Q then becomes

$$Q = \frac{\pi r_w^4 \, \Delta P}{8 \mu L} \tag{3.8}$$

This is the Hagen-Poiseuille equation used in capillary extrusion rheometer for the determination of viscosity from Q and ΔP.

Using equation (3.1),

$$\tau = \frac{\Delta P}{2L} \, r$$

or

$$\tau_w = \frac{\Delta P}{2L} \, r_w$$

and $\tau = \mu \dot{\gamma}$, equation (3.8) can be written as

$$\frac{\Delta P r_w}{2L} = \mu \, \frac{4Q}{\pi r_w^3} \tag{3.9}$$

The quantity $4Q/\pi r_w^3$ is the shear rate at the wall, and η is the symbol used for nonnewtonian liquids.

Equations for shear rate calculations

The following expressions based on flow data enable die designers to estimate shear rates as well as the dimensions of the appropriate flow channels required by the dimensions of the finished rigid PVC products. The equation

$$\dot{\gamma} = \frac{4Q}{\pi r^3} \qquad (3.10)$$

can be used to compute operating shear rates for round holes, such as rod dies, manifolds, injection molding nozzles, and mold gates. For thin sheet and slit dies, this equation becomes

$$\dot{\gamma} = \frac{6Q}{wh^2} \qquad (3.11)$$

which is the apparent shear rate at the wall of the slit, where

w = width of die
h = die opening

Since $\dot{\gamma}$ is related to the mean lineal speed of the material flowing through the die at the exit v_d and to the minimum gap in the die exit, the (3.10) and (3.11) expressions can be rewritten in the following manner. *Note*: $v_d = Q/A$, where A = cross-sectional area.

$$\dot{\gamma}_{circ} = \frac{4v_d}{r_d} \qquad (3.12)$$

$$\dot{\gamma}_{slot} = \frac{6v_d}{h_d} \qquad (3.13)$$

The shear rate for an annular die is given by

$$\dot{\gamma} = \frac{6Q}{\pi(r_o + r_i)(r_o - r_i)^2} \qquad (3.14)$$

where

r_o = outer die radius

r_i = inner die radius

Equations (3.10) through (3.14) are approximate equations for non-newtonian fluids. Since annular gaps may be treated as an infinitely wide slot die, provided that the ratio of die gap to circumference is small, the simplified expression (3.13) can also be used for annular dies. When these conditions occur, h_d is treated like $r_o - r_i$.

Shear rates for complex shapes can be calculated by computing the shear rate for each single geometric component part separately and adding these to obtain the total. These equations are the most frequently used for computing the operating shear rate for rigid PVC compounds at the die, delivery tube, mold gates, and injection-molding nozzle.

Viscosity calculations and corrections

In the standard capillary extrusion rheometer, the shear stress is
determined directly from the pressure drop, which in turn is calculated as a function of extrusion force F, plunger diameter D_p, capillary diameter D_c, and capillary length L_c, as follows:

$$\tau_w = \frac{FD_c}{\pi L_c D_p^2} \tag{3.15}$$

The shear rate at the capillary wall γ_w can be calculated as a function
of the crosshead speed U, the diameter of the plunger D_p, and the
diameter of the capillary D_c, as follows:

$$\dot{\gamma}_w = \frac{2UD_p^2}{15D_c^3} \tag{3.16}$$

These equations may be analyzed dimensionally as follows:

$$\text{Shear rate} = \frac{cm}{sec} \frac{cm^2}{cm^3} = sec^{-1}$$

$$\text{Shear stress} = \frac{dyn}{cm^2} \frac{cm}{cm} = \frac{dyn}{cm^2}$$

or

$$\text{Shear stress} = \frac{lb}{in.^2} \frac{in.}{in.} = \frac{lb}{in.^2} = 6894.8 \text{ Pa}$$

The viscosity is then

$$\eta = \frac{dyn\ sec}{cm^2} = \text{poise (P)}$$

or

$$\eta = \frac{lb\ sec}{in.^2} = 68,948 \frac{dyn\ sec}{cm^2}$$

or 68,948 P, or 6894.8 Pa sec.

As stated before, the calculation of the melt viscosity is conducted
at the capillary walls where shear rate and shear stress reach their
maximum values. The expression for the melt viscosity then becomes

$$\eta = \frac{\tau_w}{\dot{\gamma}_w}$$

Here, τ_w is the shear stress at the wall of the capillary, and $\dot{\gamma}_w$ is the corresponding shear rate. Since rigid PVC melts do not behave as newtonian fluids, one must use the form of $\dot{\gamma}_w$ valid for nonnewtonian fluids. Rabinowitsch has shown that a true shear rate can be computed by taking into account the degree of nonnewtonian behavior of the melt. For a power law model (see Sec. 3.5), this is given by the equation

$$\dot{\gamma}_{tw} = \dot{\gamma}_w \left(\frac{3n + 1}{4n} \right)$$

where n, as shown before, is the slope of a logarithmic plot of shear stress versus shear rate, or flow-behavior index, or power law flow index.

Substituting

$$\dot{\gamma}_w = \frac{4Q}{\pi r^3}$$

in the previous equation,

$$\dot{\gamma}_{tw} = \frac{4Q}{\pi r^3} \left(\frac{3n + 1}{4n} \right) = \left(3 + \frac{1}{n} \right) \frac{Q}{\pi r^3} \qquad (3.17)$$

Using this true shear rate, a true melt viscosity can be found from

$$\eta_t = \frac{\tau_t}{\dot{\gamma}_t}$$

Since the melt viscosity varies with shear rate, temperature, pressure, and other factors, melt viscosity is normally referred to as apparent melt viscosity.

As shown in equations (3.10) and (3.17), the shear rate is calculated from the flow rate and corrected using the slope of the line of a log-log plot of τ_w versus $\dot{\gamma}_w$, the n value.

In many cases, one can determine the viscosity of rigid PVC melts in the region of processing, for example, pipe extrusion, by drawing a straight line through the τ_w versus $\dot{\gamma}_w$ data and obtain approximately values for η and n suitable for engineering calculations.

In addition to the Rabinowitsch correction, there are a number of other corrections that can be applied to the raw data obtained in a rheometer. For example, there is a significant error (up to 50%)

present in much capillary data because of the pressure effect on viscosity. Also, flow activation energy increases with pressure. At
high hydrostatic pressures of >10,000 psi, low shear rate of 1 sec^{-1},
and rather low melt temperature, 365°F, the melt viscosity of some
rigid PVC compounds may be raised by a factor as high as 3, which
decreases when the shear rate increases to 100 sec^{-1}. The methods
that have been worked out for correcting these errors, however, are
not discussed here.

Another phenomenon that contributes to some error in the capillary
data is that of the capillary wall slip. It occurs only at high rates of
shear. At lower rates, alternate slipping and sticking occur at the
capillary wall. At still lower rates, there is no slip. Slip is more a
function of stress than shear rate, occurring generally above 80 psi
for PVC. Stick-slip appears in the range 40 to 80 psi. It is in this
region that melt fracture is more likely to appear. There is seldom
any slip below 40 psi for most of the rigid PVC melts at the lowest
stock temperatures that produce acceptable product. A typical force
trace, illustrating the wall slip-stick, is presented in Fig. 3.5. The
regularly fluctuating force is accompanied by an inversely fluctuating
flow rate, indicating increasing sticking to the capillary wall as the
force builds up, followed by a sudden release, during which the force
decreases and the flow of material from the capillary accelerates rapidly.

3.5 POWER LAW

Throughout this chapter, we have already seen that the viscosity η
of rigid PVC melts decreases with increasing shear rate $\dot{\gamma}$. We have
been also exposed to the concept of flow-behavior index n, also known
as the power law flow index.

The viscosity relations for rigid PVC melts are taken from the viscosity relations applicable to nonnewtonian fluids in isothermal simple
shear, as presented by McKelvey [2]. The power law:

$$\eta = \eta_0 \left| \frac{\dot{\gamma}}{\dot{\gamma}_0} \right|^{n-1} = \eta_0 \left| \frac{\tau}{\tau_0} \right|^{(n-1)/n} \tag{3.18}$$

can also be written as

$$\left| \frac{\tau}{\tau_0} \right|^{(n-1)/n} = \left| \frac{\dot{\gamma}}{\dot{\gamma}_0} \right|^{n-1}$$

or

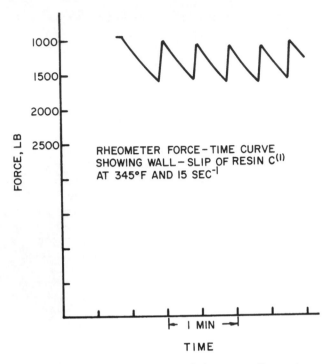

RHEOMETER FORCE–TIME CURVE
SHOWING WALL–SLIP OF RESIN C[1]
AT 345°F AND 15 SEC[-1]

FORCE, LB

1000
1500
2000
2500

├─ I MIN ─┤

TIME

(I) Low – Medium M_w, (K Value~63), Suspension Homopolymer

Fig. 3.5 Rheometer force-time curve showing wall-slip of resin C at
345°F and 15 sec[-1].

$$\left| \frac{\tau}{\tau_0} \right|^{1/n} = \left| \frac{\dot{\gamma}}{\dot{\gamma}_0} \right|$$

or

$$\tau = \tau_0 \left| \frac{\dot{\gamma}}{\dot{\gamma}_0} \right|^n \tag{3.19}$$

For convenience, $\dot{\gamma}_0$ is usually chosen as 1 sec[-1] or τ_0 as 1 dyn/cm^2
The parameter n is the flow-behavior index of the fluid. As shown,
absolute values are used in the power law because $\dot{\gamma}$ (or τ) can be
either postive or negative, but n must always be positive. When n is
unity, the power law becomes Newton's law. When n < 1, the viscosity
of the fluid decreases with increasing $\dot{\gamma}$ or τ, which is the case of
rigid PVC melts.

Temperature and shear rate dependence of the nonnewtonian viscosity of rigid PVC polymers can be expressed approximately with an equation similar to the Colwell and Nickolls equation [3],

$$\eta = \eta_0 \left| \frac{\dot{\gamma}}{\dot{\gamma}_0} \right|^{n-1} e^{-b(T-T_0)} \qquad (3.20)$$

where b, n, and η_0 are parameters characteristic of the material and T_0 and $\dot{\gamma}_0$ define a standard reference state. The $e^{-b(T-T0)}$ part of equation (3.20) is taken from an equation also used to correlate viscosity temperature data:

$$\eta = ae^{-bT}$$

which at the standard-state shear rate $\dot{\gamma} = 1$ sec^{-1} and temperature, which is arbitrarily chosen here as 180°C, becomes

$$\eta_0 = ae^{-b\dot{\gamma}T}{}_0$$

At the same shear rate $\dot{\gamma}$ but at temperature T_1, this equation becomes

$$\eta_1 = ae^{-b\dot{\gamma}T}{}_1 \qquad (3.21)$$

which when combined with the previous equation becomes

$$\frac{\eta_0}{\eta_1} = \frac{e^{-b\dot{\gamma}T}{}_0}{e^{-b\dot{\gamma}T}{}_1}$$

or

$$\eta_1 = \eta_0 e^{b\dot{\gamma}(T_0 - T_1)} \qquad (3.22)$$

Values of the material parameters n and b are established from capillary viscosity measurements as follows. To determine n, values for the apparent melt viscosity at the standard reference state $\eta_0(\dot{\gamma}_0 = 1$ sec^{-1} and $T_0 = 180$°C) and at some other condition on $\eta_1(\dot{\gamma} = 10$ sec^{-1} and $T_0 = 180$°C) are substituted in equation (3.18). For example, if from a viscosity versus shear rate curve of a medium-to-high MW PVC resin, $\eta_0 = 1.5 \times 10^6$ poise and $\eta_1 = 4.7 \times 10^5$ poise, then substituting these values, $4.7 \times 10^5 = 1.5 \times 10^6 \times 10^{n-1}$, n ~ 0.5. Keeping the temperature constant, this flow index value decreases first to ~0.36 when the shear rate increases to 100 sec^{-1} and then to ~0.33 when the shear rate becomes 1000 sec^{-1}. This 0.36 value is commonly used for engineering calculations such as to determine flow in extruders, screws, die adaptors, dies, etc.

Table 3.1 $b\dot{\gamma}$ at 100 sec^{-1} Shear Rate

MW PVC resin	$b_{\dot\gamma}(°C^{-1})$	$1/b\dot\gamma(°C)$
60 X 10^3	0.060	16.7
105 X 10^3	0.040	25.0
207 X 10^3	0.028	36.0

Since the flow index decreases when the molecular weight of the resin increases*, it is advisable to determine this value experimentally every time a change in resin molecular weight is made. As mentioned before, the flow index can also be obtained by measuring the slope of a log-log plot of the shear stress versus shear rate curves. For higher accuracy, the slope should be measured within the range of shear rates of the melt processing under consideration.

For b, the equation $\eta_1 = \eta_0 e^{b\dot\gamma(T_0 - T_1)}$ is then used when apparent melt viscosity η, at a reference temperature T_0, 180°C, and at the temperature T_1, this time 220°C, for example, are substituted into equation (3.22), which is now written as

$$\eta_1 = \eta_0 e^{b\dot\gamma\Delta T} \tag{3.23}$$

Note that when $\Delta T = 1/b_{\dot\gamma}$, $\eta_1 = \eta_0 e$ or $\eta_0 = \eta_1$ (1/e).

Using the previous equation, $b\dot\gamma$ values were determined for the viscosity-shear rate (at various temperatures) data presented by Collins and Metzger [4] for PVC resins representing various molecular weight (MW). Specifically, the $b\dot\gamma$ values were estimated at 100 sec^{-1} for PVC resins representing the following MW: 60 X 10^3, 105 X 10^3, and 207 X 10^3. The results obtained are tabulated in Table 3.1.

As seen in Table 3.1, the $b\dot\gamma$ decreases with the MW of the resin. Since the quantity $1/b\dot\gamma$ represents the number of degrees that the temperature of the PVC resin or compound must be raised at constant shear rate to decrease the viscosity by a constant factor 1/e, these data indicate that the lowest MW resin has the most temperature sensitive viscosity, and the highest MW resin has the least. The $b\dot\gamma$ values, as expected, are affected by the shear rate, decreasing when the $\dot\gamma$ increases and increasing when the $\dot\gamma$ decreases. For the PVC resin of MW 105,000, for example, this $b\dot\gamma$ value at 1 sec^{-1} shear rate and the temperature range used above (180 to 220°C) becomes 0.08 or $1/b_{\dot\gamma}$ = 12.5°C.

For comparative purposes, flow parameters of a PVC resin of MW 105 X 10^3 and a copolymer of methacrylonitrile (MAN; 90%) and

---

*Higher molecular weight resins have lower flow index values because they are less newtonian.

Table 3.2 Flow Parameters

Parameter	PVC resin, 105,000 MW	MAN (90%)-Styrene (10%) copolymer[a], 150,000 MW
n (dimensionless)	0.50	0.41
$b_{\dot{\gamma}0}$ ($^\circ$C^{-1})	0.080	0.022
η_0 (poise)	360×10^3	800×10^3

[a]Highly polar polymer with very high cohesive energy density.

styrene (10%) are shown in Table 3.2 at standard state $\dot{\gamma} = 1$ sec^{-1} and at T = 200°C. These data indicate that the viscous nature of the MAN-styrene copolymer is far more pronounced than that of this PVC resin and that rigid PVC processing shows a more temperature-sensitive viscosity than the MAN-styrene copolymer. As also shown by the flow index value, this copolymer is significantly less newtonian than the PVC resin under consideration.

Knowing the flow parameters for a PVC resin or compound, the equation

$$\eta = \eta_0 e^{-b_{\dot{\gamma}}(T-T_0)}\left|\frac{\dot{\gamma}}{\dot{\gamma}_0}\right|^{n-1}$$

can be used to calculate the η value at temperatures other than the experimental, e.g., to estimate the viscosity of the rigid PVC melts at decomposition temperatures or slightly above their decomposition temperatures. This is based on the assumption that the shear rates are the same, or fall within a narrow range of shear rates (10 sec$^{-1} \leqslant \dot{\gamma} \leqslant$ 50 sec^{-1}). It is also assumed that the temperature coefficient $b_{\dot{\gamma}}$ has been determined at temperatures near the decomposition temperature and the calculation is conducted within a range of 50°C.

Another expression, $[\exp(b_{\dot{\gamma}} \Delta T)]^{1/n}$, (3.24), which is valid for power law fluids, can help to estimate the effect of changing the PVC melt temperature on the flow rate Q of PVC melts in forming dies, particularly those of annular shape. From the experimental data, an average curve of flow rate (pounds per hour) versus head pressure at a temperature T is drawn. The effect of increasing the temperature T, 10°F, on the flow rate Q at constant pressure is then determined using the above equation and the known values of $b_{\dot{\gamma}}$ and n. For example, when $b_{\dot{\gamma}}$ and n have the values of 0.023°C^{-1} and 0.36, respectively, the factor for the flow increase is 1.42. Hence, curves for this particular PVC melt at temperatures 10 degrees Fahrenheit above and below the experimental curve can be established.

3.6 FACTORS AFFECTING MELT VISCOSITY, POSTEXTRUSION SWELLING, AND MELT QUALITY

In this section, we discuss how the factors having significant influence on postextrusion swelling, melt viscosity, and melt quality affect these properties. Knowledge in this area enables process engineers to assess the processability of rigid PVC compounds. Before we proceed, however, it is pertinent to introduce, at this point, some discussion of the so-called postextrusion swelling, or puff-up.

A piece of valuable information obtained from the rheometers is postextrusion swelling, also known as percentage memory, puff-up, percentage puff-up, and the Barus effect. These values, which are normally presented as a function of the shear rate, are determined by measuring the outside diameter of the extrudate and comparing these numbers with the inside diameter of the capillary. These values are reported as the ratio between the two diameters (the diameter of extrudate per diameter of capillary), or as a percentage,

$$\% \, \text{memory} = \frac{E_{o.d.} - C_{i.d.}}{C_{i.d.}} \times 100$$

where

$E_{o.d.}$ = extrudate outside diameter
$C_{i.d.}$ = capillary inside diameter

Die swell is related to the elastic behavior of the rigid PVC melt. Inside the die the melt is under stress, and therefore deformed. Outside the die, upon removal of the stress, the elastic component of the deformation is recovered and this causes an increase in diameter greater than that obtained with newtonian fluids.

The processability of rigid PVC compounds can be compared by measuring their percentage memory versus shear rate. The lower the memory at a given shear rate, the more easily that material is formed. Better control of the wall thickness of the extrudate is also obtained. A high degree of memory, however, is desirable for a biaxially oriented bottle process since, in general, the higher the memory, the higher this material will stretch and blow.

The influence of PVC resin molecular weight on apparent melt viscosity and postextrusion swelling, or puff-up, is presented in Table 3.3. As shown from these experimental values, we can draw the following tentative conclusions:

Rigid PVC resins are very shear sensitive; an increase in shear rate of 90 sec^{-1} significantly decreases the apparent melt viscosity.
Although only modestly, the puff-up increases as the molecular weight decreases.

Table 3.3 $\eta_{ac}{}^a$ at 350°F

Identification[b] PVC resin MW (GPC)	10 sec^{-1} (psi sec)	100 sec^{-1} (psi sec)	Puff-up at 30 sec^{-1}
71,000-75,000 (Resin C)	8.0	1.2	1.15
50,000-58,000 (Resin D)	4.9	0.85	1.17
50,400 (Resin E)	3.3	0.65	1.23
45,000	2.35	0.51	1.25
40,000	—	0.24	1.26

[a]Capillary: 0.390 X 0.0495 in. X 90°
[b]All resins containing 1.0% of an alkyltin mercaptoacetate stabilizer milled prior to their extrusion in the capillary rheometer. These resins intended for injection-molding applications.

Values of apparent melt viscosity for the top three PVC resins of Table 3.3 (resin C = 71,000 to 75,000, resin D = 50,000 to 58,000, and resin E = 50,400 $\overline{\text{MW}}$) and two higher $\overline{\text{MW}}$ resins (resin A and resin B) are plotted in Figs. 3.6, 3.7, and 3.8. As shown, at the lowest processing temperatures, these values converge with increasing shear rate; thus, the large advantage in reduced melt viscosity for low-molecular-weight PVC is partially lost at increasing shear rate. At injection-molding processing conditions (high temperature and shear), however, the effect of the lower $\overline{\text{MW}}$ on melt viscosity is quite significant (Fig. 3.8).

If the values of PVC weight average molecular weight ($\overline{\text{M}}\overline{\text{W}}$) presented in Table 3.3 are plotted logarithmically versus apparent melt viscosity (psi sec) at a fixed shear rate (10^{-1} sec), good correlation is found between these parameters (see Fig. 3.9). The line drawn among the experimental data has a slope of about 3.5. These results indicate that, in this case, the melt viscosity of these low-MW resins can be related exponentially with the weight average molecular weight by the 3.5 power, as follows:

$$\log \eta_0 = 3.5 \log \overline{\text{MW}} + K$$

which is the well-known Bueche relation [5] expressing the 3.5 power law dependence of melt viscosity on molecular weight. Although this formula is for amorphous polymers, it seems also to work for low-molecular-weight (low-crystallinity) PVC resins.

In Table 3.4 the postextrusion (percentage puff-up) of a typical Type I rigid PVC compound is plotted versus temperature and versus shear rate. As shown, postextrusion swelling increases with the shear rate and also with the temperature.

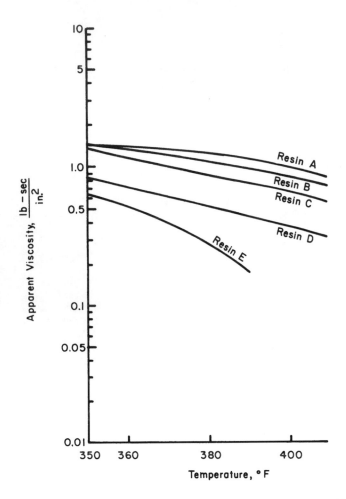

Fig. 3.6 Apparent viscosity versus temperature. Shear rate = 100 sec^{-1}; capillary D = 0.05 in., L/D = 8.

3.6.1 Effect of Crystallinity

Figure 3.10 presents the puff-up ratio and the apparent melt viscosity of a suspension resin ($\overline{\text{MW}} \sim 74,000$) and a partly crystalline low-density powder PE. As shown in this figure, PVC puff-up increases with increasing temperature, and the levels of puff-up are quite small.

For amorphous polymers, like polystyrene and most of the low-density polyethylenes (PE) in the molten state, the capillary extrudate swell is usually large and has an inverse dependence on temperature at constant rates of shear after its melting point is reached. Figure

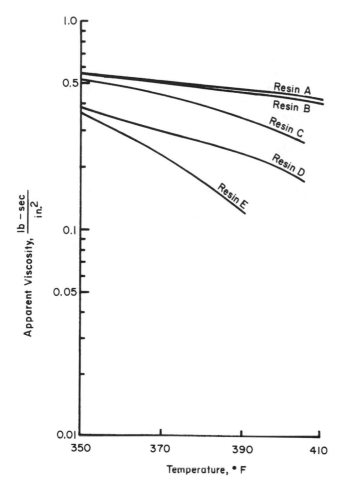

Fig. 3.7 Apparent viscosity versus temperature. Shear rate = 275 sec^{-1}, capillary D = 0.05 in., L/D = 8.

3.10 shows that this partly crystalline PE has a well-defined T_m and the puff-up indeed increases with increasing temperature until T_m is reached. As shown, the extrudate swell or puff-up of rigid PVC at extrusion temperatures is somewhat uncommon.

Formulation components and, specifically, the concentration of inorganic fillers and pigments, the presence of small amount of plasticizers, and the type and degree of dispersibility of the rubber used as the impact modifiers have a very decisive effect on the magnitude of the rheological properties mentioned so far. Rubber grafting and some additives like cetyl vinyl ether (CVE), for example, contribute

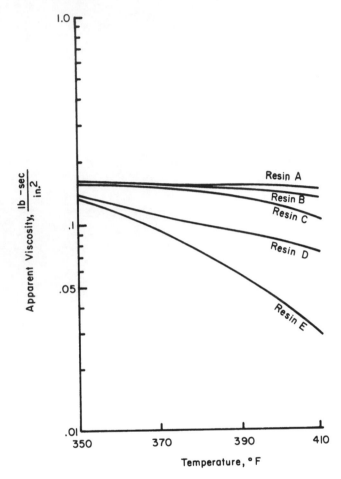

Fig. 3.8 Apparent viscosity versus temperature. Shear rate = 1000 sec^{-1}, capillary D = 0.05 in., L/D = 8.

greatly to lower melt viscosity, lower postextrusion swelling, and excellent strand smoothness. Also, PVC grafting (PVC blends produced by polymerization of vinyl chloride monomer, VCM, in the presence of a rubber) is very helpful in producing blends capable of preserving toughness under simulated injection-molding conditions. The presence of small amounts of primary plasticizers in rigid PVC formulations tends to increase viscosity, which is often attributed to an increase in crystallinity. This in turn produces a decrease in impact resistance.

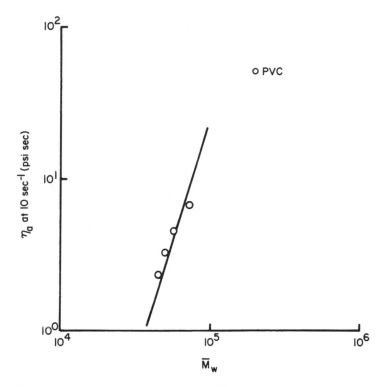

Fig. 3.9 Rheology of PVC at 175°C.

Table 3.4 % Puff-up (Swell) at Shear Rate, sec^{-1}

Temperature, °C	8.77	43.8	87.7	438	877	1754
170	5	7	9	10	12	14
180	5	9	9	14	14	15
190	7	7	14	17	19	17

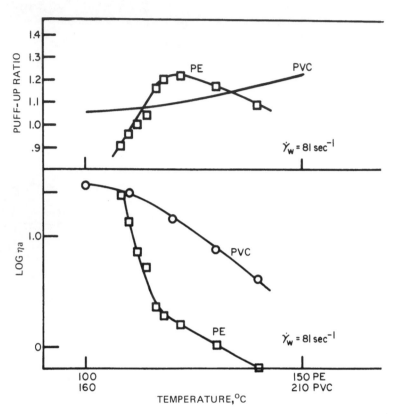

Fig. 3.10 Extrudate swell and apparent melt viscosity of (a) an unplasticized PVC suspension resin; (b) a partly crystalline low density
PE.

The explanation of the anomalous behavior of PVC is thought to lie
in the fact that PVC is still crystalline at the processing temperatures
(PVC $T_m \sim 210°C$), and because of the high melting point of the
crystalline phase, fusion for the most part does not appear to take
place at the highest temperatures compatible with thermal stability.
Munstedt [6] has reported the existance of a critical melt temperature below which PVC melt samples show rheological behavior similar
to that of slightly cross-linked polymers and above which the behavior of PVC is similar to that of amorphous polymers. Collins and Krier
[7] have shown that PVC has two distinct flow-activation energies on a
capillary rheometer: the transition from one flow energy to the other
one was found to occur at about the T_m of the PVC crystals. At temperatures below T_m, primary PVC resin particles (~ 1 μm) seem to re-

tain their integrity and crystallites behave as pseudo-cross-links in a
three-dimensional network in the melt, thus hampering flow. Thus,
the poor flow of the unmodified rigid PVC melts can be attributed to
the incomplete melting of PVC particles due to residual crystalline
structure. The residual crystallinity maintains the primary particle's
integrity and is responsible for the unusual velocity profile and stiffness of the melt exiting the die, which contributes to the excellent
die-forming properties of rigid PVC in complex dies [8], such as some
window lineals.

In the melt processing of rigid PVC compounds, it seems that we are
dealing with at least two flow mechanisms: a particulate flow mechanism consisting of partially interfused large deformable particles acting as supermolecular units, which occurs at lower temperature, and a
flow mechanism at higher temperature corresponding to the melting
of PVC crystallites and characterized by a high activation energy due
to the interaction and fusion between the particles. Berens and Volt
[9,10] have presented some evidence that indicates the PVC "melts"
consist of large flow units (supermolecular). Also, Mooney [11,12]
has claimed that even amorphous melts consist of deformable flow
units. He even developed equations to describe the flow of such systems; his equations, although containing many experimentally difficult
to determine parameters, do give the functional relation that η_a should
depend on G', the deformability of the flow units, the real part or elastic component of the complex modulus, also called the storage modulus.

As the temperature increases and approaches 200°C, the crystallinity decreases and the deformability and breakdown of the particle
increases. At the lowest temperatures, one might visualize that, in
the extrusion process, for example, one is mainly shifting around
rigid spheres. Thus, a high viscosity and low puff-up are expected,
which is the actual case.

As the deformability of the particles increases (by decreasing crystallinity with increasing temperature), the particles become elongated
and partially fused in the extrusion process, freeing polymer chains
for interdiffusion between the primary particles. This degree of fusion, however, varies with the extrusion process. PVC melts in twin-
screw extruders, which normally are at temperatures below T_m, are
more crystalline than the melt produced in single-screw extruders,
where the melt, or at least part of the melt, is at a temperature near
or over T_m. As shown in Fig. 3.10, the puff-up increases, and the
viscosity decreases quite sharply as T_m is approached.

The above-stated can be of help not only when designing the optimum melt processing for a particular product but also when analyzing
existing processes. In pipe extrusion, for example, it is known that
low puff-up and higher melt viscosity (stiffness) are desirable for the
die and, also, postforming operations. These properties contribute to
an easier pipe sizing. Since dimensions, mainly in pipe extrusion,

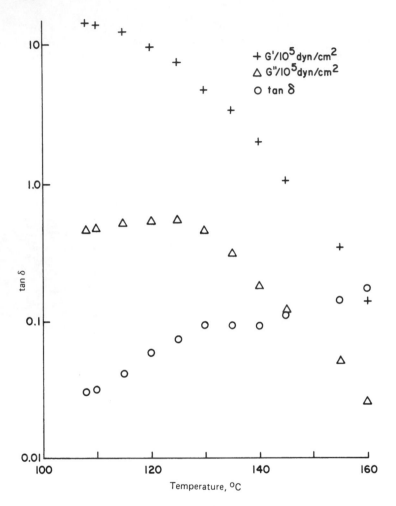

Fig. 3.11 Weissenberg rheogoniometer complex modulus: temperature behavior at 1 Hz. $\overline{MW}$ (GPC) = 13,500.

can be held much closer to the minimum wall thickness, these properties also contribute to significant costs savings. Thus, twin-screw ex-extruders, which deliver a more uniform melt (radial melt temperature is very small) at lower temperatures, for the most part below 210°C, are more suitable than the single-screw extruders. In blow-molding applications, however, high puff-up and lower melt viscosity are more desirable to achieve higher orientation and consequently better mech-anical properties and a better definition of mold details. For trans-parent applications, low crystallinity or its absence contributes to a

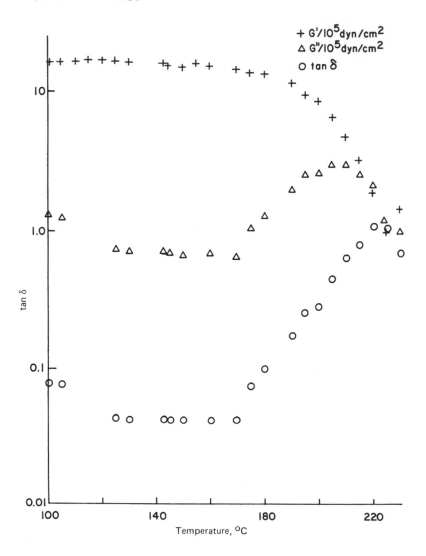

Fig. 3.12 Weissenberg rheogoniometer complex modulus: temperature behavior at 1 Hz. $\overline{MW}$ (GPC) = 57,000.

more sparkling container. In this area, single-screw extruders, which can deliver the melt, or at least part of the melt, at T_m or higher, are more suitable than the multiscrew extruders.

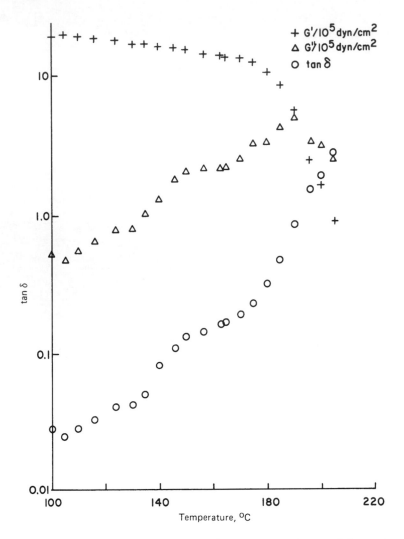

Fig. 3.13 Weissenberg rheogoniometer complex modulus: temperature behavior at 1 Hz. $\overline{MW}$ (GPC) = 67,000.

3.6.2 Complex Modulus Behavior of PVC

Since, as stated by Mooney, modulus has an effect on viscosity, complex modulus behavior was measured for three selected PVC fractions using the Weissenberg rheogoniometer. The measurements at the various temperatures were made at 1 Hz.

Figures 3.11, 3.12, and 3.13 show actual storage and loss moduli data on three fractions of a model PVC polymer with the following characteristics.

Fraction	$\overline{MW}$ (GPC)	Room temperature % crystallinity[a]
I	13,500	22.5
II	57,000	26.0
III	67,000	23.5

[a]Samples heated 0.5 hr at 150°C, and slowly cooled.

We see that, for fraction I, the storage modulus G' drops markedly with temperature, whereas for the fraction III material the modulus is flat to about 180°C and then begins to drop. For the much more crystalline fraction II material the modulus is flat to about 200°C. These results suggest that the modulus G' is influenced by crystallinity and molecular weight. Furthermore, as presented by Nielsen [13], crystallinity increases modulus, and also increasing molecular weight increases modulus [14], except that after certain levels of $\overline{MW}$, modulus values well above T_g reach a plateau.

In the above, we have presented simple examples to illustrate some of the uses of dynamic mechanical data G' and G" for obtaining viscoelastic behavior of PVC. This technique can be of help also in determining differences in degree of crystallinity between two polymers of about the same molecular weight.

Dynamic mechanical properties measurement provides a better approximation of low shear limiting viscosity of rigid PVC melts without the problems inherent in capillary extrusion measurements of a partially crystalline fluid. Furthermore, dynamic mechanical properties measurements also yield the normal stress difference, which is a measure of melt elasticity.

3.6.3 Effect of Fillers

The concentration of inorganic pigments and fillers seems to have a very pronounced effect on the puff-up, post extrusion swelling. Table 3.5 presents a plot of the percentage puff-up $[(De/Dc)-1] \times 100$ versus shear rate at three temperatures, 170, 180, and 190°C, for two type I rigid PVC compounds containing, respectively, 5 and 15% of TiO_2. As shown in this table, we can conclude the following.

1. For both compounds, the percentage puff-up increases with the shear rate.
2. For two rigid PVC compounds, differing mainly in the concentration of TiO_2, the higher the filler content, the lower the percentage puff-up. Also, the higher the filler load, the lower is the percentage increase in postextrusion swelling when the temperature increases from 170 to 190°C.

Table 3.5 % Puff-up $[(De/Dc)-1]$ 100^a Shear Rate, sec^{-1}

Temperature °C	8.77	43.8	87.7	438	877	1754
170						
Type I rigid PVC with 5% TiO_2	7	7	5	15	17	19
Type I rigid PVC with 15% TiO_2	5	7	9	10	12	14
180						
Type I rigid PVC with 5% TiO_2	9	12	17	20	20	20
Type I rigid PVC with 15% TiO_2	5	9	9	14	14	15
190						
Type I rigid PVC with 5% TiO_2	15	20	24	29	29	27
Type I rigid PVC with 15% TiO_2	7	7	14	17	19	17

[a]De = extrudate outside diameter; Dc = capillary opening.

3. At the lower temperature, 170°C, postextrusion swelling is about the same for both compounds.

As mentioned before, (Sec. 2.2), emulsion resins are added quite frequently to suspension resins to modify some of the resin compound properties. Specifically, these emulsion resins reduce the puff-up and the puff-up sensitivity to stock temperature, and because of reduced elastic deformation, improve extrudate surface. If the molecular weight of the emulsion resin is high enough, these emulsion resins produce a modest increase in impact resistance. Also, some gain in power economy has been occasionally observed, due to the lower apparent melt viscosity of the resin blend.

A possible explanation for the improved behavior of the blend of resins is that fine particle size paste resins, 0.2 to ~ 1 µm, act as relatively inert fillers in a continuous matrix. This is more so if there is a significant difference in molecular structure between the paste resin and the matrix. It seems that below its crystalline melting point, T_m, the original emulsion resin particles preserve their shape, size, and particularity, hence contributing to slippage of resin particles and to lower apparent melt viscosity and low elastic deformation and recovery.

In addition, the surfactants on the emulsion resins are partly responsible for the lower viscosity.

The use of emulsion PVC in pressure-rated pipe, however, has been
restricted because traces of water-sensitive emulsifiers still remaining
in the resin shorten the life of pipe exposed to water.

3.6.4 Melt Fracture

Irregularities in the surface of extruded rigid PVC polymer are caused
by the release of elastic energy, stored by the polymer as it is processed through the extruder die [15]. These elastic responses and
melt fracture tendencies of polymer formulations can be quantitatively perceived as a function of shear stress, shear rate, temperature,
and compounding additives.

When operating an extruder at a high throughput rate, the extrudate sometimes takes on a rough irregular appearance that cannot be
attributed to any other cause than the physical breakdown of the surface of the melt or "melt fracture." It appears to be a stick-and-slip
phenomenon, i.e., the plastic melt sticks and builds up to a certain
volume, which then releases or slips, resulting in a steady pattern of
irregular flow.

This phenomenon occurs when the shear stress of the melt exceeds
its shear strength. An example of this is in an extrusion die where,
due to a substantial reduction in channel width, a sudden increase
occurs in the shear rate. Original investigations by Tordella [16,17]
showed that melt fracture occurred at a critical pressure point that
varied with the viscosity of the melt, the die pressure point, and the
die design or geometry. Special attention should be given to the last
factor, since it is the only one that can be modified without causing a
loss of extrusion output.

For a given rigid PVC compound being processed at a given temperature through a standard die, melt fracture is found at some definite
output rate, or shear rate through the die. Shear stress-shear rate
curves developed using capillary extrusion rheometers can be used to
predict shear rates when melt or surface fracture will occur. This information, used in conjunction with the computed shear factors for
any die system, can estimate the maximum output for the production
of quality material.

Paradis [18] has presented a series of practical examples showing
how the rheological characteristics of rigid PVC compounds can be
used to estimate performance in terms of output and the suitability
of a compound for any extrusion and/or injection-molding fabrication
process. Paradis conducts his technique in three steps, as follows.
In step 1 a capillary rheometer is used to determine rheological properties, like shear rate-shear stress curves, as well as the shear
rate at which melt or surface fracture occurs (inflection point) on
typical compounds (e.g., Type I, Type II, or blow-molding) at typical
processing temperatures. Data are then plotted on log-log paper as

a family of curves. In step 2, the operating shear rates are then
determined for each extruder die size and extrusion rate and the data
are plotted in a family of shear rate (sec^{-1}) versus extrusion output
(pounds per hour) curves. Equations (3.10, (3.11), and (3.14) are
used to determine the operating shear rates. The number of curves,
however, can be significantly reduced if these curves are developed
at the lowest stock temperature that produce acceptable product. In
this section we use only the annular die shape to illustrate Paradis'
technique.

In step 3, from the plots generated in step 1, the shear rate data at
which melt fracture occurs is taken to the shear rate versus extrusion rate plot generated in step 2, and the corresponding extrusion
output (pounds per hour) reading is the maximum output for the production of quality material for the requested product (pipe or tubing
e.g., see Figs. 3.14 and 3.15).

A simple example taken from Paradis' presentation follows: If we
want to know the optimum output of an extruder producing 1 in.
schedule 40, Type II pipe and the average extrudate melt temperature
is 415°F, we use a plot similar to that in Fig. 3.14 to obtain the shear
rate at which melt fracture probably occurs, in this case at 100 sec^{-1}.
This number when applied to Fig. 3.15 gives about 180 lb/hr. Of
prime importance to the success of this technique is an accurate measurement of the extrudate stock temperature and the availability of a
series of shear stress versus shear rate curves at typical extrudate
stock temperatures.

The same technique can then be applied for blow-molding operation
via parison extrusion, which also uses an annular die.

The problems associated with the occurrence of melt fracture during
the extrusion of rigid PVC blow-molding compoundings are more criti-

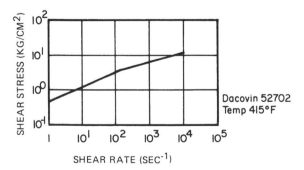

Fig. 3.14 Shear stress versus shear rate for Type II pipe compound.
(Courtesy of SPE, Brookfield Center, Connecticut.)

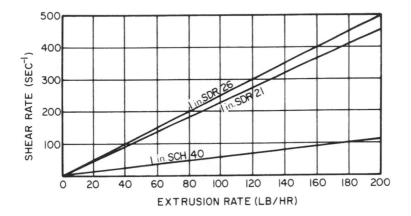

Fig. 3.15 Extrusion rate versus shear rate for 1 in. pipe. (Courtesy of SPE, Brookfield Center, Connecticut.)

cal in the case of glass-clear containers, since the presence of melt fracture can ruin the surface appearance and the clarity of the container. In some other applications of only functional value, such as PVC pipes or tubing, some melt fracture is more tolerable. Even in the case of some opaque profiles, like siding and its accessories, some microroughness, due to a mild case of melt fracture, is not objectionable.

We have already seen the existence of a value of shear rate at which melt fracture occurs. Schulken and Boy [19], who have studied the theoretical aspects of melt fracture, call it a critical shear rate, which depends mainly on the strength of the melt, the rate of increase of shear rate, and the viscosity breakdown factors for the melt.

Following the pioneering work of Schulken and Boy [19], Fisher [20] presented very illustrative example on how this information can be used not only to analyze performance of existing dies but also to provide die designers with techniques to estimate optimum die entry geometry. This technique could also be applied to estimate the optimum extruder die adaptor dimensions, such as entry angle and the length of the entry section. As stated before, these dimensions, when optimized, contribute to higher outputs.

For a given rigid PVC compound, the critical shear rate is determined with a set of capillaries having, except for different entrance angles, the same capillary length and diameter. By varying the flow, Q, the critical shear rate, the shear rate at which melt fracture occurs, is then determined and plotted against the half-angle of entry (degrees) for each capillary. This is the critical shear rate curve for the rigid PVC under consideration (see Fig. 3.16).

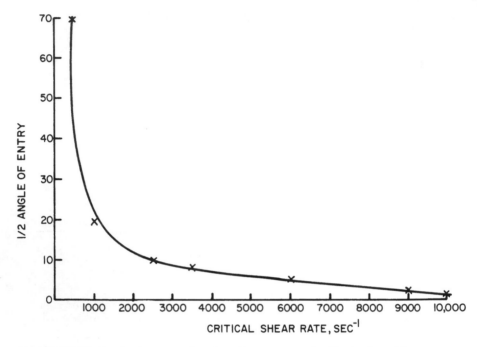

Fig. 3.16 Estimated curve showing the expected effect of capillary entry angle on critical shear rate for a rigid PVC Type I compound.

For a desired output Q, this curve can be used as a reference for calculating the optimum entrance angle for the rigid PVC compound under consideration. The equation for shear rate

$$\dot{\gamma} = \frac{4Q}{\pi r^3}$$

where

    Q = the rate of flow
    r = capillary radius
    $\dot{\gamma}$ = shear rate

is used to determine the shear rate corresponding to the desired output. This $\dot{\gamma}$ value is then taken to the curve, and, from the plot, the optimum entry half-angle is obtained.

If, for example, we have a die adaptor or a die with one taper angle, the previous equation in combination with the critical shear rate curve can be used to determine the maximum flow rate Q, free from surface fracture, that one can obtain extruding the given compounds through the die.

$$Q = \frac{\dot{\gamma}\pi r^3}{4}$$

where

r = exit radius
$\dot{\gamma}$ = critical shear rate taken from the curve

Powell [21] offers the following approximate equation for half-angle α of entry: tan α = $\sqrt{2\eta/\lambda}$, where η is the apparent melt viscosity corresponding to the shear rate $\dot{\gamma}$ at the die entry, and λ = extensional viscosity corresponding to the tensile rate, also at the entry. Tensile flow data are obtained using a constant stress melt tensile rheometer like the one used and discussed by Cogswell in Ref. 22 or the Goettfert Rheotens melt extensiometer as discussed in Ref. 23.

These are simple cases. This technique, however, can be applied to geometrically complex dies or adaptors. It is a common practice to use tapered flow channels to achieve maximum output rate under conditions of laminar flow. (Dies of constant cross section are rarely used alone.) Tapers are used in the body of the die to reduce the overall pressure drop or at the die entry to eliminate dead spots. They also contribute to a more streamlined flow, which is desirable for heat- and shear-sensitive material like rigid PVC.

As shown by Powell's expression, the choice of taper angle depends on the extensional properties of the melt as well as on the shear properties: in converging flow in a uniform tapered channel the tensile stress in the flow direction increases and reaches a maximum value at the narrow end of the taper, where shear rate and pressure drop also become maximum. As stated before, Tordella [16,17] has shown that, if this tensile stress exceeds a critical value, a critical pressure point, rupture of the extrudate surface can occur.

Having said all this, we should note that basic differences exist in dies designed for single-screw and twin-screw extrusion. Stiffer melt, lower temperature twin-screw extrusion responds better to higher inventory dies with a higher compression ratio; single-screw extrusion at higher melt temperatures benefits from lower inventory dies with lower compression rates.

3.7 MELT STABILITY

Melt stability is a measure of a material's ability to maintain a constant melt flow or melt viscosity with time at extrusion temperature. This information can be easily obtained in some rheometers like the Dayton and Instron rheometers, which are equipped to extrude automatically four charges of a constant volume of molten polymer at various residence times in the rheometer barrel. Specifically, the test is run as follows. Using a tamping rod, the polymer is compacted in the barrel

while the barrel is in forward position. Once the barrel is slid into
position under the ram, the polymer is molten and kept at a temperature T. The four clocks controlling the residence dwell time are set
at, i.e., 5, 10, 15, and 20 min. Under a constant pressure P, four
extrusions representing the various rheometer dwell times are then
recorded.

There are many ways of expressing the melt stability. One common
way is the change in apparent melt viscosity with time, $\Delta \eta_a$/time at
the various temperatures. A positive $\Delta \eta_a$ indicates that the PVC compound increased in viscosity due to cross-linking. Unsaturated rubbers used as impact modifiers tend to cross-link. A PVC compound
with 0 $\Delta \eta_a$/time is more ideal for production, since the melt viscosity
remains unchanged as a function of time, and long production runs
can be made with minor processing changes. On the other side, a
negative value of $\Delta \eta_a$/time indicates that the PVC compound under
evaluation is decreasing in molecular weight (chain scission) due,
very probably, to degradation (see Fig. 3.17).

Perhaps the easiest way of analyzing the melt stability data is by
plotting the extrusion time in seconds versus the residence time of
the molten polymer in the rheometer barrel in minutes. An increase
in extrusion time with dwell time indicates an increase in viscosity or
cross-link, but a decrease in viscosity or cross-link during a decrease

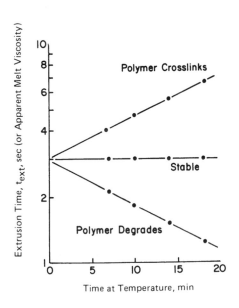

Fig. 3.17 Determination of melt stability factor.

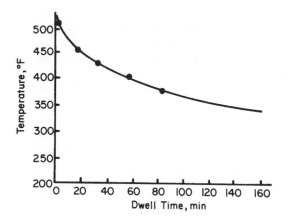

Fig. 3.18 Time for a copolymer of acrylonitrile (70.0%) and styrene (30.0%) to reach + b = 17.5.

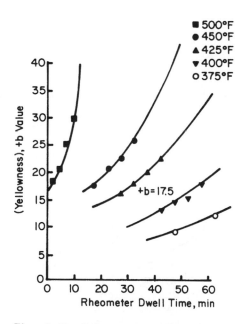

Fig. 3.19 Color propagation curve for a copolymer of acrylonitrile (70.0%) and styrene (30.0%).

in extrusion time indicates that the polymer is degrading. This technique can be used very effectively for materials comparison.

In another version of this test, the Instron rheometer is used to obtain the data, which are obtained at standard melt flow conditions as follows. Zones 1 and 4 are run at the same load (psi) with a 40 min period between the two, and the shear rate (sec^{-1}) versus the viscosity (kilopoise) is determined and plotted through zone 3, as per standard. Then the viscosity corresponding to zone 4 is obtained and compared to that of zone 1. A ratio >1.0 of the 45 min viscosity (zone 4) versus the initial viscosity (zone 1) indicates cross-linking, and <1.0 indicates chain scission has occurred.

3.8 COLOR PROPAGATION CURVES

Capillary rheometers can also be used to measure the time-temperature effect on polymer color stability. The procedure, which is rather simple, follows. The barrel is loaded with polymer at a fixed constant temperature and at various dwell times. The polymer melt is then manually extruded under constant pressure through the same capillary. The extrudate is then flattened to a constant thickness and yellowness index, +b value, is measured in a HunterLab colorimeter. Yellowness, +b value, is then plotted versus dwell time at the various temperatures evaluated. This technique works very well with glass-clear PVC formulations, specifically with the food-grade compounds containing calcium-zinc stabilizers or low levels of the di-n-octyltin maleate or di-n-octyltin-S, S'-bis(isooctyl mercaptoacetate) stabilizers. Examples of color propagation curves are presented in Fig. 3.19 to illustrate this technique. As shown, the polymer is not PVC but an unmodified, glass-clear acrylonitrile (70.0%) and styrene (30%) copolymer that, although with higher melt viscosity than PVC, can be processed in similar equipment.

From these curves we can derive a time-temperature plot (time at temperature to produce a fixed yellowness, +b value) (see Fig. 3.18). These curves can then be used to give an overall idea of the color stability of the rigid PVC compound under evaluation.

3.9 GLASS TRANSITION TEMPERATURE T_g

Most of the commercial rigid PVC compounds and the resins used to manufacture them have T_g between 74 and 82°C. Below these temperatures, the longer molecular segments are not free to rotate and there is little room for Brownian movement. Motions of shorter segments, however, are likely to continue, and other transition temperatures below T_g may be found. The high degree of chain-to-chain attraction reinforced by the high polarity of the pendant Cl group produces high cohesive energy density. These interconnecting factors are re-

Table 3.6 Effect of $\overline{MW}$ on T_g

PVC resins $\overline{MW}$ (GPC)	T_g, °C[a] by DTA	Specific viscosity[b]
95,000	81	0.47
75,000	80	0.39
58,000	77	0.30
57,000	77	0.29
50,000	74	0.25
VC/VA polymer 6.1% VA	65	0.43

[a]Differential thermal analyzer.
[b]0.4 g per 100 ml cyclohexanone (at 25°C).

sponsible for the rather high T_g values of rigid PVC compounds. The pendant Cl also contributes to improved barrier properties and chemical inertness.

At T_g the cohesive forces yield drastically and the polymer expands so that there is room for segmental motion. The polymer then starts to flow. Above T_g, a small unit increase in temperature produces a much larger increase in volume. This explains why, in the dry blend preparation of rigid PVC compounds in high-intensity mixers, the PVC resins are normally sheared up to about their T_g value: above that temperature the intermolecular forces are rather weak, favoring the mutual dispersability of resins and additives.

Table 3.6 presents the effect of molecular weight on T_g for various PVC resins. As shown in Table 3.6, T_g increases with the $\overline{MW}$ of the resin. As presented by Daniels and Collins [24], T_g also increases with decreasing polymerization temperature, and since syndiotacticity normally shows the same dependence on temperature of polymerization, the increase in T_g could be due to the increase in syndiotacticity. As reported by the same authors [24], the melting point T_m of PVC is also strongly affected by syndiotacticity (crystallinity) and to some extent by molecular weight and chain length of the polymer. The higher are the tacticity and molecular weight, the higher the T_m. At normally encountered molecular weights, however, it is not the chain length but the tacticity that controls the melting point of the crystallites. As is also shown in Table 3.6, copolymers of VC/VA when compared with PVC of equivalent specific viscosity show a sharp loss in T_g. It is apparent that the insertion of VA units at regular intervals in the PVC chain results in less hindered mobility of the chain because of reduced steric factors (possibly reduced interchain hydrogen bonding of dipole interactions) or because of the reduced crystallinity. Copolymerization is expected to result in reduced crys-

tallinity because of disruption of structural and compositional uniformity along the chain.

The T_g of a PVC homopolymer resin and a chlorinated PVC resin,
$\sim 65.6\%$ chlorine, are $\sim 80\,^\circ C$ and $\sim 118\,^\circ C$, respectively. As seen, a
significant increase in T_g is achieved by the chlorination of the PVC,
that is, by the significant increase in polarity, which stiffens the
chains and reduces segmental motion, resulting in reduced flow because of the reduced mobility of chains in the melt.

Some additives like high molecular weight poly(methylmethacrylate)
(PMMA) polymers or the AN-AMS-S (acrylonitrile-α-methylstyrene-
styrene-terpolymer) when polyblended with PVC increase the T_g
of the compound. They also act as modifiers to improve the deflection
temperature under load (DTUL) of PVC. For this reason they are
called heat distortion improvers. Some other additives, like the AN-
AMS-tert-butylacrylamide terpolymers, tend to depress to T_g. In this
case the amide group may compete for the strong dipole interactions in
PVC that contribute to its rigidity.

It is evident that, in order to obtain PVC blends of high T_g, one
must use a large amount of diluent polymer or a diluent polymer of
very high T_g containing no strong polar or hydrogen bonding groups.
As already discussed in this chapter, T_g increases with decreasing
polymerization temperature. This approach, however, simultaneously
produces a more dramatic increase in molecular weight and syndiotacticity of the PVC chain, which have a significant effect on polymer
processability. Collins and Daniels [24] have also demonstrated that
the effects of molecular weight and syndiotacticity can be independently
manipulated, which suggests the possibility of selectively increasing
the T_g values of the basic resin with no major change in polymer processability.

3.10 MELTING POINT T_m

As mentioned before, PVC is a partly crystalline polymer, which in itself is not unusual, but its crystalline melting point is well over $210\,^\circ C$,
where thermal degradation becomes serious. Processing must be done,
therefore, well below the melting point. During processing, PVC
grains break down to primary particles, which fuse when the crystallites acting as pseudo-crosslinks start to melt, freeing polymer chains
(supermolecular units) for interdiffusion between the primary particles: hence, the importance of crystallinity on PVC melt processing.
This crystallinity accounts for most of the major differences between
PVC melt flow and normal amorphous flow, such as is found in polystyrene at similar temperatures. As it has been also shown before,
the puff-up behavior of PVC is similar to that of low-density partly
crystalline powder polyethylene (PE), below its crystalline melting
point, increasing with temperature. Above its T_m, this PE puff-up
decreases with increasing temperature. No similar decrease has been

observed with ordinary PVC, indicating that they are still below their
crystalline melting points, about 215°C.

The x-ray diffraction technique for measuring crystallinity in PVC
at room temperature is quite good. Adaptation to high-temperature
measurements, however, has not yet provided quantitative values, for
the most part. Thus, we must rely on the premise that, as crystallinity increases, T_m also increases, as with PE and polypropylene materials. Crystallinity, furthermore, increases with the molecular
weight. High-molecular-weight PVC, being more crystalline and more
viscous, may not mix to as nearly an ideal extent as low-molecular-
weight PVC. Also, wetting of the resin particles by heat stabilizers,
processing aids, and elastomer particles is probably adversely affected by high molecular weight and crystallinity. Consequently, the
melt processing of unplasticized PVC compounds based on high-molecular-weight resins, when compared with the ones based on lower MW
resins, may not be as thermostable at their respective T_m values.
Determination of T_m for high-molecular-weight resins thus becomes
even more difficult. The above-stated, however, does not mean that
high-molecular-weight PVC resins should necessarily be less heat
stable than lower MW resins. Due to the longer chains and fewer
structural irregularities, e.g., double bonds, carbonyl groups, and
hydroperoxides, associated with lower polymerization temperatures,
they should be, at least in theory, more stable [25,26]. Recent
studies conducted in a C. W. Brabender Plasticorder, which provides
a continuous mixing of the melt [27], show a shift of the degradation
peak to a higher temperature for the higher molecular weight PVC
resins.

Unplasticized PVC compounds with a T_g/T_m ratio ~ 0.40 do not fit
the well-known ratio range between 0.50 and 0.75 shown by most of
the thermoplastic polymers.

3.11 TEMPERATURE DEPENDENCE OF VISCOSITY

The influence of temperature on the viscosity of polymer melts is usually expressed in terms of an Arrhenius-type equation,

$$\eta = Ae^{E/RT}$$

where

    η = melt viscosity
    A = constant
    e = base of natural logarithm
    E = activation energy
    R = gas constant
    T = absolute temperature

From the measured values of melt viscosity at several temperatures,
log η versus $1/T$ can be plotted for a given PVC compound. The

slope of the line represents the activation energy of flow for the rigid
PVC compound under consideration. Since unplasticized PVC melts
do not behave as a newtonian fluid, determination of the flow-activation energy becomes rather complicated. Several factors affect the
flow-activation energy of unplasticized PVC melts, mainly, resin
molecular weight, melt temperature, shear rate, and particulate nature of the polymer. Also certain modifiers appear to alter the activation energy at the lower temperatures.

Collins and Metzger [4] have shown that the influence of temperature on viscosity and flow-activation energy depends on the temperature range; that is, the flow-activation energy is not consistent irrespective of whether this flow-activation energy is based on viscosities
at constant τ or $\dot{\gamma}$ but can be approximated by two values, one for the
low temperatures and the other for high temperatures. The change in
activation energy occurs near the temperature region associated with
the melting point of PVC, and since the melting point increases with
the MW, and the activation energy is associated with melting, the activation energy should increase as the molecular weight increases. This
is the case at constant shear stress. As presented by the same authors, however, this is not the case at constant shear rate where the
flow-activation energy based on viscosity decreases as the $\overline{MW}$ increases. These conclusions are drawn from the experimental data
tabulated in Table 3.7.

A plot of the effect of temperature on a medium-MW PVC resin melt
viscosity is shown on Fig. 3.20. This plot seems to establish that,
in the temperature range 177 to 191°C and at the values tested, PVC
melt viscosity does not show as great temperature dependence as it
does either above or below this range. Fig. 3.20 also shows that
the effect of temperature on viscosity decreases as the shear rate increases to 300 sec^{-1}. Since this plot represents conditions similar to
those encountered in extrusion operations, this type of information is
of great value to process engineers. For example, an increase in extruder barrel heat from about 177 to 190°C has a limited effect on lowering the melt viscosity. An increase in screw rpm, however, provides a more effective approach. Furthermore, if the log of the viscosity is plotted versus the processing melt temperatures at a constant shear similar to that of the melt-processing system under consideration, very valuable information can be obtained. The slope of
the plot, which is the flow-activation energy, indicates the temperature dependence of viscosity: the greater the slope, the more sensitive is the viscosity to temperature. This information can guide process engineers in dealing with processing problems. For example,
if the slope of an injection-molding compound at the processing temperature is small, raising the temperature has a small effect on mold
cavity filling. Increasing the screw rpm (shear), injection and holding pressures, and injection speed have a more significant effect.

The flow-activation energy is reported as kilocalories per mole and
must be calculated at a constant shear rate. The higher the activa-

Table 3.7 Flow-Activation Energy

Resin[a] designation	Shear rate, 10 sec^{-1} kcal/mol	Shear stress, 3 X 10^6 dynes/cm^2, kcal/mol
A	36.0	31.2
B	29.3	42.1
C	26.9	50.3
D	22.5	34.5
E	14.1	58.5

[a]PVC samples listed in order of increasing molecular weight.
Source: Data courtesy of the Society of Plastics Engineers.

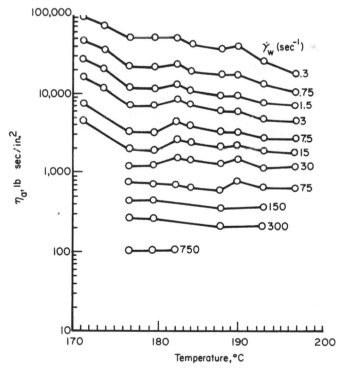

Fig. 3.20 Temperature dependence of PVC viscosity resin C.

tion energy, the more dependent is the viscosity on the temperature. At a shear rate of 10 sec^{-1}, common values of activation energy are in the range of 15 to 35 kcal/mol. Figures 3.6, 3.7, and 3.8 show η_a versus the sample temperatures at three different shear rate values (100, 275, and 1000 sec^{-1}) for five commercial PVC resins with average molecular weights measured by GPC of 112, 90, 74, 56, and 50 $\times 10^3$. As shown, little difference in η_a versus temperatures at the three shear rates evaluated is seen for the higher molecular weight materials. As crystallinity and molecular weight drop, considerable differences in η_a emerge. The greater the drop, the more pronounced are the differences. Comparing these three figures, one can see that the η_a-T dependence is less as the shear rate $\dot{\gamma}$ increases.

Since the shear rate as well as the temperature ranges presented cover extrusion as well as some low-shear injection-molding processes, the information presented is of value for resin selection. For example, a resin molecular weight of about 56 $\times 10^3$ appears to be about the minimum that provides a controllable amount of η_a-T drop. For an extrusion and die forming process, however, an even higher MW resin (MW $\geqslant$ 74 $\times 10^3$) is more desirable.

To allow the separate assessment of the effects of crystallinity and molecular weight on the temperature dependence on viscosity of PVC, model polymers were fractioned, using established procedures. This study is presented in Table 3.8 and Fig. 3.21. Figure 3.21 shows a rather flat η_a-T behavior for the high-molecular-weight and high-

Table 3.8   Data on Fractions of Model Polymers[a]

Sample	X	$\overline{MW}$ (GPC)	Log $\overline{MW}$	145°C	160°C	175°C	190°C
				\multicolumn{4}{c}{η_a lb sec/in²}			
1	30.0	109,000	5.037	3.30	2.76	2.30	1.91
2	26.5	117,000	5.068	3.00	2.48	2.05	1.70
3	23.5	67,000	4.826	3.64	2.36	1.54	1.01
4	23.5	25,500	4.406	1.22	.57	.27	.13
5	23.5	39,000	4.591	2.77	1.41	.72	.37
(40 pt4 60 pt3)							
6	22.5	39,500	4.597	2.66	1.07	.44	.18
7	22.5	13,500	4.130	.30	.09	.02	.009

[a]X = room temperature percentage crystallinity of samples heated 0.5 hr at 150°C and slowly cooled. $\overline{MW}$ = weight average molecular weight from gel permeation chromatography (GPC).
η_a = extrusion rheometer apparent viscosity ($\dot{\gamma}$ = 82.5 sec^{-1} L/D = 8).

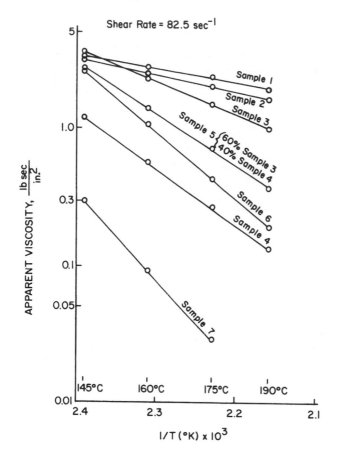

Fig. 3.21 Table 3.6, plot of apparent viscosity versus 1/T (K) and versus T(°C).

crystallinity fraction, whereas the lower crystallinity and lower molecular weight have considerable variation of η_a with T. The blend of the two fractions with the same percentage of crystallinity but different molecular weights behaves like a monolithic system, with a $\overline{MW}$ almost predictable from the $\overline{MW}$ of the two resin components.

3.12 PRESSURE DEPENDENCE OF VISCOSITY

To determine the effect of higher pressures on the temperature dependence of viscosity, the Doolittle viscosity free-space relationship was combined with the Spencer and Gilmore [28,29] equation of state, as follows. Doolittle shows [30] that if the viscosity of many low-molecular-weight hydrocarbons is plotted against reciprocal relative free-space, V_0/Vf, straight lines are very precisely determined for each compound. Thus, this viscosity free-space relationship can be represented by (A and B are constants)

$$\eta = Ae^{B/(Vf/V_0)} \tag{3.25}$$

where

Vf = volume of free space per 1 g of liquid at any temperature

V_0 = volume of 1 g liquid extrapolated to absolute zero without change of phase, which is also known as the specific volume W at 0 K.

V = volume of 1 g of liquid at any temperature, also known as the specific volume at temperature T

The Vf/V_0 relationship is also written as $Vf/V_0 = (V - V_0)/V_0$ or $Vf/W = (V - W)/W$. Equation (3.25) can be then written

$$\eta = Ae^{BW/(V-W)} \tag{3.26}$$

If this equation is combined with the well-known Spencer and Gilmore [28,29] equation of state for polymers:

$$\frac{P + \pi}{V - W} = nRT$$

where π is internal pressure, n is the number of moles, R is the gas constant, and T is the absolute temperature, we obtain the following expression:

$$\eta = Ae^{BW[(P+\pi)/nRT]} \tag{3.27}$$

This compares with the usual

$$\eta = Ae^{\Delta E/RT}$$

so that

$$\Delta E = \frac{BW}{n}(P + \pi) \tag{3.28}$$

which predicts that the temperature dependence of viscosity, ΔE increases with pressure. Also of interest in this equation is the dependence of ΔE on the internal pressures π. This property is also known as the cohesive energy density and is a measure of molecular polarity. The increase of ΔE with polymer polarity has long been recognized.

Using an Instron capillary rheometer modified with a double plunger to allow measurements at high hydrostatic pressures, the pressure dependence of viscosity was studied for a Type I, rigid PVC compound.

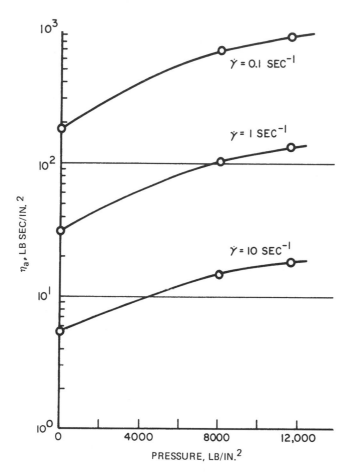

Fig. 3.22 Hydrostatic pressure dependence of Type I, rigid PVC compound viscosity at 345°F.

Some results are presented in Fig. 3.22. Under the conditions of this study, the influence of pressure on the melt viscosity of the rigid PVC compound under consideration was as follows. At the lowest shear rate evaluated, $\dot{\gamma} = 0.1 \ sec^{-1}$, and at 345°F, an increase of 10,000 psi caused a three- to fourfold viscosity increase. At 100 sec^{-1}, and at 365°F (see Fig. 3.23), the same pressure causes only a two- to threefold viscosity increase. The results at 365°F show this compound viscosity at this temperature to be less pressure dependent than at 345°F, so that a further temperature increase might be expected to continue this trend.

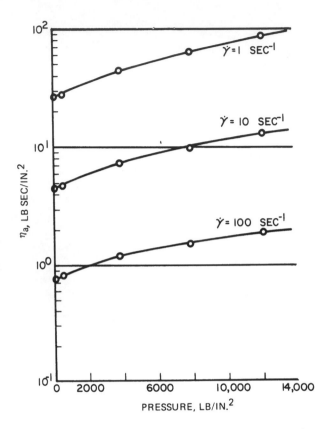

Fig. 3.23 Hydrostatic pressure dependence of Type I, rigid PVC compound viscosity at 365°F.

Table 3.9 Some Activation Energies

Type I, rigid PVC compound, pressure, (lb/in.^{2})	Shear rate, sec^{-1}	ΔE, kcal/mol
0	1	5.6
0	10	6.3
5000	1	14.5
5000	10	11.4
10000	1	16.7
10000	10	13.1

It must be pointed out that the above results and, in fact, all capillary rheometer data, are somewhat in error because of the effect of pressure in increasing the viscosity of the fluid in the capillary, particularly near the entrance region. Correction of this error, however, is not discussed here.

Table 3.9 lists some activation energies at two shear rates, 1 and 10 sec^{-1}, and three hydrostatic pressures, at increasing order. These experimental data confirm that the temperature dependence of viscosity ΔE increases at higher pressures. Also, it seems that, at the higher pressures, it decreases when the shear rate increases from 1 to 10 sec^{-1}.

REFERENCES

1. L. A. Utracki, *Polymer Eng. Sci* 14(4):308–314, 1974.
2. J. M. McKelvey, *Polymer Processing*, John Wiley & Sons, New York, 1962.
3. R. E. Colwell and K. R. Nickolls, *Ind. Eng. Chem.* 51:841, 1959.
4. E. A. Collins and A. P. Metzger, *Polymer Eng. Sci.* Vol. 10(2):1970.
5. F. Bueche, *Physical Properties of Polymers*, Interscience, New York, 1962.
6. J. Munstedt, *J. Macromol. Sci. Phys. B-14,* 2:195, 1977.
7. E. A. Collins and C. A. Krier, Trans. Soc. Rheol. 11(2):225, 1967.
8. R. J. Brown, *SPE ANTEC 40th*, 1982, pp. 507-508.
9. A. R. Berens and V. L. Folt, Trans. Soc. Rheol. 11:95, 1967.
10. A. R. Berens and V. L. Folt, Polymer Eng. Sci. 8:5, 1968.
11. M. Mooney and W. E. Wolstenholme, J. Appl. Phys. 25:1098, 1954.
12. M. Mooney, J. Appl. Phys. 27:1956.
13. L. E. Nielsen, *Mechanical Properties of Polymers*, Reinhold, New York, 1962.
14. J. D. Ferry, *Viscoelastic Properties of Polymers*, John Wiley & Sons, New York, 1961.
15. C. L. Sieglaff and C. G. Vinson, Polymer Eng. Sci. 9(1): 1969.
16. J. P. Tordella, *J. Appl. Phys.* 27(5):454, 1956.
17. J. P. Tordella, *Trans. Soc. Rheol.* 1:203, 1957. (Twenty-fifth Anniversary Technical Conference of the Society of Plastics Engineers, Inc., May 15-18, Detroit.)
18. R. A. Paradis, Rheology of rigid PVC defined in terms of commercial production conditions, *SPE ANTEC 25th* May 1967, pp. 1073–1077. (Sixteenth Anniversary Technical Conference of the Society of Plastics Engineers, Inc., Jan. 1960, Chicago.)

19. R. M. Schulken and R. E. Boy, Cause of melt fracture and its relation to extrusion behavior, *SPE Technical Papers, 16th,* 1960, p. 6 (Session 23, Paper 82).

20. E. G. Fisher, *Extrusion of Plastics*, Interscience, New York, 1964.

21. P. G. Powell, Polymer Eng. Sci. Vol. *14*(4):198—306, 1974.

22. F. N. Cogswell, *Rheol. Acta 8*(2):187, 1969.

23. A. S. Pazur and L. C. Uitenham, *SPE ANTEC 39th* May 4—7, 1981, pp. 573—576.

24. A. C. Daniels and E. A. Collins, Polymer Eng. Sci. Vol. *19*(8): 1979, pp. 585—588.

25. G. Scott, ACS Symposium Series 25:1976, 340—365.

26. E. D. Owen, ACS Symposium Series *25*: 1976, 208—217.

27. E. A. Rabinovitch, *SPE ANTEC 40th*, 1982, pp. 498-499.

28. R. S. Spencer and G. D. Gilmore, *J. Appl. Phys.* 20:502, 1949.

29. R. S. Spencer and G. D. Gilmore, J. Appl. Phys. *21*:523, 1950.

30. A. K. Doolittle, J. Appl. Phys. *22*(12): 1951.

4

Rigid PVC Extrusion

I. LUIS GOMEZ/Monsanto Company, Springfield, Massachusetts

4.1   Introduction 153

4.2   Extrusion of Rigid PVC 153

4.3   Modern Single-Screw Machines 155

4.4   Polymer Residence Time in Modern Single-Screw Machines 156

4.5   Instrumentation-Application to Monitor Operating Variables 157

4.6   Extrudate Temperature T_* 157

4.7   Melt Pressure 159

4.8   Flow Function ϕ 160

4.9   Mechanical Power p 161

4.10 Analysis of Screw Performance 161

4.11 Energy Factor 162

4.12 Mechanical Energy Dissipated as Heat in the Polymer 164

4.13 Scale-up Rules for Single Screws: Analysis and Discussions 166

4.14 Rules for Semigeometric Scaling 167

4.15 Flaws of the Semigeometric Scaling Rules 170

4.16 Drag Capacity of the Metering Section of the Screw 171

4.17 Analysis of Experimentally Determined Performance Data 172

4.18 Analysis of Two-Stage Single-Screw Extrusion in Vented Barrel Extruders 174

4.19 Characteristic Curves 178

4.20   Processing Rigid PVC Dry Blends in Two-Stage Single-Screw
       Vented-Barrel Extruders 182

4.21   Two-Stage Single-Screw with a Barrier or Double-Channel
       Section 187

4.22   Vent-Zone Operation 191

4.23   Pumping-Zone Operation 192

4.24   Screw Nose Design 192

4.25   Screw Temperature Control 193

4.26   Extruder Temperature Control 194

4.27   Combined Screw and Die Performance 196

4.28   Flow-Restricting Devices 200

4.29   Plate-out 201

4.30   Twin-Screw Extrusion Process 203

4.31   Analysis of Twin-Screw Extrusion Process 208

4.32   Matching the Formulation to the Extrusion Process and to the
       Finished Products 208
       4.32.1 Two-Stage Single-Screw Vented Extrusion 209
       4.32.2 Conical Twin-Screw Extrusion 210

4.33   Extrusion of Pipes and Conduits: Single Versus Conical Twin-
       Screw Dry Blend Extrusion Process 212
       4.33.1 Two-Stage Single-Screw Vented Extrusion 212
       4.33.2 Conical Twin-Screw Extrusion 214
       4.33.3 Pipe Die Design: Matching Dies to Single and Twin-
               Screw Extrusion 215
       4.33.4 Pipe and Conduit Takeoff (Calibration, Cooling Train,
               and Haul-off Units) 221
       4.33.5 Cooling Bath Requirements 221
       4.33.6 Processing and Quality Problems That Can Be Remedied
               with Temperature Adjustments 228

4.34   Rigid PVC Extrusion of Siding and Accessories 230

4.35   Rigid PVC Extrusion of Window and Door Profiles 236

References 242

Bibliography 244

4.1 INTRODUCTION

Throughout this work we have dealt with (1) the preparation of rigid
PVC dry blends, mainly the equipment, the process, the resin, and the
dry blend properties; (2) the various melt-processing techniques used
to pelletize these dry blends; and (3) the rheological properties of
typical rigid PVC compounds and how these properties affect the melt-
processing characteristics of these compounds.

The process of extrusion immediately follows those of polymer preparation and/or melt compounding. It consists of continuously converting, through an external heat source and applied mechanical energy, a
suitable resin compound into a profile of a specific cross section by
forcing the material as a fluid through an orifice or die under controlled conditions. In order that this concept can have practical value,
however, certain requirements must be satisfied, concerning both the
raw material and the equipment.

Rigid PVC compounds are based on resins that when unmodified,
are processed with great difficulty. Through modification via additives,
however, their processability window is widened, but still the thermal
and shear sensitivities of these compounds remain as a latent threat
to any extrusion operation. As already shown, melt rheology provides
a valuable tool to assess whether the rigid PVC compound under consideration satisfies the requirements of the extrusion process.

The equipment must have a means of softening the resin compound
and also be capable of providing sufficient pressure continuously and
uniformly on the material, in order to convey the molten compound
to and through the die in as uniform manner as possible to achieve dimensional stability of the finished product.

There are so many factors involved in the areas of screw and die
fabrication that to do justice to such specialized areas, a separate
book would be necessary. Consequently, these areas will not be
covered here. Furthermore, no attempt is made here to discuss the
basic principles of the extrusion process and the theory of the various
screw zones, as well as the various components of the extrusion machines, which through the years have been, for the most part, extensively covered in the literature (see Bibliography).

In this chapter, emphasis is placed on process, process operation,
instrumentation, experimental measurements, and analysis of the quantitative data, as well as some of the theoretical framework that should
be useful in the design of the extrusion process.

4.2 EXTRUSION OF RIGID PVC

Extrusion is the most important forming technique of rigid PVC polymer
processing. More rigid PVC is converted into useful products by extrusion than by any other process. It is also a process common to other
rigid PVC melt-processing operations. Some specific examples of these

processing operations are (1) the modern reciprocating injection molding machines, which are built around an intermittent extrusion process;
(2) some new calendering lines, which use compounding extruders like
the planetary extruder or single-screw kneaders as the source of the
PVC melt, and (3) some of the older calendering installations, which
have short strainer-extruders located between the mill rolls and the
calendering rolls to protect the calender from metal.

Because rigid PVC polymers are inherently difficult to process,
standard extrusion technology cannot be applied directly to the processing of these compounds. The combination of their rather high
viscosity with their thermal sensitivity, which increases with the temperature, makes the development of high-capacity rigid PVC extrusion processes particularly challenging. The so-called static mixers,
for example, which are successfully used in other segments of the industry, show some processing limitations when used with rigid PVC
melts, i.e., a variable degree of polymer stagnation and degradation,
which occurs at the edges of the mixing elements and, as expected,
increases with the extrusion time (see Fig. 4.1).

Mainly, for economic reasons, the trend of this segment of the industry is toward a dry blend extrusion process that requires significantly more attention and high levels of experimentation by the process
engineers. Experimentation with production-sized extruders, however,
is expensive. Ideally, experimentation should be carried out on laboratory-sized extruders and the results of such experiments used to predict the performance of scaled-up equipment. The laboratory-sized
extruder thus becomes an extremely valuable development tool. It becomes possible for such parameters as feedstock formulations, polymer
properties, and operating variables to be thoroughly evaluated in the
laboratory and the results applied with some confidence to production-
sized equipment. The critical part here is in the development of scaling
rules applicable to quality and uniformity. As expected, these scaling
rules are not absolute and deviations between the predicted and the
experimentally determined performance data are somewhat common.

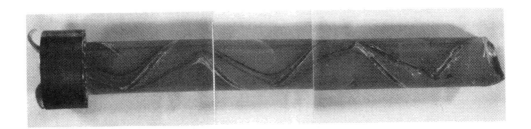

Figure 4.1 Static mixer.

4.3 MODERN SINGLE-SCREW MACHINES

The two-stage single-screw extruder designs discussed in Sec. 2.5.6
as well as those single-stage screws with an L/D ratio of up to 24/1
may be arbitrarily considered a sort of "first-generation" design. Single-screw extrusion machines with screws with L/D ratios higher than
24/1 (30/1 to 34/1), higher outputs and power economy (pounds per
horsepower hour), lower degree of extrusion surging, a more uniform
melt quality, and quite often special nose design and no breaker plate,
may be considered "higher temperature-second generation" designs.
These units normally find application in the dry blend extrusion of
thin-walled and small-diameter pipes and conduits, siding, window profiles, and panels. Through a combination of proprietary screw and die
designs, one of the largest pipe producers in this country extrudes
large-diameter pipe (up to 12 in. diameter), from powder via single-
screw extrusion. Because of the greater thrust capability of single-
screw extruders, they handle the back pressure of smaller pipe sizes
very effectively. This area is dominated by two-stage, single-screw,
long-barrel, vented extruders. The combination of the same extruders
in a nonvented mode with a compression relief screw and vacuum hopper
provides an alternative to the previous system. The potential polymer stagnation and degradation problems at the vent are traded for
potential air leaks in the screw vacuum seal system, which by fluidizing
the dry blend in the feed zones, causes serious feed problems.

It has been reported [1] that vacuum hoppers are not effective at a
moisture level higher than 0.065%. It is difficult, however, to produce
PVC compounds and keep them under this moisture level. By heating
the vacuum hopper and/or using a vacuum hopper feed system as explained in Ref. 2, removal of volatiles is significantly helped. This
method, however, is not as effective as the use of a vented barrel extruder in combination with a high level of vacuum.

This section is not intended to be a comprehensive review and discussion of all the modern two-stage single-screw extrusion systems;
rather, it attempts to interpret the significance of the extrusion processing work in terms of the following issues.

Screw performance characteristics (rate, pressure, and uniformity)
    as a function of the operating variables
Performance of the vent
Quality of the melt that can be generated by the extrusion process

Many factors contribute to the great variability among the extrusion
plants throughout the world. In this book we have already seen most
of these factors, and no further attempt to cover them will be made.
Instead, typical extrusion processes will be described. Selection of
these processes, although somewhat arbitrary, is based on the industry
trend which, in turn, is influenced by processing efficiency and power
economy. Dry blend extrusion in those machines offering high power
i.e., E > 8 lb/hp-hr, seems to fit this trend.

Although we concentrate on dry blend extrusion using two-stage single-screws in vented barrel extruders, and specifically on those designs with barrier-type sections, the discussion that follows in this chapter is applicable to any single-screw extrusion.

4.4 POLYMER RESIDENCE TIME IN MODERN SINGLE-SCREW MACHINES

Before discussing modern single-screw extruder machines, let us discuss the series of events that seem to have taken place during the early days of rigid vinyl extrusion and that led to the present trend.

Initially, emphasis was placed on product development and on die and takeoff equipment. For the most part, conventional vinyl screws were used. There were so many problems shaping, cooling, and forming the rigid vinyl profiles that for years no special attention was given to the output of the extruders.

Improved equipment and more experienced process engineers and extrusion technicians allowed longer production runs, which in turn, led to a switch in research activities toward higher extrusion outputs. At first, screw speed was increased, leading to shorter residence time and reduced melt homogeneity. Extrusion surging and subsequent forming problems were intensified.

To improve this situation, the length of the extruder barrel and screw was increased, which in turn increased the residence time and polymer thermal history. For some pelletized compounds, however, the heat history imposed on the polymer was excessive. The efforts spent in solving these processing problems, aided by the advantageous economic implications, seem to be the reasons behind the present trend in single-screw extrusion, which is dry blend extrusion in two-stage, long-barrel, vented extruders. Obviously, longer screws increase the power and torque requirements. On the other hand, the increased barrel length with its attendant increase in residence time permits complete fusion and physical property development of rigid PVC dry blends under less severe temperature profiles than shorter L/D machines. The net result is a step toward twin-screw extrusion philosophy and—at least for pipe—lower cost formulations with lower stabilizer levels.

This modern trend is dominated by the designs of the barrier screws. All these double-flighted screws act very efficiently, transferring heat from the barrel to the unmelted polymer. Hence, they are excellent polymer melters. Most of these designs also produce excellent pressure stability. The presence of a double-flight section in the first stage of a two-stage screw tend to increase the required power and the torque even further. This, however, is compensated by the higher outputs and improved extrusion stability, which produces significantly higher power economy.

4.5 INSTRUMENTATION-APPLICATION TO MONITOR OPERATING VARIABLES

Thus far, frequent references have been made to radial temperature gradient and pressure changes in the melt at the various screw zones. This section discusses the use of instrumentation in determining temperature as well as pressure profiles and to monitor such factors as extrusion uniformity and degree of channel filling. Ideally, every critical extruder zone such as the first metering, second metering zone, and extruder adaptor, should be provided with pressure transducers. Also, a traversing melt thermocouple at the delivery tube is indispensable for determining the temperature gradient of the melt delivered to the die. This, however, constitutes an expensive proposition, hard to maintain in production lines. At least at the early stages of the extrusion development, instrumentation helps to identify the significant screw performance characteristics and operating variables and then measures them over a sufficiently wide range of values. Quantitative relationships among these values can then be established.

Conceptually, for each experimental run a relationship like the following can be established.

Extrudate quality and power economy = $f(N, T_B, \phi)$

where

N = screw rotational frequency (rpm)

ϕ = flow function for screw, which is defined by the equation

$$\phi = \frac{Q_p}{Q_d}$$

or

$$\phi = \frac{Q_d - Q}{Q_d} = 1 - \frac{Q}{Q_d}$$

where Q_d is the volumetric drag flow rate, Q_p is the volumetric pressure flow rate, which for the most part oppose the drag flow rate, Q is the actual volumetric extrusion rate, and T_B is the temperature profile imposed in the barrel. The extrudate quality, which should be judged once the extruder has achieved a steady state, is closely related to extrudate temperature and melt pressure.

4.6   EXTRUDATE TEMPERATURE T*

The melt temperature leaving the end of the screw varies with both position and time. A traverse thermocouple in the flange at the extruder head measures the temperature profile in the extrudate stream, which shows the schematical shapes shown in Fig. 4.2.

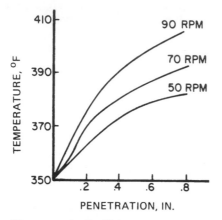

Figure 4.2 Radial temperature profiles at various rpm measured at the delivery tube of a 2.5 in. 32/1 L/D extruder. Band heater temperature, 350°F.

Although conduction and shear heating effects are significant in radial melt temperature measurements, corrections for these effects are not presented here. These correction factors, however, are needed to convert experimental measurements into true radial temperature profiles. A method to systematically correct for these factors for a moving polymer systems has been presented by Kim and Collins [3].

Each point on these profiles represents the average temperature for the given penetration and for that layer of the molten polymer. Upon analysis of the data, a decision can be made on how to define the average value $\overline{T}_*$ to best reflect the performance of the extruder at a given rpm. An additional use of this temperature profile is to establish ΔT, the maximum temperature variation in the extrudate stream. This ΔT value may be the single most important variable affecting extrudate color and yellow streaking in the extrudate, which is the early warning of degradation. Shallow screws in the metering zone, low h_M values, by generating high shear,

$$\dot{\gamma} = \frac{\pi D N}{h_M}$$

tend to produce high melt temperatures and ΔT values. As seen in this equation and Fig. 4.2, high screw speed N also produces high-shear, high melt temperature, and high ΔT values. The thermal life of rigid PVC melts at those high T and ΔT values, however, is relatively short (the higher N values reduce the residence time, and thus the degree of thermal abuse).

Extrudate temperature versus screw speed, as well as radial temperature profile and ΔT at the delivery tube, are parameters of great

value for screw performance comparison. For example, if we compare
two screws with the same overall L/D and the same channel depth h_M in
the metering zone but different L_M/L ratios (L_M is the length of the
metering zone), the analysis of the above parameters helps to determine if the extrudate temperature is governed by the channel depth
or the screw metering zone length. This answer has some significance
for design and scale-up purposes. This information also helps to decide if the extra length of the metering zone is needed to achieve a
steady-state temperature profile. As expected, a supply of heat transfer fluid to the screw is vital for maintaining a stable thermal balance
of the melt in most extrusion operations.

4.7   MELT PRESSURE

Pressure at the end of screw P_* and pressure at the extruder barrel,
preferably in the metering zone area, should be measured with strain-
gauge pressure transducers and their outputs recorded. This output
is directly proportional to the force acting on the transducer. Note
that the alternatives to having an indicating gauge rather than a record-
ing transducer or having no pressure measurement in the metering
zone make it more difficult to optimize extrusion conditions or troubleshoot problems. These transducer traces provide perhaps the single
most important parameter to assess extrusion stability and, consequently, degree of surging. Because the pressure P_* at the end of
screw does not have an appreciable variation with radial position, a
single transducer position is sufficient and $\bar{P}_*$ can be defined as the
average of the P_* readings made during steady-state operation. The
transducers penetrating the extruder barrel generate, on a moving-
strip chart, the well-known sawtooth shapes, at a frequency equal to
the rotational frequency of the screw. Analysis of the maximum and
minimum values of the sawtooth as well as its general shape yield useful information about the performance of the screw. Maximum values
are usually obtained when the leading or pushing edge of the flight
reaches the transducer. Minimum values usually occur when the
trailing edge of the channel is under the transducer. Figure 4.3
shows a typical sawtooth pressure cycle at low chart speed. The number of sawteeth in a 1' period could be used to measure the screw
revolution per minute (rpm). Perhaps the most sensitive measure of
extrusion surging is the stability of the head pressure P_*. Extrusion
stability could be defined as

$$ES = \frac{\Delta P_*}{\bar{P}_*}$$

ΔP_* represents the maximum variation in head pressure observed during a steady-state run and $\bar{P}_*$ the average pressure.

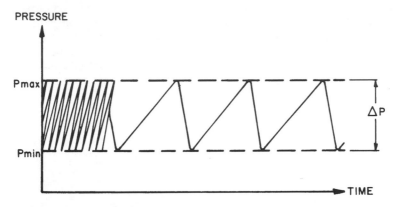

Figure 4.3 Sawtooth pressure cycle at high and low chart speeds.

For perfect uniformity the sawtooth wave determined at the metering zone repeats itself without deviation, and at low chart speed it generates a perfectly smooth wave envelope (see Fig. 4.3).

With many screws, a high degree of uniformity can be achieved at low screw speeds. (The real test of a screw design occurs at higher speeds; with increasing speed, all screws tend to show a loss of uniformity.) The better screw designs permit operation at high speeds. At high speed, the difference in uniformity can be striking.

Screw speed N is the primary operating variable. Melting rate, radial temperature gradient, and melt pressure are direct functions of the screw speed.

Barrel temperature profile T_B is, in general, a secondary operating variable with rigid PVC polymers because of temperature build-up due to the shear heating characteristic of single-screw processing. For every compound and screw geometry combination, some answers to the question of how much effect T_B has on T and ΔT should be determined.

4.8. FLOW FUNCTION ϕ

As shown before, and as presented by McKelvey and Wheeler [4], the flow function ϕ is defined by the expression

$$\phi = 1 - \frac{Q}{Q_d} \tag{4.1}$$

This expression if converted to mass rates becomes

$$\phi = 1 - \frac{W}{W_d} \tag{4.2}$$

where

W = mass extrusion rate (lb/rev)

W_d = drag flow capacity of the screw (lb/rev)

The maximum value of ϕ occurs at closed discharge when the helical pressure gradient $\partial P/\partial z$ reaches its maximum value, which happens when there is no flow of the melt down the channel, W = 0. The positive drag flow is neutralized by the negative pressure flow, and $\phi = 1$. This W_d factor, as we see later, is mainly a function of the screw geometry (diameter D; helix angle θ, flight depth h, and some correction factors). At a given screw speed, ϕ changes if the die resistance changes. The importance of ϕ is that the most useful theoretical approaches to scale-up require ϕ to be the same for both the model and the large extruder. Hence, this explains why ϕ is frequently treated as an independent variable.

It is apparent that, for a given die head, if W is independent of screw speed N [see equation (4.2) and (4.28)], then ϕ is also independent of N. Under these conditions, the head resistance becomes the direct control of ϕ. Even if W does show some variation with N, no attempt should be made to correct the die resistance in order to maintain a constant value of ϕ.

4.9 MECHANICAL POWER p

The mechanical power transmitted from the screw to the polymer is a key performance variable. If one had a direct measurement in the screw at the feed hopper, then p could be calculated simply by multiplying the torque by the rotational frequency. Unfortunately, direct screw torque measurements are not that feasible. Usually, one must measure the electrical power input to the drive mechanism and apply appropriate factors for the drive and transmission efficiencies. These are often not known as a function of speed and load, and hence considerable error can be introduced, depending upon the nature of the drive.

The mechanical power requirement p and the screw torque T_r are related by the equation

$$p = 2\pi N T_r \qquad\qquad (4.3)$$

If T_r has units foot-pound and N is expressed in revolutions per minute, then p will be expressed in horsepower if the right side of this equation is divided by 33,000. *Note:* 1 hp = 550 ft-lbf/sec.

4.10 ANALYSIS OF SCREW PERFORMANCE

The performance of a given screw is a function of the following factors.

1. *Screw design*, which includes screw dimensions, such as screw diameter D, screw length L, flight helix angle θ, flight depth h, in the feed

zone h_F and in the metering zone h_M, and their relationship h_F/h_M, the length of the various screw sections, and the so-called capacity factors, which are obtained from some of the previous parameters. In worn-out screws, the clearance between the top of the screw flight and barrel, the c value, also has a significant effect on screw performance. In fact, as a rule of thumb, screws are replaced when c values become 15% of h_M.

2. *Feed properties*, which include the following feed variables: particle size, particle surface and porosity, percentage fines, amount of reworked material, resin molecular weight, molecular weight distribution, thermal history, additives, stabilizers, and volatiles. Most of these variables have already been covered. *Force feeders*, which are mainly used to increase the density of the powder and to pack the screw channel more effectively to get higher output. By densifying the dry blend, not only more material is forced into each flight of the screw, but the dry blend is also forced against the heated barrel surface, thus promoting conveying and melting. They contribute to higher pressures in the feed zone.

3. Operating variables, mainly N, screw speed, T, temperature profile, and T_{core}, screw core temperature.

The screw performance is also judged by the following variables, which are mainly functions of the screw speed, the only true independent variable: Average mass extrusion rate G (pounds per hour); energy factor E (pounds per horsepower-hour); T_{ex}, extrudate temperature, melt temperature gradient ΔT, and melt quality: color, bubbles, melt fracture, postextrusion swelling, extrudate surface, surface gloss, among others.

4.11 ENERGY FACTOR

The energy factor E, also known as the energy economy of an extruder, is defined by the expression

$$E = \frac{G}{p} \tag{4.4}$$

where G is the mass extrusion rate in pounds per hour and p is the mechanical power input in horsepower.

One can gain some understanding of the significance of energy economy by considering an overall energy balance for the extruder. For a steady-state process, the first law of thermodynamics states that

$$G \Delta H = q + p \tag{4.5}$$

where

   p = rate of mechanical energy input
   q = rate of heat energy input

G = mass extrusion rate (lb/hr)

ΔH = change in specific enthalpy of the polymer in the extruder

This equation when combined with the energy factor equation

$$E = \frac{G}{p}$$

results in

$$G = \frac{1}{\Delta H} \ (p + q)$$

or

$$E = \frac{1}{\Delta H} \left(1 + \frac{q}{p}\right) \tag{4.6}$$

If q = 0, the process is said to be adiabatic, and the above equation becomes

$$E = \frac{1}{\Delta H}$$

Figure 4.4a is a typical plot of enthalpy (Btu per pound) versus temperature (°F) for a rigid PVC compound. These data were taken from the enthalpy versus temperature data at 3000 psi presented by Griskey and colleagues [5,6]. Since the heat added to or removed from an extrusion process at constant pressure is equal to the change

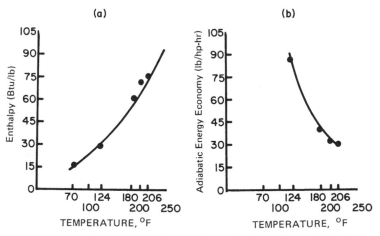

Figure 4.4 (a) Enthalpy temperature curve for type I rigid PVC compound. (b) Adiabatic energy economy for type I rigid PVC compound.

in enthalpy of the system and is given by the equation dH = Cp dT,
ΔH can be calculated if Cp polymer-specific heat data as a function of
temperature are available. The pressure and temperature behavior of
the polymer, however, is needed for the computation of the effect of
pressure on enthalpy H at constant temperature T.

The extrusion trend, which consists of axial grooves in an inten-
sively cooled feed section (sometimes referred to as forced-feed extruders) produces a very early melt and may be designed to run adia-
batically thereafter. A pressure profile like the one in Fig. 4.5, with
a pronounced peak at the end of the grooved feed section, reveals an
early melt and a desirable axial negative pressure gradient. Negative
pressure gradients at the metering zone, by providing a forward pressure flow work with the drag flow, thus enhances the pumping rate
($\phi < 0$ and $Q > Q_d$).

Figure 4.4b is a plot of $1/\Delta H$ versus temperature, which is derived
from previous data. It represents the adiabatic energy economy
(pounds per horsepower-hour) versus temperature curve. This type
of relationship is of value in determining the amount of power that
must be delivered to the screw to extrude at a certain rate. This is
based on the assumption of perfect adiabatic conditions. If, for example, the adiabatic energy economy is 13.5 lb/hp-hr, one would have
to deliver 74 hp to the screw to extrude at a rate of 1000 lb/hr.

The preceding discussion is presented to illustrate the value of
the enthalpy versus temperature relationship to establish the adiabatic
energy economy of an extrusion process; however, it is recognized
that the temperatures used in these examples are way below the normal
processing temperatures for rigid PVC.

4.12 MECHANICAL ENERGY DISSIPATED AS HEAT IN THE POLYMER

The equation G ΔH = p + q can also be used to establish an energy
ratio relationship e between the mechanical energy dissipated as heat
in the polymer to the total energy increase of the polymer. This can

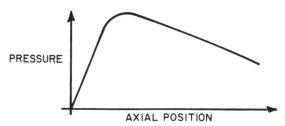

Figure 4.5 Pressure profile showing an axial negative pressure gradient.

be described by the expression

$$e = \frac{p}{G \ \Delta H}$$

which, when combined with a rearrangement of the previous equation, results in the following:

$$q = G \ \Delta H - p$$

and substituting

$$\frac{q}{G \ \Delta H} = 1 - \frac{p}{G \ \Delta H}$$

yields

$$q = G \ \Delta H \ (1 - e)$$

Hence, we see that if $e = 1$, then $q = 0$ (adiabatic operation). The mechanical energy dissipated as heat in the polymer is equal to the total energy increase of the polymer. If $e < 1$ then $q > 0$; heat input is required. If $e > 1$ then $q < 0$; heat removal is required.

The energy ratio $e = p/G \ \Delta H$ is accessible from the experimental data and the temperature-enthalpy curve of the polymer. Note that p is the mechanical power input in horsepower and G is the mass extrusion rate in pounds per hour.

Figure 4.6 illustrates two approximate plots of e, energy ratio, versus screw speed N for a rigid PVC compound prepared from a high-molecular-weight (MW) resin and at two die openings. As shown, we can note that, for both die restrictions and the rigid PVC compound tested, the value of e is >1 and q < 0. This means that a portion of the mechanical energy p put into the process must be removed by heat transfer through the barrel wall and discarded in the cooling water.

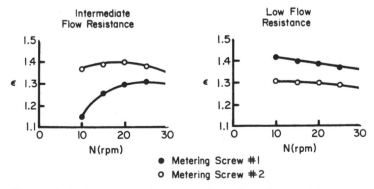

Figure 4.6 Energy ratio ε as a function of speed N.

This is characteristic of extrusion processes, involving high-viscosity melts, in which melt viscosity is in control. The above-stated is more so for rigid PVC compounds based on high-molecular-weight resins containing a low level of processing aids for single-screw extrusion, where the melt at the metering section exhibits high values of radial temperature gradient, which increase with the screw speed.

4.13 SCALE-UP RULES FOR SINGLE SCREWS: ANALYSIS AND DISCUSSIONS

Over the past 25 years, extensive extrusion studies have been carried out in the field with all sorts of screw designs, rigid PVC compounds, and extruder sizes. As expected, many different screw design approaches have been developed, built, and used commercially in this progressive industry.

Experimentation with production-sized extruders is not only very expensive but also, for obvious reasons, impractical. Ideally, experimentation should be carried out on laboratory-sized, fully instrumented extruders, and the results of such experiments should be used to predict the performance of scaled extruders. Hence, it is important to establish, experimentally, scaling rules between the extruders using the same feedstock formulations that will be used in the production-sized equipment. Throughout this work, a semigeometric scaling approach is followed, and since it is not possible to maintain exact similarity of all conditions, the rules must be considered approximate only.

In dealing with the rather high viscous nature of most commercially available PVC compounds, mainly those for applications in which high heat distortion temperatures are of prime importance (e.g., residential siding, window frame profiles, pipes, and gutters), the extrusion process is dominated by viscosity and, hence, internal heat generation. This statement is valid even in the case of well-balanced formulations. If, for example, one goes to larger extruders, lower screw speed is required. Performance similarity is maintained if the speed of the larger extruder differs from that of the small one by the factor 1/X, where

$$X = \frac{D_{large}}{D_{model}} \tag{4.7}$$

Scaling rules for output G and screw torque T_r can also be established, based on the drag capacity of the screw and the semigeometric scale-up rules. The significance of these rules is that they make it possible to examine experimentally new screw designs and concepts at the smaller scale and to make reasonable estimates of the performance of scaled versions of the new designs. It is pertinent to state at this point that the degree of accuracy of these rules is higher when scaling single-stage metering screws and, specifically, when predicting screw performance in the metering zone.

4.14 RULES FOR SEMIGEOMETRIC SCALING

We have already defined the scale factor X by the expression

$$X = \frac{D_{large}}{D_{model}} \qquad (4.7)$$

Semigeometric scaling requires that all linear dimensions of the larger extruder be increased over those of the smaller one by the factor X. The exception is the channel depth h, which is normally increased by the factor, X^m, hence the term "semigeometric."

The exponent m normally varies between 0.5 and 0.9; the higher the molecular weight and the shear sensitivity of the PVC compound, the closer to unity this value is, since these materials need deeper channels to minimize temperature generation. For low-temperature single-screw extrusion, the 0.5 exponent value approximates more accurately the scaling performance. A summary of the above follows.

$$h_{large} = h_{model} X^m \qquad (4.8)$$

$$LD_{large} = LD_{model} X \qquad (4.9)$$

$$\text{Helix angle } \theta_{large} = \text{Helix angle } \theta_{model} \qquad (4.10)$$

where

LD = linear dimensions, $0.5 \leqslant m \leqslant 0.9$.

The channel depth h normally is a function of position along the screw. The rules require that equation (4.8) be applied to channel depths at all corresponding points between the two screws. Hence, for a metering screw, the previous equation is applied to h_{feed} as well as $h_{metering}$.

Semigeometric scaling also requires that the following constraints be imposed upon the operating variables.

$$E_{large\ extruder} = E_{model\ extruder} \qquad (4.11)$$

$$T_{barrel\ large\ extruder} = T_{barrel\ model\ extruder} \qquad (4.12)$$

$$\Delta P_{head(large\ extruder)} = \Delta P_{head(model\ extruder)} \qquad (4.13)$$

$$N_{large} = N_{model}\ \frac{1}{X} \qquad (4.14)$$

To simplify these scaling rules, a prime (') notation is used for the dimensions of the extruder model.

In summary, the larger extruder must be operated at lower screw speed N, which differs from that of the small extruder by the factor

1/X. Ideally, flow function ϕ; barrel temperature profile T_B; power economy E; and head pressure gradient ΔP of the two extruders should be identical.

Before pursuing the discussion of the performance of the two extruders and the rationale for the relationship, let us state the two major goals of the scaling process, which can be summarized as follows.

1. To design the larger extruder so that it has similar performance characteristics to that of the model extruder
2. To relate the performance characteristics of output, power, energy balance, and torque to the scale factor X

A simplified equation for mass extrusion rate (pounds per revolution) derived from equations (4.2) and (4.28),

$$W \propto D^2 h_M (1-\phi) \tag{4.15}$$

and the well-known equation for mechanical power requirement p as a function of the screw torque T_R and screw rpm N,

$$p = 2\pi N T_R$$

or

$$T_R = \frac{p}{2\pi N} \tag{4.3}$$

are the starting points for the rationale of the scale-up rules.

Combining equation (4.15) with the pertinent scaling rules, (4.8) through (4.14), one can derive the following intermediate expression for the mass extrusion rate of the screw. This intermediate expression when simplified, yields expression (4.16) below.

$$\frac{W}{W'} = \frac{D^2 h_M (1-\phi)}{D'^2 h'_M (1-\phi)}$$

or

$$\frac{W}{W'} = \frac{D'^2 X^2 h'_M X^m}{D'^2 h'_M} = X^{2+m}$$

or

$$\frac{W}{W'} = X^{2+m} \tag{4.16}$$

Therefore, the output per revolution increases roughly in proportion to the scale factor raised to the 2 + m power.

If one now combines equation (4.3) with the equations for power economy, $E = G/p$ (G = output rate and p = mechanical power) and output rate $G \propto WN$, and the simplified expression (4.16), one comes up with the following simple relationship:

$$\frac{WN}{2\pi N T_R} = \frac{W'N'}{2\pi N'T_R'}$$

or

$$\frac{W}{T_R} = \frac{W'}{T_R'} \qquad \frac{W}{W'} = \frac{T_R}{T_R'} = X^{2+m} \tag{4.17}$$

which states that the screw torque is directly proportional to the mass extrusion rate and increases roughly in proportion to the scale factor raised to the $2 + m$ power.

The previous rules can also be applied to the overall energy balance of the extruder, the expression $G\Delta H = p + q$, which can be written as

$$\frac{G\,\Delta H}{G'\,\Delta H'} = \frac{p+q}{p'+q'} \tag{4.18}$$

is the starting point of this demonstration.

From the expression $G \propto WN$, which can also be written as

$$\frac{G}{G'} = \frac{WN}{W'N'} \qquad \frac{G}{G'} = X^{2+m}\,\frac{N'X^{-1}}{N'} = X^{1+m} \tag{4.19}$$

one obtains

$$\frac{G}{G'} = X^{1+m}$$

and combining equations (4.3) and (4.17), one has

$$\frac{p}{p'} = \frac{NT_R}{N'T_R'}$$

or

$$\frac{p}{p'} = X^{2+m}\,\frac{N'X^{-1}}{N'} = X^{1+m} \tag{4.20}$$

We now consider the heat transfer rate q, given by the expression $q = UA\,\Delta T$. Assuming that the heat transfer coefficient U is proportional to N and that $\Delta T = \Delta T'$ and $A = A'X^2$ (A is the surface area of the extruder barrel), the previous equation becomes

$$\frac{q}{q'} = \frac{\sqrt{N}A}{\sqrt{N'}A'} = \frac{\sqrt{N'\,X^{-1}}\,A'X^2}{\sqrt{N'}A'} = X^{1.5}$$

From the previous relationships, we can assume that the heat transfer rate q is given by the equation

$$\frac{q}{q'} = X^{1.5} \tag{4.21}$$

Comparing equations (4.18) through (4.21) we have

$$\frac{G \ \Delta H}{G' \ \Delta H'} = \frac{p + q}{p' + q'}$$

$$X^{1+m} \ \frac{\Delta H}{\Delta H'} = \frac{p'X^{1+m} + q'X^{1.5}}{p' + q'} \tag{5.22}$$

which indicates that the change in specific enthalpy of the polymer in both extruders, model and large, is the same only when m = 0.5. This means that $T* = T'_*$ for low-temperature extrusion. For high molecular-weight polymer, however, m ≠ 0.5 and $T* ≠ T'_*$, which is the actual case.

As expected, the preceding scaling rules have some flaws. When adapted by process engineers to their particular case (e.g., resins, compounds, or extruder), however, they are of great value: (1) to verify performance of the scaled extruder and (2) to continue their extrusion research efforts. Furthermore, these simple relationships, established among the most common process variables that process engineers must face in their daily work, will help their troubleshooting efforts enormously.

4.15 FLAWS OF THE SEMIGEOMETRIC SCALING RULES

If the previous scaling rules are applied to the superficial shear rate $\dot{\gamma}$ defined by the equation

$$\dot{\gamma} = \frac{\pi D N}{h}$$

assuming m = 0.5, we have

$$\frac{\dot{\gamma}}{\dot{\gamma}'} = \frac{\pi D'X(N'/X)/h'X^{0.5}}{\pi D'N'/h'} = \frac{1}{X^{0.5}} = X^{-0.5} \tag{4.23}$$

which indicates that the actual shear rate in the larger extruder is less than that of the model.

Let us consider now the effect of shear rate on the average melt viscosity. This, in fact, is a main defect in the rationale for the semigeometric scale-up procedure. Since rigid PVC melts are power law fluids, we have the following expression:

$$\eta \propto \dot{\gamma}^{n-1} \tag{4.24}$$

which, when combined with equation (4.23),

$$\frac{\dot{\gamma}}{\dot{\gamma}'} = X^{-0.5}$$

gives the following equations:

$$\frac{\dot{\gamma}^{n-1}}{\dot{\gamma}'^{n-1}} = X^{-0.5(n-1)}$$

or

$$\eta = \eta' X^{0.5-0.5n} \tag{4.25}$$

where n is the flow index, ~0.5 for some rigid PVC melts.* Hence, in scaling from small to large extruders, the effective viscosity of the polymer melt in the large extruder at the corresponding screw speed increases. This also increases the torque requirement. This is also what is expected from expression (4.23), since lower shear rate values produce lower melt temperatures and consequently higher melt viscosity and torque.

4.16 DRAG CAPACITY OF THE METERING SECTION OF THE SCREW

The scale-up rules outlined above indicate that, if all the constraints stated before are satisfied, then the performance of the two extruders is given by the following simple relationships [note, from expression (4.2), that $W = W_d(1 - \phi)$]:

$$\frac{W}{W'} = \frac{W_d}{W'_d} \qquad \frac{T_R}{T'_R} = \frac{W_d}{W'_d} \tag{4.26}$$

The quantity W_d is the drag capacity for the metering section of the screw and is given approximately by the relationship

$$W_d \simeq \frac{1}{2} \rho \pi^2 D^2 h_M \sin \theta \cos \theta \tag{4.28}$$

Hence, from equation (4.17), we see that the drag flow capacity of the metering zone increases in proportion to the scale factor raised to the 2 + m power:

$$\frac{W_d}{W'_d} = X^{2+m}$$

Without the various correction factors for screw channel geometry (shape), leakage flow, and channel curvature, for example, which can have a significant effect on the numerical value of W_d, equation

---

*At the standard reference state, 180°C and 1 sec^{-1}.

(4.28) should be treated as an approximate expression for drag capacity only. Assuming that screws have been built according to the scaling rules, the next step is to analyze if their performance is similar to that of the model screws.

4.17 ANALYSIS OF EXPERIMENTALLY DETERMINED PERFORMANCE DATA

The experimental data, specifically, output per screw revolution W (pounds per revolution), screw torque T_R (foot pound), melt temperature T_*, and temperature gradient ΔT are first tabulated versus screw rpm values, N. It is assumed that the model and the scaled-up extruders, used to generate the experimental data were equipped with adjustable dies to provide for low, intermediate, and high melt flow restrictions and that the pressure drop ΔP for the large extruder-die system is the same as that of the model extruder.*

These experimental data are then examined. These data show that, as expected when dealing with high-viscosity polymer, the primary operating variable is screw speed. They clearly show that N has a dominating first-order effect on performance, particularly the performance characteristics of rate and power. Die resistance and, if evaluated, barrel temperature show a relatively small second-order effect on rate and power.

The experimental data are then compared with the predicted scale-up performance, using the semigeometric scale-up theory. This can be done more objectively with a series of graphs (Figs. 4.7 to 4.10).

Since the output per revolution of the large extruder increases roughly in proportion to the scale factor raised to the 2 + m power, a log-log paper should be used to show plots of output per revolution as a function of screw speed N for the two extruders. Figure 4.7 represents such a plot. The deviations between the predicted and the observed values should be examined together with the corresponding pressure profiles. (The reason is that the scale-up rules for output are based upon flow in the metering zone, and the pressure profile in the metering section of the screw has a very significant effect on output).

The predicted torque of the large extruder is also related to that of the small one by the same exponential function, X^{2+m} [equation (4.17)]; consequently, it should be treated like the output per revolution (see Fig. 4.8).

Semilog paper is used to plot melt temperature and temperature gradient as a function of screw rpm. The larger extruder has to be run at lower screw revolution to maintain the same temperature profile and about the same temperature gradient (see Fig. 4.9).

---

*The approach followed in the forthcoming discussion has been modeled after J. M. McKelvey's extrusion studies at Monsanto, Bloomfield, CT [7,8].

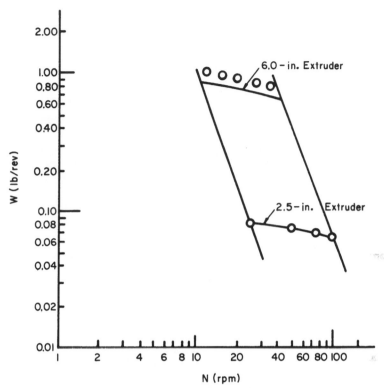

Figure 4.7 Scale-up of output data. Circles represent experimental data. Solid line for 6.0 in. extruder is predicted from scale-up rules.

For the pressure profile data, linear graph paper can be used. Pressure can be plotted versus the relative position of the pressure transducers or mixing devices, like mixing rings in the metering section of the screw (Fig. 4.10). The shape of these pressure profiles provides very valuable information. For example, a negative pressure gradient in the metering section indicates that, for this screw, the pressure flow is in the forward direction, thus enhancing the drag flow. Consequently, the observed values for outputs most probably exceed the predicted values. For screws with a positive pressure gradient in their metering sections, the pressure flow opposes the drag flow. The outputs of these screws should be less than predicted. The effect of pressure in the screw metering on output is another flaw of the semi-geometric scale-up rules, since implicit in these rules is the assumption that ΔP at the head in the model and the large extruder are similar and this similarity quite frequently does not exist.

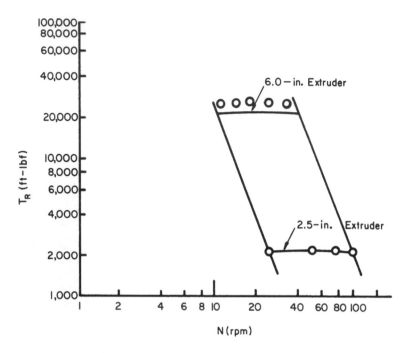

Figure 4.8 Scale-up of torque data. Circles represent experimental data. Solid line for 6.0 in. extruder is predicted from scale-up rules.

4.18 ANALYSIS OF TWO-STAGE SINGLE-SCREW EXTRUSION IN VENTED BARREL EXTRUDERS

Figure 4.11 is a schematic representation of a two-stage single screw in a vented-barrel extruder. The function of the first stage is to heat, melt, and compress the dry blend and deliver it to the vent section for removal of gases and volatiles. A high level of vacuum is normally used at the vent to aid in the removal of volatiles. Typical polymer resi-dence time at the vent is 4 to 10 sec. Once the melt passes through the vent it is decompressed in an almost explosion-type operation. As seen, the vent section is quite deep and, by design, is partially filled only. Moisture (absorbed and adsorbed) and trace monomer from poly-merization are the vapors most commonly removed. The vent section, where removal of the volatiles takes place, is also called the extraction section, and these two-stage extruders are also called extractor-ex-truders. Entrapped air is also removed at the vent section of these extruders. At this section some cooling of the melt always takes place.

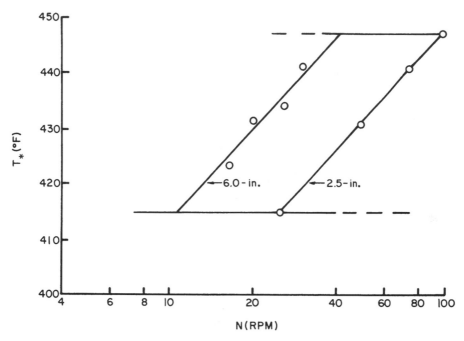

Figure 4.9 Scale-up of extrudate temperature data. Circles represent experimental data. Solid line for 6.0 in. extruder is the expected temperature.

From the vent, the melt enters the second stage, where further shearing and heating occur. The function of the second stage is to generate sufficient pressure to force the melt through the head-die system at the required rate.

In the steady state, the melting rate in the first stage R_1, the pumping rate in the second stage R_2, and the flow rate through head-die system R_3, must all be equal: $R_1 = R_2 = R_3$. If for any reason this condition is not satisfied, the operation of the extruder fails. For example, if the PVC compound is melted in the first stage at a higher rate than it is pumped by the second stage, $R_1 > R_2$, melt begins to flow out of the vent. A sudden flow restriction in the head-die or second stage could produce the same effect, $R_2 > R_3$.

At first glance it could seem to be an extremely difficult proposition to design screws with first and second stages having melting and pumping rates exactly equal over a reasonable range of operating conditions. However, a two-stage screw has a self-regulation feature that greatly simplifies the design problem.

Normally, not all the screw channels in the second stage are filled. In Fig. 4.12, L_* designates the filled length. It is this portion only

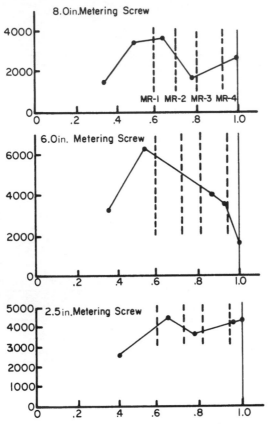

Figure 4.10 Typical pressure profile data versus mixing rings relative positions for three metering screws containing four mixing rings in the metering zone.

that has pressure-building capabilities. Thus, variations in L_* affect the pressure-building capability of the second stage. In general, as L_* increases, the pressure-building capability increases. The range of variation possible for L_* is between zero and L_M (length of second metering zone). Within these limits the second stage regulates itself to accommodate to changes in all operating conditions. The operation undoubtedly fails when a pressure-building capability is needed that requires an L_* greater than L_M.

Regardless of the specific screw design, each of the various components of an extractor-extruder should perform a similar function for processing both pelleted or powder compounds. The only exception is the extruder hopper, which in the case of powder should be of a forced-feed nature. Gravity-fed powder is frequently erratic,

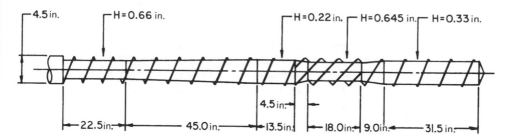

Figure 4.11 Two-stage single screw in a vented barrel extruder.

resulting in surging and inconsistent melt. As mentioned before, a
well-designed compactor in a dry-blend extrusion can improve extrusion
output and extrusion stability. As expected, a great number of two-
stage screws have been developed through the years, and every major
extruder manufacturer is offering its own design. In this section, we
will follow a qualitative and functional approach, which for the most
part is valid in describing the operation of any two-stage screw designed for vented-barrel extruders. The preceding and forthcoming

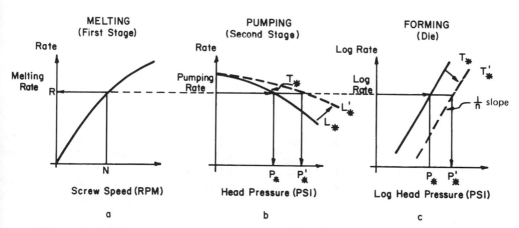

Figure 4.12 Characteristic curves for two-stage single-screw vented
barrel extrusion process in steady-state operation.

presentations have been strongly influenced by McKelvey's analysis of polymer processing in two-stage single-screw vented-barrel extrusion [7,8]. This qualitative-functional approach uses characteristic curves to describe the performance of each of the components of the screw. As shown in Fig. 4.12a,b,c the performance of the first and second stage of the screw, as well as that of the die, is described by a characteristic curve. The ordinate of these curves is a rate R: the melting rate, the pumping rate, and the die flow rate. For reasons mentioned before, these three rates must be equal for steady-state operation of the process.

4.19 CHARACTERISTIC CURVES

Consider now the first-stage melting curve. Typically, such curves are nearly linear at low speed and develop a nonlinearity as screw speed N increases. As already shown, screw speed is a primary variable affecting melting rate for a given screw and polymer. It is also responsible for the radial temperature gradient in the melt.

The characteristic pumping rate curve shows the typical convexity associated with nonnewtonian nonisothermal flow. If $L'_* > L_*$, as shown in Fig. 4.12b, the head pressure increases, as does the pressure stability, measured by individual pressure transducer traces in the pumping stage and delivery tube, which tends to show less variability. As already discussed, pressure stability is a sensitive indicator of extrusion uniformity. For perfect uniformity (no surging), the pressure transducer trace (sawtooth wave) repeats itself without deviation.

As shown in Fig. 4.13, at a given screw speed N a family of these curves is generated, if one considers the filled length L_* of the screw to be a parameter. Similarly, for a given filled length L_*, another family of curves is generated by considering the screw speed N as a parameter. As is shown in Fig. 4.13, at a given screw speed and rate the pressure-building capability of the screw increases as the filled length increases. Similarly, if the filled length is constant, the pressure-building capability at a given extrusion rate increases as the screw speed increases, since the melting rate increases with N, and the amount of melt delivery to the second stage must be higher. As reported by Squires [9], gains of up to 30% in pressure-generating capability of this pumping zone could be achieved via design optimization, mainly channel depth and helix angle, without enlarging the total length of this zone. This author highly recommends Ref. 9 as a thorough analysis of polymer processing in two-stage extractor screws.

In a standard screw, material transport depends on the helix angle of the screw flights as well as flight depth. As helix angle decreases, material transport increases and frictional shear (slippage) decreases. It would seem, then, that efficient material transport leading to high output rates and low shear can be readily achieved by reducing the

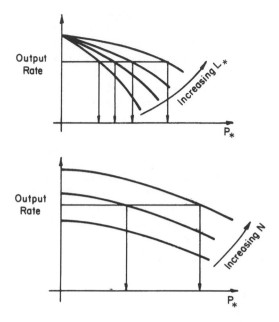

Figure 4.13 (a) Change in characteristics at constant speed N as L∗ changes. (b) Change in characteristics at constant L∗ as N changes.

helix angle.* However, the need to achieve a uniform, homogeneous melt with minimal adverse time history requires certain constraints on screw geometry.

The higher the pressure-generating capability of the pumping zone of a two-stage single screw, the better is the overall performance of this screw. Specifically, this design shows less tendency to produce vent flow problems and greater ability to pump against the die pressure.

According to Squires [9], the optimum channel depth, offering a wider range of acceptable performance, is obtained when this channel depth is 2.0 times the drag-flow depth, h_D, which is the channel depth when only drag flow is involved (zero helical pressure gradient along the channel), $\partial P/\partial z = 0$. If the helix angle is optimized simultaneously with channel depth, the optimum angle will be close to 30°. This 30° optimum angle is independent of polymer properties and rate but is a

---

*Most single-screw extruders have been built with a so-called square pitch, i.e., helix angle is 17.7° (screw pitch = diameter). This represents a design that has reasonably good solid-conveying performance for a variety of materials and adequate pressure-developing potential.

function of the number of flights and the ratio of flight width to diameter. The smaller the flight width becomes, the closer to 30° is the helix angle. In polymer melts, however, the above is never the case and the optimum helix angle depends rather strongly on the power law index n.

For an isothermal (or constant viscosity) screw pump, the 30° angle also produces the maximum output at a given pressure and screw speed. But in the extrusion of highly viscous, shear-sensitive power law fluids, such as rigid PVC melts, and with minimal adverse time-temperature history, the optimum helix angle is less than 30°. For example, with the aid of any suitable computer programs, an optimum helix angle can be established if the potential polymer melt residence time in the second metering zone is measured as a function of helix angles and the corresponding optimum channel depths for maximum pressure generation. These computations, when carried out, show that a fairly broad optima exists in the region around 22°. Figure 4.14 represents the results of some of these computations for various pumping zone designs of a two-stage single-screw 3.5 in. extractor-extruder.

As is shown throughout this book, for a given rigid PVC compound, die and design output rate, screw characteristics curves are of great value in selecting the optimum screw design pumping section for maximum pressure-development capability. Other factors that have to be considered in this selection are the so-called head pressure stability and the temperature T_* of the melt delivered by the screw. For a given screw and PVC compound, these factors vary as a function of screw speed N, filled length L_*, and pumping rate R. The analysis of flow, pressure, and temperature development in the pumping sections of extruders continues to be the object of intensive study, and computer programs for calculating the theoretical flow rate characteristic of the pumping section, as well as for scaling purposes, are readily available. This explains why general features of screw characteristics curves are reasonably understood.

Let us now analyze the head-die system. The polymer melt passes through head and die under the driving force of the hydrostatic pressure, developed by the pumping section of the screw. If one assumes the validity of the power law model for the polymer melt, isothermal flow rate versus head pressure P data for dies can be plotted as a straight line on log-log paper. The slope is the reciprocal of n, the flow-behavior index, a characteristic property of the polymer. For unfilled PVC melts, n is usually in the vicinity of 0.5.*

Consider the process operating in the steady state as represented in Fig. 4.12a,b,c. Here the melting rate is R, the filled length of the second stage L_*, the head pressure P_*, and the melt temperature in the head T_*. Suppose now that some change is made in the forming die causing increases in its resistance to flow. When the process again

---

*At 1 sec^{-1} and 180°C and 0.36 at 100 sec^{-1} and 180°C.

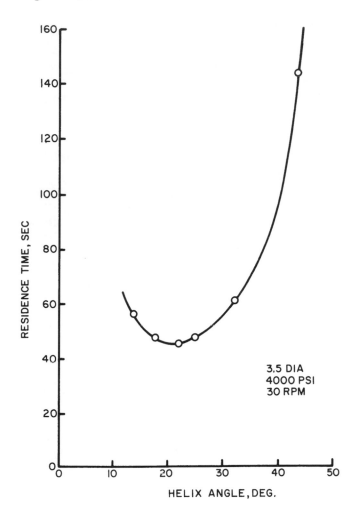

Figure 4.14 Polymer melt residence time (estimated) in a metering
zone as a function of various helix angles.

reaches a new steady state, the melting rate will still be R (screw
speed and feedstock remain unchanged); but the characteristic curves
for pumping and die flow will have changed, and temperature, pressure, and filled length assume new values designated $T_*^!$, $L_*^!$, and $P_*^!$.
Clearly, the higher die resistance requires a higher head pressure,
which in turn requires a greater pumping capacity and larger L_*.
This results in increased working and, hence, a higher melt temperature at the die and a lower polymer melt viscosity. This kind of self-

adjustment can work only as long as the required pumping length L_* is equal to or less than the actual length L_M of this second metering zone length.

If for any reason the process conditions are such that an L_* greater than the length of this second metering zone is needed, the process will fail. Hence, one can establish upper bounds for a process by examining the pumping characteristics of the second stage when $L_* = L_M$.

4.20 PROCESSING RIGID PVC DRY BLENDS IN TWO-STAGE SINGLE-SCREW VENTED-BARREL EXTRUDERS

In this section, we discuss some of the techniques used to analyze the experimental data and how these data are used to determine the performance capability of the screw or screws under evaluation. This is done within the conceptual framework of two-stage extrusion as described in the previous section. First, for a typical rigid PVC compound, the characteristic melting rate curve, output rate (pounds per hour) versus screw revolution N are established. Figure 4.15 is a typical melting rate curve for a rigid PVC compound in a 4.5 in. 32/1 L/D vented extruder. This melting capacity is, essentially, independent of the performance characteristics of the second stage and the head-die.

To determine whether the second stage of the screw has the required pressure-building capability, one must first establish the required head pressure, which, as shown in the previous section, is obtainable from the characteristic curve for the head-die system. To determine the characteristic curve for the head-die system experimental work is carried out using standard adjustable die systems in which the polymer flows through an annulus or through a slit orifice. Figure 4.16 is a typical head-die characteristic curve (log-log paper) for a rigid PVC compound at a fixed die opening and at the same N values used in Fig. 4.15; as shown, this characteristic curve is a straight line. The position of the inner portion of the die annulus, a mandrel or tip with a tapered section, as well as the flow through the slit, are adjustable. Moving the mandrel up or down or adjusting the lip or restrictor bar changes the resistance to flow. Hence, the position of these movable parts becomes a process variable.

Other head-die characteristic curves are then determined by regulating the flow through the die and measuring the rate (pounds per hour) versus the head pressure for every restriction and N value tested. These data, when plotted on log-log paper, take the form of parallel straight lines. If the flow-behavior index n for a given rigid PVC compound is 0.36, the slope of its head-die characteristic is 2.78 (reciprocal n) and the angle that these curves form with the abscissa is 70.2°.

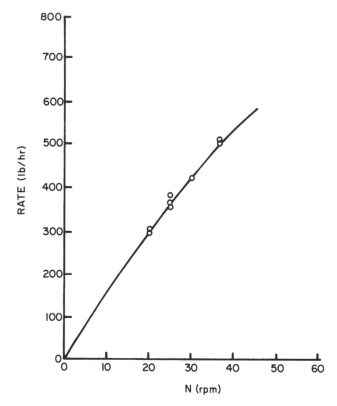

Figure 4.15 Melting rate curve for a 4.5 in. 32/1 L/D extruder with a rigid PVC type I compound.

For each die opening, an estimate of the effect of changing flow temperature on melt flow can now be made using the following approximate expression at constant pressure.

$$[\exp(b\dot{\gamma}\ \Delta T)]^{1/n}$$

See equation (3.22). As stated before, $b\dot{\gamma}$ is a parameter characteristic of the material, the quantity $1/b\dot{\gamma}$ representing the number of degrees that the polymer temperature must be raised at constant shear rate $(\dot{\gamma})$ in order to decrease the viscosity (η) by the factor $(1/e)$. To illustrate this technique, we use the value determined by Westover [10], which is 51°C at a shear rate of 40 sec^{-1}. Consequently, $b\dot{\gamma}$ = 0.0196°C^{-1}. For example, for a flow temperature increase of 10°F when $b\dot{\gamma}$ and n have the values of 0.0196°C^{-1} and 0.36, respectively, the factor for the flow increase is +1.35. For a melt temperature of

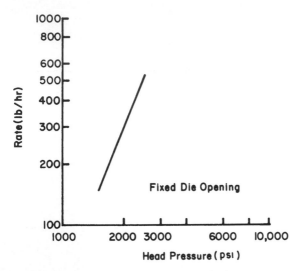

Figure 4.16 Head-die characteristic curve at a fixed die opening.

380°F, a given rigid PVC compound and a given die restriction, curves
at temperatures 10 degrees Fahrenheit above and below the average
curve can be established. This is summarized in Fig. 4.17. This
figure can be used to illustrate the following example. If a fixed rate of

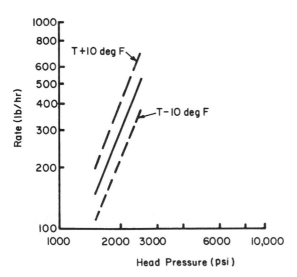

Figure 4.17 Head-die characteristic curves at various temperatures,
at a fixed die opening.

400 lb.hr is required, we can expect the following head pressures at
the various temperatures:

Temp, °F	Head pressure, psi
370	2600
380	2250
390	2000

In summary, from the established melting rates and for a given die restriction, the required head pressures for these established rates can
be determined by the die characteristic curves. Furthermore, for a
given output, the effect of raising or lowering the melt temperature
10 degrees Fahrenheit on the head pressure can be estimated. Conversely, if one wants to keep the head pressure for a given die restriction constant, these curves can be used to obtain an approximation of the output rates as a function of the melt temperature.

The experimental determination of the second-stage pumping rate
characteristic curves, however, is complicated by the fact that the
pumping rate versus head pressure curves are direct functions of the
filled length L_* of the second metering zone of the screw (see Fig.
4.13a). To establish these curves, at least three pressure transducers
should be provided. Ideally, there shall be a transducer at the beginning, middle, and end of the second-stage metering zone. A pressure
transducer trace fluctuating between 0 and P indicates that the corresponding channel is partially filled. Distance L_* is counted from the
first channel that becomes completely filled. As expected, the average
pressure value increases with the distance from the position when the
channel becomes completely filled and is proportional to this distance.
By plotting the average pressure transducer reading when the channel
becomes completely filled as a function of the transducer relative position and expressing this distance as a fraction of the total length of
the second-stage metering zone, one can determine the approximate
filled length of the second pumping section of the screw, which should
be treated as the parameter L_*.

If now we want to apply all the previous discussions to determine
experimentally the approximate pressure-building capability of the
second stage, we can proceed as follows.

First, at each operational screw speed, the head-die is restricted
until the melt starts to flow through the vent. At this point, one
starts opening the die until a point is reached at which no more vent

flow is observed and the pressure transducer near the vent shows complete channel filling. The output (pounds per hour) versus head pressure for L = 1 or f = 1 is then measured and plotted. The die-opening process is then continued until this transducer indicates a partially filled channel. All these intermediate head-pressure data may also be plotted in the same graph identified by the screw rpm and L = 1. This head-die opening process continues, focusing the attention

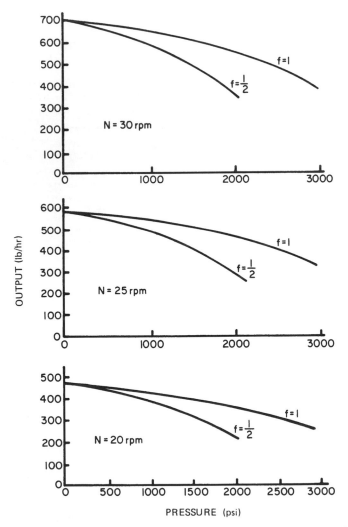

Figure 4.18 Screw characteristics for second stage of 4.5 - inch PVC screw.

around the transducer located near the $L = \frac{1}{2}$ point. The data obtained in the vicinity of $L = \frac{1}{2}$ are then plotted, this time in the $f = \frac{1}{2}$ graph. These data correspond to the $L = \frac{1}{2}$ or $f = \frac{1}{2}$ curve. By extrapolating these two series of data points to zero pressure, while maintaining the typical convexity associated with nonnewtonian nonisothermal flow we have approximate pumping-rate characteristic curves of the second stage of the screw. This same procedure is then repeated for other screw rpm. Figure 4.18 represents a family of these approximate curves. As shown, this figure is similar to Fig. 4.12b. -

To determine whether the pressure-building capability of the second stage of the screw is adequate, we consider the overall process operation of the screw at typical screw rpm. From Fig. 4.15 the melting rates are established. Then, from the head-die characteristics (output rate versus head pressure) for the given compound and the tip position and/or die opening required for the given extrudate, we determine the required head pressures for the rates established. These points are then taken to Fig. 4.18, and if they fall within the capacity of the second stage of the screw, (within the $f = 1$ and $f = 1/2$ curves) one can conclude that the pressure-building capability of the screw is adequate. It is the author's experience that, quite frequently when full channel filling or pressure development starts in the tapered compression section of the second stage, severe pressure and rate surging usually occur at the die. It appears that PVC melts are incapable of developing steady pressure and uniform flow in a converging channel at low hydrostatic pressures, and an unsteady slip-stick flow situation results.

4.21 TWO-STAGE SINGLE-SCREW WITH A BARRIER OR DOUBLE-CHANNEL SECTION

The first so-called barrier screws appeared early in 1960 [11]. In this prototype early design, two separate channels were cut in the same screw shaft, one of which connected with the feed hopper, and the other channel supplied the melt to the metering zone of the screw. The feed channel width and depth decreased progressively until it disappeared, but the output channel increased at a balanced rate until it attained full width and depth at a position adjacent to that at which the feed channel width became zero. At this point melting efficiency also diminished. The following simple sketch representing the unwrapped channels of the Maillefer screw summarizes this description:

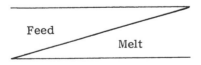

The only connection between the two channels is over the secondary flight land, which is also called dam flight.

The main function of the secondary channel is to drain off the freshly made melt pool and prevent it from interfering with the desirable shear action at the barrel surface. As the solid bed of resin is pushed forward down the screw channel, the shear at the solid bed/hot barrel surface generates more heat, which melts the resin at the bed surface nearest the barrel. This creates a film of melt between the solid bed and the barrel, which tends to adhere to the barrel surface. As the screw flight passes by it wipes or scrapes this film off the barrel and, having no place else to put it, deposits it in the reservoir channel. The barrier-screw principle tends to control the melting mechanism by not allowing the solid bed to float or be isolated by the melt pool but by forcing it into intimate contact with the barrel. Most of the energy (p + q) put into the plastic is concentrated in the unmelted polymer when it is most needed, and better melting takes place. In addition, the feed characteristics of the screw are more uniform because pressures have been reduced by the draining off of the early melt. As shown in Fig. 4.19, which represents a 32/1 L/D metering screw with a barrier section in its compression zone, which was ejected from an unvented extruder barrel after a melt containing regrind was frozen, another function of the dam flight is to create a second melt pool at the forward edge of this intermediate flight that acts as a solid-bed breaker. This technique of freezing the melt and ejecting the screw provides very revealing information on the progressive formation of the melt pool at the expense of the solid bed. Also, one can observe how the reworked material surrounded by the melt behaves like late melters. Figure 4.20 presents another screw, a metering screw this time, which was ejected, as explained before, while extruding the same feedstock used with the barrier screw. As shown one can detect unmelted regrind as far as the middle of the metering zone of this screw.

The melting mechanism explained above improves heat transfer tremendously, resulting in extrusion systems with high power economy. Some of the most recent designs also produce excellent head-pressure stability with only a modest increase in pressure variation when output increases significantly. Figure 4.21 represents head-pressure variation versus output for a 3.5 in. 34/1 L/D two-stage single-screw with

Figure 4.19 A 32/1 L/D metering screw with a barrier section in its compression zone which was ejected from an unvented extruder barrel after the melt from a feedstock containing regrind was frozen.

Figure 4.20 A 32/1 L/D metering screw (no barrier section) ejected
from the extruder of Fig. 19, while using the same feedstock.

a barrier secton in the first stage. As shown, head-pressure stability
is excellent.

In summary, the improved head-pressure stability, by producing
less extrusion surging, contributes to an overall better control of extrudate wall thickness and weight and consequently leads to substantial
savings. In this area, the barrier-type, two-stage single-screw,
vented-barrel extruders approach the desirable extrusion performance
common to most of the twin-screw extruders.

Inherent to these barrier-type screws is a higher torque. This,
however, is highly compensated for by the higher output rates and
head-pressure stability, which contribute to higher power economy.

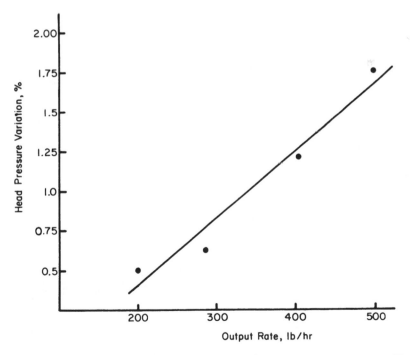

Figure 4.21 Head pressure variation (%) versus output rate (lb/hr),
3.5 in. 34/1 L/D.

   As expected, the early design approaches have been extensively
modified. The main objective of these modifications are to maximize
melting while minimizing energy input and maximizing extrusion uniformity and to improve circulation of the melt at the end of the melting
section. In the United States, well-known barrier-type screw designs
are the Hartig-Barr, the Efficient, the Davis-Standard barrier, the
Uniroyal, the Barr-2, and the NRM barrier.
   In one of the two best known designs (U.S. patent 3,698,541),
the width of the channel containing the solid bed is held constant to
avoid deflection of the solid bed while the depth diminishes. The
constant solids-channel width allows maximum contact of solid polymer
with the barrel surface for the most efficient melting. This design ac-
commodates the changing solids-melt ratio by gradually decreasing
the depth of the solids channel and increasing the depth of the melt
channel. This results in a very deep melt channel toward the end of
the melting section. In another well-known design (see Fig. 4.22),
the solids and melt channels gradually change, but at rates that vary
in accordance with melt conditions. The melting section is divided into at least two lengthwise zones, an upstream zone and a downstream
zone. The solids channel never ends completely, but once it reaches
a greatly reduced depth it again increases so that no sudden pressure
buildup can occur.

Figure 4.22 DSB barrier screw. (Courtesy of Davis-Standard.)

Quite frequently when a barrier-type section is used in the first stage of a two-stage screw and for a feed containing a high level of reworked material, mixing devices, such as mixing pins, are incorporated in the screw after the barrier section and before the decompression zone, leading the vent area. These mixing devices reduce the presence of partially fused reworked material in the vent area and improve the overall quality of the melt.

4.22 VENT-ZONE OPERATION

Vent-zone channels are designed to be partially filled only. Pressure gradient in the axial direction is zero. There is, however, a small cross-channel pressure gradient with the pressure maximum occurring at the forward side of the flight wall. Any positive melt pressure at the vent causes plugging of the vent opening and loss of venting effectiveness.

The channel depth h can be computed from the nominal drag flow equation, which can be written as follows: $Q_d = F_d \alpha' h N$. This equation, however, is modified for the fractional filling required. For example, if a 25% channel filling is desired for maximum bubble removal, the above equation is modified so that Q_d becomes $4Q_d$, because we want the extra capacity in this area to create a partial channel filling. The typical value of F_d, the drag flow shape factor for a partially filled channel, is about 0.95. α' is a function of the channel geometry only and is given by the equation $\alpha' = \alpha h$, where

$$\alpha' = \frac{1}{2} \pi^2 D^2 \left(1 - \frac{ne}{t} \right) \sin\theta \cos\theta$$

where

D = screw diameter
n = number of parallel channels
e = flight width
t = $\pi D \tan\theta$
θ = helix angle

The efficiency of devolatilization is influenced by both polymer melt temperature and vacuum level. A vacuum of 28 in. mercury is recommended at the higher screw speeds. At higher screw speeds the residence time of the melt at the vent area is shorter, but the temperature is higher. If at the first metering section of the screw the polymer is not properly fluxed, there will not be a good seal between the vent port and the feed opening. This limits the amount of vacuum that one can apply at the vent. Good vacuum is critical to an effective devolatilization. Since the devolatilization process occurs primarily by a process of diffusion and the rate of diffusion increases with temperature, a well-fluxed melt and a relatively high temperature when the polymer reaches the extraction section are important.

The extract obtained from the vent is a complex mixture. The principal component is water, which when evaporating causes some beneficial cooling of the polymer at the vent zone. Along with the water some vinyl chloride monomer is also removed in a mild steam-stripping operation

There exists some problem of polymer stagnation, however, in the partially filled vent flights. Such material may slowly make its way back into the active stream as additional color or other degradation products. As expected, this situation, which becomes more acute after extruder shutdown and start-up operations, is a potential source of contamination.

The importance of a vent-zone temperature level was first quantified by Latinen [12], who defined a devolatilization number N (Dev.) as a function of D = diameter at the vent zone; M = the number of flights at the vent (M = 1 for a single-flighted vent and M = 1.6 for four-flighted vent); screw speed; L/D at the vent, Pv = vent pressure, and P ($H_2O(v)$) = volatile (H_2O) vapor pressure at melt temperature. Devolatilization is presumed to be proportional to N (Dev.).

4.23 PUMPING-ZONE OPERATION

For the most part, the operation of the pumping zone of a two-stage single-screw has been already covered. This screw zone functions as a metering pump, from which the molten plastic material is delivered to the die system at constant volume and pressure. More is known about the behavior of this component of the screw than any other. As we have seen, the polymer in the filled channels of this zone is in the fluid state, and the equations and relationships discussed earlier can be applied more accurately than in any other step of the screw extrusion process. Pressure and temperature measurements are also more accurately measured either at, or immediately after, this zone, where the polymer is in the fluid state.

Pressure gradients along the pumping zone of the screw have a significant effect on output. A pressure profile with a positive gradient works against the drag flow; a negative gradient reinforces the drag flow. These phenomena contribute to outputs different from the calculated values

4.24 SCREW NOSE DESIGN

The use of a breaker plate and screen combination for so-called high-temperature extrusion remains a trouble area for many reasons, such as,

Breaker plates are difficult to streamline.
Breaker plates provide places for polymer to stagnate and degrade.
Breaker plates cause pressure drops and temperature rises.

Some extruder manufacturers provide an extended conical screw nose of 60° or 90° with a matching adapter designed to minimize hold time and hang-up tendency. The use of a conical nose also reduces the plate-out tendency and the amount of material in the die approach, and unlike a screen pack system is almost self-cleaning.

The screw nose surface has always been a problem area: not only is the melt at its highest temperature (see typical radial temperature gradient), but also this material has the longest residence time at the highest temperature, which even under the mildest conditions causes color propagation problems. In the early days, localized nose air cooling was used to reduce ΔT.

At present, one of the most effective means of attenuating this potential problem is by designing a slightly off-center conical nose that, when rotating, tends to break up the hot melt that normally forms there. Another means of helping this situation is by chrome plating the screw nose area, specifying "hard" chrome and formulating the compound with effective internal lubricants. Reference 2 describes a breaker plate design approach that minimizes these problems.

4.25 SCREW TEMPERATURE CONTROL

In most high-temperature extrusions of rigid PVC, a temperature-controlled screw is required. The frictional heat generated in the screw must be removed. This is done not only to influence the interaction between the polymer and the surface of the screw but also to control the melt temperature, which in modern single-screw machines processing high-viscosity PVC resins can be very high. Another beneficial effect of a temperature-controlled screw is better control of the radial temperature gradient in the metering zone and, consequently, of the melt delivered to the die.

Oil heat-exchange systems provide the most precise control of screw temperatures. These units must be of sufficient heating and cooling capacity to quickly respons to heat transfer needs. Typical oil temperatures for rigid PVC extrusion are between 130 and 180°C.

The temperature-controlled oil is normally fed to the screw nose area through a tube centered in the screw core and then returned between the tube and the inner wall of the screw. If an extended conical tip is used, this tip should also be oil temperature controlled. A very hot tip can produce a very high temperature gradient with maxima in the screw nose area. This could cause polymer stagnation and discoloration and, eventually, degradation.

In so-called low-temperature single-screw extrusion, which dominated the early days of siding and panel extrusion, air cooling of the screw nose was a common practice. This practice, however, is not effective when dealing with high-shear extrusion.

The temperature-controlled oil, however, is not free from problems. Even in the presence of a very efficient force-feeder hopper, there is always a risk of losing the feed due, among other things, to a very early melting of the polymer in the feed zone area.

4.26 EXTRUDER TEMPERATURE CONTROL

Figures 4.19 and 4.20 presented in Sec. 4.21 give an objective idea of
the melting mechanism of rigid PVC compounds; in this case compounds
containing high levels of reworked material. The presence of reworked
particles that are late melters, however, contributes to a better definition of the solid bed/melt interface. As shown, within a screw channel,
the solid bed or solid plug is surrounded by melt. This solid bed
melts primarily at the solid bed/melt interface as it is conveyed down.
Consequently, the temperature of the melt films and melt pool surrounding the solid bed has a very decisive influence on the rate at which the
solid bed melts. Furthermore, the melt film formed over the solid bed
exchanges heat with barrel by conduction. This melt, which is scraped
off into a melt pool, contributes to the temperature of the melt pool.
The heat generated within the interlayers of this melt film by viscous
dissipation, however, is far more than the heat conducted from the
barrel and primarily controls the melting mechanism. This heat generation, in turn, is mainly a function of the melt viscosity of the rigid
PVC compound being extruded. Because of screw rotation, the solid
bed as well as the melt pool are pumped down along the screw channel.
The melt pool increases progressively at the expense of the solid bed
as the channel depth decreases. This is accelerated by the gradual decrease of the volume of the channels, which increases the overall compression. During this conveying process, the solid bed receives heat
continuously from the melt film, which increases its average temperature.

   Because of the combined effects of increased temperature and compression, the solid bed should be completely melted at the metering
zone. Quite often when using feedstocks containing a high concentration of reworked polymer or coarse particle size regrind, the presence
of partially melted regrind, which looks like gels floating in the melt
pool, can be seen in the metering zone. This situation, however, can
be easily improved with the use of some mixing devices, such as a
series of a double row of mixing pins in the metering zone with decreasing gaps.

   In Sec. 4.6, we discussed some of the characteristics of the melt
in the metering zone, mainly the existence of a radial temperature
gradient ΔT, which increases from the barrel inner surface to the
screw surface. As already shown, this ΔT also increases with the
compound melt viscosity and with the screw rpm, N. The existence
of this ΔT value is a fact process engineers learn very early in their
careers to live with and accept.

   For a given feedstock composition and extrusion system running in
a steady-state operation, however, process engineers are in a better
position of controlling the process if this ΔT remains as constant as
possible throughout the extrusion run. Furthermore, since the barrel
temperature at the feed and compression zones of the extruder in-

fluences the melting mechanism, close control of the temperature in
these zones results in a more uniform melting mechanism, which in
turn produces a more uniform and controllable process.

This can be achieved by an accurate and quick means to both add
and extract heat from the screw and the various barrel zones. We
have already discussed the importance of screw cooling. In this section, we discuss barrel temperature control systems.

Both water- and air-cooled extruders are used in the extrusion of
rigid PVC. For air-cooled extruders, heating rate and cooling rate are
relatively equal, and time-proportioning controllers provide a very
satisfactory control system. This system works as follows. When the
measured temperature is below the lower edge of the band, the heater
power is fully on. As the temperature starts to rise and moves into
the set temperature band, the controller starts to decrease its output
automatically and proportionally (reverse action). This eliminates
short-term oscillations, maintaining a more accurate thermal balance.
When the sensed temperature matches the set temperature, output to
the heater is supplied 50% of the time. In general, the narrower the
bandwidth, the less the stability of the control process. If the bandwidth is opened to improve stability, however, the temperature drift
within the band also becomes wider. Wider bands are also more prone
to temperature "offset" (droop), which is the temperature difference
between the setpoint and the temperature at which process stabilizes.
Automatic reset compensates for offset problems by shifting proportional band around the set point. This is not a quick-response approach
that can cope with constant system upsets. The heater is not continuously on, but it cycles between the on and off positions, and the
time average power input is proportional to the deviation from the
set point.

This situation is improved, however, by the inclusion of "rate action", which provides more anticipation by sensing process rate of
change. What it does is start reducing the heat as the rising temperature falls within the proportional band. It makes stepwise changes in
power input to produce a faster response to sudden temperature upsets. Most controller manufacturers recommend and use these three-
mode combinations, known as the PID-controller, a combination of proportional plus reset (integral) plus rate (derivative). This combination bring the process back to the set point automatically, faster than
the reset operation alone.

Water-cooled extruders present an additional problems because the
water-cooling rate is faster than the heating rate. To alleviate this
problem, the flow of water is automatically controlled by solenoids sit-
uated in each zone. The heaters and solenoids are operated from an
automatic, three-mode controller, usually a solid-state, PID system.

Two basic types of temperature controllers are offered: (1) an
analog system containing a certain number of discrete components, and
(2) a digital system that has a microprocessor replacing the discrete

components. Also available are analog-indicating, digital-set point controllers, as well as an analog controller with an analog-to-digital converter.

Another important factor to be considered in dealing with temperature control is the actual location (depth) and the number of thermocouple probes in the extruder barrel. As far as the die is concerned, every process engineer feels comfortable with at least one thermocouple probe at the die and, preferably, another one at the delivery tube. The reading or readings can then be used to change zone temperatures with one of the several commercial melt-temperature controllers. For a single-sensor system, process engineers place the probe midway between the heater band and the barrel liner. This system is adequate for a smaller extruder running noncritical profiles.

In the newer extrusion lines, however, the use of dual thermocouple probes per zone, one close to the barrel liner (melt) and the other near the heater-cooler band, is becoming more and more common. A further development of the dual probes, which is offered by Davis-Standard, one of the largest U.S. extruders manufacturers, consists of reading the thermocouples separately and giving a weighting to each thermocouple reading, taking into account the thermal lag of the barrel and feeding these readings to the controller's input circuit, which brings the inner barrel surface temperature to stability quickly and automatically. The controller connected to the deep probe monitors the temperature close to the process, and its output provides the set point for the controller that operates the heaters, which also receive input from the shallower probe. This system works on an anticipatory basis without any adjustments required by the operator and is covered by U.S. patents 3,751,014 and 3,866,669.

In the processing of shear- and heat-sensitive materials such as rigid PVC, the control of extruder barrel and die temperatures is even more critical than when extruding some other commodity polymers. In the extrusion of food-grade clear, rigid PVC compounds, which for the most part are mildly stabilized only, the control of processing temperatures within a narrow band becomes even more critical and the best anticipatory control system should be used. This will assure not only a more stable extrusion but also greater control of the optical as well as the organoleptic properties of the containers. It is up to process engineers to decide what temperature-control instrumentation is best for their needs. Fortunately, the selection has never been wider and better. For a more detailed presentation of the several types of temperature controllers available, this author recommends Refs. 13 to 17.

4.27 COMBINED SCREW AND DIE PERFORMANCE

Through the years, extrusion research has placed more of its emphasis on screw design than on die design optimization. Today, the search for screws delivering not only the highest output and energy economy (pounds per horsepower-hour), but also the best temperature and pressure uniformity, continues. Several reasons appear to be behind this trend.

1. The die exit geometry is fixed and dictated by the profile to be manufactured.

2. Since postextrusion swelling of most rigid PVC compounds for building products applications is rather low, no major correction of the die exit cross section is necessary. Die exit dimensions are rather close to those of the finished product.

3. Since energy economy is a major goal and the power required for forming through the die is only about 5% of the power required for the melt processing of the polymer [18], the emphasis is obviously on screw design, looking for the most efficient polymer melters.

For the most part, the design engineers are left with few die areas to be optimized. These are die entry geometry to accommodate the higher flow rates delivered by the newer screw designs; the die land or channel length to minimize melt residence time and to balance the flow of the melt in the die; the die compression ratios to balance pressure drops and temperature rises; and the streamlining of flow, which is of extreme importance for heat-sensitive materials such as rigid PVC. Althougl. no standard exists for the angle of streamlining, it is known that angles of 45° or less tend to minimize the chances of dead spots.

The quality of an extruded PVC product, however, also depends on the efficiency of the die and takeoff equipment, which are part of the process. The main functions of the head-die are to shape the PVC melt pumped by the screw into the required cross section and to build up enough back pressure to ensure complete mixing of the melt in the screw and to streamline the polymer melt so that dimensionally stable products can be extruded.

This is done by channeling the melt from the extruder bore to the exit, which profile is that of the required form. As expected from materials exhibiting laminar flow, some of the characteristics of the melt leaving the extruder, velocity and temperature gradients mainly, are also maintained when the melt moves through the die. The velocity of the layer in the center is faster than that at the die walls. The combined effect of pressure drops and heat generated in the polymer due to viscous dissipation, however, tends to invert the temperature gradient, and the layer near the wall may be at a higher temperature than the center. For most annular dies, the changes in melt temperature gradients at the die tend to reduce the ΔT values. The existence of this equalizing effect can be verified by measuring with an infrared camera, for example, the inner (profile has to be opened) and outer surfaces of an annular profile and comparing this value (steady-state operation) with the ΔT at the delivery tube. The pressure characteristics of the melt leaving the screw also show some changes when the

the melt travels through the die. Melt pressure, for example, drops due to the restrictions in die cross section and temperature rises. Too great a head restriction results in too high a stock temperature and too low a restriction may result in lumps or nonhomogeneous extrudate.

This section deals with the overall relationship of screw and die performance. Expected from the die used in this study is some adjustable melt flow restriction, which could be imposed by a restrictor bar, adjustable mandrel, and adjustable lips, among others. As shown before, very valuable research tools are the screw pumping and the die characteristic curves. In both cases, the ordinate is always output rate (pounds per hour), while the abscissa is the head pressure (psi). These curves, if properly manipulated, can be used to study the interactions between die and screw pumping zone and also to study second-stage screw optimization. To do so, characteristic curves for a given screw are determined for three die flow restrictions, i.e., full opening, intermediate, and high flow restrictions at a constant screw rpm. The output rate versus head pressure for this screw and the three die openings are plotted. Using the same die settting and screw rpm, the same evaluation is repeated for another screw design candidate and the data plotted in the same graph for performance comparison. Obviously, this kind of extrusion research requires a sort of screw library that could be a very expensive proposition. Some of the well-known extrusion research laboratories offer, however, not only a library of screw designs but also various barrel lengths, fully instrumented extruders, and screw ejection facilities, which allow quick screw ejection for melt formation studies, for example. These laboratories are ideal for conducting any extrusion research program.

Figure 4.23 summarizes a parallel evaluation of two screws differing in the second-stage configuration, mainly second metering zone depth. Screw 2 has a small compression angle due to a deeper metering zone. As shown in Fig. 4.23, the points at which the screw characteristic curves cut the die characteristics curves represent the pressure at the screw tip and the output of the machine for a particular die opening at a given screw speed. To obtain the maximum information from any experimental run, the extruder should be stopped at any of the experimental points, while quickly cooling screw and barrel to freeze the melt. Once the melt is frozen, the screw is then ejected and replaced with the other screw candidates.

If screw 1 is replaced by a screw with a deeper metering zone but otherwise identical geometry, the screw characteristic curve is the line with steeper slope. As shown in Fig. 4.23, screw 2 gives a much lower output when die restriction increases. The difference in output between the two screws diminishes as the die opening increases. As shown, for a fully opened die one can obtain a higher output with a deep-flighted screw.

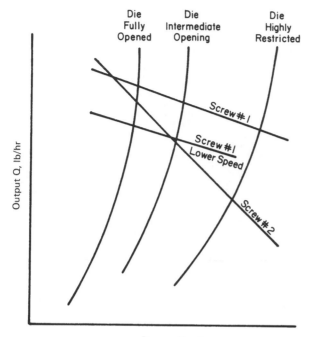

Figure 4.23 Parison die: constant sleeve clearance and three tip positions.

Additional information of great value for the analysis of combined screw and die performance are head pressure variation and radial temperature gradient at the flange and also, if possible, near the land section of the die. Thus, these measurements should be made at the entrance to the flow head and also at the die exit.

The axial die wall temperature profile is also of great value in measuring the temperature rise due to the heat transferred from the melt, which increases in temperature due to the viscous shear heat generation and/or pressure drops. Figure 4.24 is an axial die wall temperature profile measured in a tubular die taken from Ref. 19. This rather smooth profile suggests that the streamlining of this die is quite adequate. Power economy data of the screws evaluated are also needed to make a judgment.

All the previous data, when plotted for each screw evaluated, are of great value for screw selection and also in discriminating between screws with similar characteristics curves, e.g., screws having the same metering zone depth but different lengths.

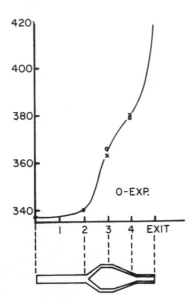

Figure 4.24 Plot of axial die wall temperature profile.

The information compiled above could be extrapolated to similar dies
of different sizes, from wide cross-sectional areas to narrow ones.
For example, a die with a small cross-sectional area, in which there is
a great deal of restriction, requires a screw with a shallower metering
zone. In this case, screw 1 produces higher output. On the other
hand, dies with large cross-sectional areas perform quite acceptably
with deeper metering zone screws (or lower overall compression ratios).

In this study of the two-stage screw and die interactions, the values
of head pressure play an important role in overall extrusion performance. For example, some adjustable dies—restricted or partially restricted—produce sufficient back pressure to fill completely most of
the channels of the second metering zones. As shown before, this
situation contributes to better extrusion stability. In summary, a
properly designed head and the correct balance between the head and
the screw are as important as having a properly designed screw for
the overall success of the extrusion process [19].

4.28 FLOW-RESTRICTING DEVICES

The existence of temperature and velocity gradients across the melt,
as well as pressure and temperature gradients along the extrusion dies,
dictate the need for flow-controlling devices that are mainly used to

correct for any localized variation in the amount of melt shaped and delivered by the die. This, for the most part, is manifested by uneven profile wall thickness, which when too severe could produce some surface rippling of the extrudate. If the wall thickness variation is uniform across the profile, screw rpm (pumping capacity) provides a more efficient way of controlling this problem. For dies of complex shape, like window lineals and gutters, and some simple dies for siding accessories, flow-restricting devices are impractical.

Adjustable restrictor bars, die lips, and mandrels, for example, as well as fixed restrictions in die cross sections (mostly in annular dies containing spiders), are normally used for flow adjustment and to increase back pressure. In annular dies with spider legs, the restrictions in flow cross section imposed on the melt also helps the self-diffusion process of the polymer molecules and hence helps the healing or "knitting" process at the weld lines.

Adjustable flow restrictors that operate by regulating die resistance and consequently back pressure also influence mixing of the melt at the screw. The die resistance can also be increased if the land length of the discharge orifice is increased. This approach, however, increases polymer residence time at high shear, which could have a detrimental effect on extrudate color.

4.29 PLATE-OUT

One perplexing processing problem that occasionally arises during long extrusion runs, especially with opaque, highly pigmented compounds such as those used for residential siding and window frames, and with opaque, highly filled compounds, like the one used to extrude some conduits and drain, waste, vent (DWV) pipes, is the "plate-out." This term is used to describe the buildup or deposit of incompatible material that accumulates on internal tooling surfaces, such as on the screw nose and at the die areas where melt pressure suddenly drops, e.g., at restrictor bars, die lips, spiders, or mandrel.

These deposits also occur on roll take-up or haul-off units, supporting stations, sizing sleeves, corrugated shoes of panel lines, and, overall, units downstream from the extruder.

As these deposits start to build up at the internal die surfaces, the extrudate begins to show die lines, blemishes, variable degree of gloss, and other defects, which for residential siding or window frame applications are objectionable. In time, this deposit provides sites for sticking, leading to degradation and, consequently, machine shutdowns. When extruding window lineals or siding, for example, the appearance of edge roughness, which does not disappear when raising die temperature, is the first warning that some plate-out accumulation is taking place in the die.

In extrusion dies containing spiders (extruded tubular and blow-molded articles), the polymer melt stream splits at the spiders and

recombines there after forming weld lines, which are normally the
weakest points in the extruded articles. As expected, the presence
of plate-out in the spiders area aggravates this weld line problem,
even if this plate-out is at an incipient state. Low-molecular-weight
additives, by diffusing out to the surface of polymer melt stream (at
an early stage of plate-out), interfere with the self-diffusion process
of the polymer molecules in the weld line.

As Chung [20], has stated, these low-molecular-weight additives
increase the time period for which the polymer melt stream remains
split, thus impairing the strength of the extruded articles at the weld
lines.

All the factors responsible for causing plate-out are not fully understood, since this phenomenon may also occur in clear and translucent formulations based on Ca-Zn stabilizers as well as those based
on di-octyltin maleates. Both stabilizer systems are commonly used to
formulate food-grade rigid PVC compounds for container applications.
The appearance of progressive die lines in the bottles is the first in-
dication that some plate-out accumulation is taking place in the die.

As reported by Lippoldt [21], plate-out is the result of material
transfer by a fluid system incompatible with the plastic matrix. This
fluid carries both organic and inorganic components, taken from the
rigid PVC compounds being processed, and is released as a conse-
quence of pressure drops. The previous statement implies that this
phenomenon is related to both formulation and process.

As it has been successively reported by Nass [22] and Thacker
[23], an invariable combination of two classes of metals are usually
present in plate-out-prone opaque compounds—alkaline earths (calcium, magnesium from the filler, or barium from the stabilizer) and
heavy metals (most commonly titanium from titanium dioxide pigment
and lead from stabilizers).

Lippoldt [21] has also reported that, for pipe formulations containing paraffin waxes, this phenomenon is caused by some solution of the
stabilizer in the molten hydrocarbon lubricant (paraffin waxes). The
hydrocarbon-stabilizer solution dissolves the alkaline earth soap, which
in turn is adsorbed on the polar surfaces of inorganic salts, such as TiO_2
or $CaCO_3$. This complex is then released from the hot melt as a con-
sequence of pressure drops in restricted die areas and collects on the
metal surface. The resultant solid film serves as a receptive surface
for further deposition. Lippoldt supports this plate-out mechanism
with experimental data.

In the case of food-grade clear formulations containing di-n-octyltin
stabilizers and epoxidized vegetable oils, the plate-out mechanism is
of a completely different nature. In this particular case, it seems that
these multifunctional tin stabilizers promote some crosslinking of the
epoxy plasticizer. This partly cross-linked plasticizer becomes insoluble in the PVC melt and then is released on the surface of the restricted die areas. That this plating accumulation is significantly re-

duced when the epoxy plasticizer is removed from these clear formulations seems to corroborate this hypothesis.

If the plate-out problem is related to stabilizer composition, great
help can be obtained from the stabilizer manufacturers, which for the
most part can adjust the composition of their highly proprietary products to minimize these problems. As expected, this approach works
well with mixed-metal stabilizer blends, like the one containing various
ratios of alkaline earth metals and heavy metals, and not so well with
stabilizers with a fixed chemical structure. The manufacturers of
mixed-metal stabilizers have some means for improving the compatibility
of the various components of their products, thus minimizing plate-
out. These include adjusting the concentration of alkaline earth metal
and that of the polyols, including surfactants and/or small concentrations of liquid zinc-based stabilizers. In summary, knowing the existence of a plate-out problem, the heat stabilizer producer, for the
most part, can supply a low plate-out version of the existing product.

As mentioned before, the practice of blending emulsion resins up
to 30 phr (parts per hundred of resin) with suspension resins reduces
plate-out tendencies. Also, inclusion of very fine silicate filler or
silica flour at very low concentration contributes to lowering plate-
out, by means of a mild scrubbing or polishing action.

The buildup of plate-out in a chilled vacuum sizing sleeve can occur
during long pipe extrusion runs. One effective technique used to
keep sizing sleeves clean is the dropwise addition of mineral spirits to
the top of the hot PVC pipe just before it enters the sleeve.

4.30 TWIN-SCREW EXTRUSION PROCESS

So far, we have discussed the performance characteristics of two-
stage single-screw vented extruders. Also, in Sec. 2.5.7., we discussed
the overall performance characteristics of some of the best-known one-
stage twin-screw extruders. Although these machines are primarily
used in the extrusion of rigid vinyl sheet, pipes, tubing, and profiles, it is a known fact that, due to their versatility, they are also
used for pelletizing. This secondary application is greately helped by
the availability of pelletizing heads and cooling equipment, which are
offered by most of the twin-screw extruders manufacturers as a package.

In this section, we discuss the conical twin-screw extruders only,
which undoubtedly represents a major trend in single-stage twin-screw
extrusion, although many parallel single-stage and two-stage twin-
screw (four-screw) extruders are also in use. Figures 4.25, 4.26,
and 4.27, for example, represent two different extrusion lines based
on two-stage twin-screw extruders extruding large-diameter rigid PVC
pipes. This is perhaps the main application of these extruders. A
major performance difference between single- and twin-screw extruders

Figure 4.25 $5\frac{1}{4}$ in. 2 + 2 counterrotating twin-screw extruder in production of PVC pipe. (Courtesy of Egan Machinery Company, Somerville, New Jersey.)

Figure 4.26 Extruder A4 100/105C Cincinnati, 326 X 13.1 mm. Extrusion head RK5 American Maplan. (Courtesy of Brasilit, Sao Paolo, Brazil.)

Figure 4.27 Takeoff equipment of extruder A 4 100/105 Cincinnati, 326 X 13.1 mm. Extrusion head RK 5 American Maplan. (Courtesy of Brasilit, Sao Paolo, Brazil.)

is the conveying mechanism. In single-screw extruders, it is based on frictional forces in the solids-conveying zones and viscous forces in the melt-conveying zone. On the other hand, all the intermeshing counterrotating twin-screw (conical and cylindrical) extruders and some of the intermeshing, parallel, corotating twin-screw extruders with decreasing three-step screw and barrel diameters are perhaps best known for their positive displacement (gear pump) conveying characteristics.

Another major difference is in the mixing mechanism, which in a single-screw machine is mostly done within the channels, but in the intermeshing twin is mostly conducted in the area where the screw intermeshes. The degree of barrel and screw wear is also quite different for both machines. As expected, twin-screw extruders are more prone to screw wear than the single-screw extruders, and more frequent replacement can occur in the field, depending on the formulation and operation of the extruder. In recent years, the proliferation of companies rebuilding screws seems to confirm the need for more frequent screw replacement.

The amount of screw-barrel wear, however, is influenced by the formulation to be processed and also by the sense of screw rotation. The corotating twin screws show, for the most part, lower rates of wear or more even wear than the counterrotating. Corotating twin screws drive the material in a figure 8 path around the screws and wear tends to appear at the 3 and 9 o'clock positions and at the peaks where the two barrel halves meet. Counterrotating screws create a 10 o'clock-2 o'clock thrust on the barrel due to accumulation of material at the bottom of the barrel—generally more severe at the compression zone just before the vent, and in the metering zone (see Fig. 4.28 (24)]. Regardless of the sense of rotation, the screw-barrel wear is minimized when lead-stabilized, rigid PVC formulations are used. This behavior is attributed to the use of the combination of lubricants dibasic lead stearate and the normal lead stearate. There is also the opinion that any of the lead salts used as heat stabilizers produce the same effect. A properly balanced external-internal lubricant system for the newer tin stabilizers, however, can be run just as long as the lead system with no greater wear. Filler loadings can aggravate the wear problem due to abrasion, but mostly it is a melt viscosity function that forces screws out to the barrel, causing wear. Actual wear is metal to metal, PVC itself is not abrasive, and $CaCO_3$ has a low hardness. TiO_2 is perhaps the most abrasive component in rigid PVC formulations.

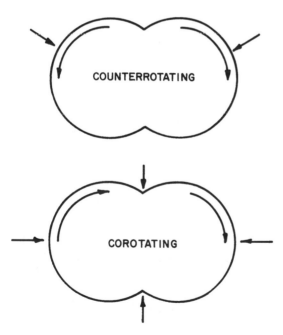

Figure 4.28 Areas of screw wear in intermeshing twin-screw extruders. Short arrows show locations of most severe screw wear.

4.31 ANALYSIS OF TWIN-SCREW EXTRUSION PROCESS

Even though twin-screw extruders have been in use since late 1930,
progress in theoretical analysis of twin-screw extrusion has been slow.
In spite of the most recent efforts [25] to describe their performance
accurately, the theoretical analysis of twin-screw extruders remains in
its infancy. There is such machine variety that it is virtually impossible to normalize discussion of twin-screw extrusion regarding scale-
up rules, temperature generation, overall performance, and so on. If
some knowledge is gained in a particular design configuration, this
knowledge cannot for the most part be used to predict performance
characteristics for different machines. Furthermore, the lack of reliable scaling rules makes it difficult to predict extruder performance
based on an extruder model. This explains, at least partly, why extrusion research in twin-screw extruders is conducted on production-
sized machines.

     However, the recent availability of laboratory-sized conical and
parallel twin-screw extruders may eventually result in developing
scale-up data to achieve predictability of compound performance in
larger extruders. (Laboratory extruders now available from C. W.
Brabender, Haake, and Krauss Maffei.)

     Some practical experience has shown that, among extruders of the
same design, processing in small, medium, and larger versions of that
design can be related qualitatively. Within the conical twin-screw
family, for example, extruding the same pipe compound at maximum output for each machine, the smallest extruder (CM 55) requires a fairly
steep barrel temperature profile and runs at the highest rpm ($\leqslant 40$),
but the intermediate CM 80 requires a less severe barrel temperature
profile and runs at a lower ($\leqslant 30$) rpm to achieve the same work input.
The (now extinct) CT-222 (the largest conical ever produced) required
a *very* gradual barrel temperature profile and ran at a low (8 to 12)
rpm. The relationship here was to residence time and radial velocity
of the screws, which determined rpm limits and the amount of heat
needed to process at optimum conditions.

     As stated before, in the following section we select one of the major
present trends in the field and base our discussion on conical, intermeshing, counterrotating twin-screw extruders only. Because of the
reasons stated before and the lack of ability to quantitatively describe
twin-screw extruders, a practical approach based on a great number
of developmental extrusion trials and extrusion runs, specifically
conducted with rigid PVC, is followed.

4.32 MATCHING THE FORMULATION TO THE EXTRUSION PROCESS
AND TO THE FINISHED PRODUCTS

For any given PVC extruded product, the first step for the process
engineer is that of matching the formulation to the required finished

product properties and to the extrusion process. The choice of resin, impact modifiers, pigments, and fillers, as well as the concentration of these additives that enhance properties, is mainly dictated by the end use and/or appearance of the finished products, but the choice and concentration of the additives enhancing the process, mainly heat stabilizers, processing aids, and lubricants, are dictated by the extrusion process.

Single- and twin-screw extruders both achieve the same goal of producing a continuous melt that is shaped out of a die. These extruders, however, produce melts showing significant differences in peak temperature and radial temperature gradient, which in turn dictate the need for substantial differences in the concentration of those additives enhancing the process, e.g., heat stabilizers, lubricants, and process aids.

A comparative view of process characteristics and formulation needs for the two types of extrusion is presented below. Most of these process characteristics have been discussed already in greater detail in other sections throughout this book. To make this a more fair comparison, let us assume that both types of extruders (with vented barrels) produce the same profile at about the same output from a dry blend feedstock.

4.32.1 Two-Stage Single-Screw Vented Extrusion

This process operates at high screw rpm that produce (1) high shear, (2) higher peak melt temperature (the melt near the screw surface is at the highest temperature), and (3) wider temperature gradient across the melt. These characteristics of the melt are maintained when the melt, pumped by the screw, travels through the delivery tube. At the die, depending on its design, the temperature of the melt contacting the die surface may rise due to pressure drops, producing a narrower ΔT, but for the most part, at higher temperatures. Higher temperatures produce lower melt viscosity and lower melt strength, which makes the sizing of the profile a more difficult task. Also, the higher temperature can create some other serious processing problems. For example, for heavier walled pipes, the residence time in the cooling tanks may be not long enough to reduce the temperature gradient, thus causing thermodegradation of the polymer and consequently unacceptable products. For thinner wall and high surface area profiles, i.e., SDR or schedule 40 pipe up to 4 in. (nominal wall thickness ~ 0.250 in.) large-diameter thin-walled pipes and conduits, panels, residential siding, gutters, and downspouts, cooling, for the most part, does not present a problem. In the first-stage, the screw also develops high compression (2.5 to 3.5:1) to ensure adequate fusion of the melt when reaching the vent.

The pressure of the melt delivered by the screw also shows some fluctuations with the screw motion. As shown in Sec. 4.7, every time

the screw rotates and the trailing edge of the flight is directly under a
pressure transducer, we have a minimum pressure value that goes to a
maximum value when the leading edge is under the transducer. This
periodical pressure fluctuation (sawtooth) produces a variable degree
of surge (flow rate fluctuation). Obviously, this problem is inherent
to single-screw extrusion and can create a greater overweight factor
in attempting to maintain minimal wall thickness.

Single-screw extrusion needs a higher concentration of internal lubricant to promote fusion, reduce internal heat generation, and protect
the localized overheating of the melt on the screw surface. Due to
higher polarity and solubility in the PVC melt, internal lubricants
lower the melt viscosity. The concentration of external lubricants,
however, should be carefully watched, since a high concentration of
external lubricants, because of their incompatibility, interfere with the
frictional forces of the polymer against the barrel surface, and hence
interfere with the conveying mechanism of single-screw extruders and
promote surging. If the concentration of these lubricants is excessive,
the conveying mechanism may fail and the extrusion can come to a
complete halt. Therefore, external lubricants are kept to a minimum
in single-screw compounds.

4.32.2 Conical Twin-Screw Extrusion

This process operates at lower screw rpm than the corresponding single
screw at the same output, which produces (1) low shear, (2) lower melt
temperature, and (3) except in the calender gap, no temperature
gradient within a screw flight for the polymer. It is pertinent to state
at this point that the main reasons for the relatively low screw rpm is
to minimize radial screw velocity, the otherwise extraordinarily high
wear stresses on the screws and barrel [26] and the high mechanical
stresses on the screw bearing. Since these extruders need longer
adaptors and bulkier dies (barrel exit is not circularly symmetrical),
which in turn produce longer polymer residence times, the lower melt
temperature and temperature gradient are very advantageous melt
features. The higher inventory of these larger dies also permits a
higher compression ratio and much of the final "work input" or final
development of physical properties is actually accomplished in the die.

Because of the positive displacement "gear pump" action of counterrotating, intermeshing screws, the flow of melt delivered to the adaptor
and die is more even. Also, the melt pressure is more uniform.

Twin-screw extrusion needs less internal and more external lubricant. A higher concentration of external lubricants has a two-fold
purpose: first, it delays fusion toward the metering zone, which helps
to control amperage at the higher screws rpm; and second, it prevents polymer from sticking to the large adaptor and die inner surfaces.
However, there is always the risk of using too much external lubricant.

Table 4.1 Example of Formulation Differences: ASTM Type I PVC
Pressure Pipe (Class 12454)-NSF

Single-screw		Twin-screw
100.00	PVC, K value 65 to 68, parts	100.00
1.00	Titanium dioxide, phr	1.00
1.0-3.0	Surface-treated (stearic acid) calcium carbonate, phr	1.0-3.0
0.7-1.2	Alkyltin mercaptide stabilizer, (methyl, butyl, or estertin)	0.25-0.4
1.5-3.0	All acrylic process aid, phr	0-1.5
1.0-1.5	Calcium stearate (internal lubricant), phr[a]	0.5-0.8
0.7-0.9	165 Paraffin wax (mostly external), phr	1.0-1.2
0.1	Low-molecular-weight PE wax, phr	0.15

[a]Calcium stearate does promote fusion, but unlike conventional internal lubricants does not reduce melt viscosity
Source: Courtesy of Cincinnati Milacron; George Thacker, Jr.

If, for example, polymer at the vent is too powdery and this situation
does not improve when the temperatures of the rear barrel zones are
increased, the formulation may have too much external lubricant. Some
external lubricants, such as the low-molecular-weight powder polyethylenes, also help to reduce melt fracture in large-diameter pipes.
    An excellent example of formulation differences between single- and
twin-screw extruders is taken from a booklet on rigid PVC extrusion
prepared by George A. Thacker, Jr., for Cincinnati Milacron Chemicals, Inc.* [23] (October 1976). This company, which with its parent,
Cincinnati Milacron, at the time offered both conical twin-screw extruders and additives for the PVC industry, pioneered significant additive improvements for these conical machines (see Table 4.1).
Keeping the lubricant balance the same, both these basic formulations
can be modified to produce a Type II (medium-impact) pipe, conduit, or
outdoor profile using the adjustments stated below.

    Type II pipe: 4 to 6 phr impact modifiers in powder form are added
                   to the basic formulations. Most commonly used im-
                   pact modifiers are CPE, ABS, and MBS.

---

*Now Carstab Division, Thiokol Corp. (since 1979).

Outdoor Profile: Add 6-9 phr impact modifiers, mainly, CPE, EVA,
and most recently all acrylic impact modifiers. Reinforce the heat
stabilizer by adding 0.5 phr more. For a white profile, add at
least 9.0 phr (preferably 12 to 13 phr) of a moderately chalking
titanium dioxide. For a pastel-colored profile, add at least 9.0
phr of a nonchalking titanium dioxide plus the ultraviolet (UV)
light stable pigment system.

Conduit: Add 4 to 6 phr of calcium carbonate (surface treated fine
particle size); 2 to 3 phr of impact modifier added. Processing
aids are used at higher levels to improve hot melt strength.

Previous dry blends are prepared following the procedures described
in Secs. 2.5.1 and 2.5.3. For applications other than NSF, the alkyltin system can be replaced with a lead system, taking care to reduce
lubricant levels to compensate for the lubricity of the lead stearates.
In lead systems, part of the tribasic lead sulfate is replaced with dibasic lead phosphite, when formulating for outdoor weathering applications.

4.33 EXTRUSION OF PIPES AND CONDUITS: SINGLE VERSUS
CONICAL TWIN-SCREW DRY BLEND EXTRUSION PROCESS

Having achieved properly balanced dry blend formulations for both
single- and twin-screw extrusion, process engineers must then concentrate their efforts on the extrusion process per se. With a good
understanding of the inherent characteristics of each process and the
effect that these characteristics have on melt quality, they are in a
better position to produce an optimum quality product at commercially
acceptable output rates. Furthermore, they can decide if the extrusion process is economically feasible.

To perform their jobs, processors must use all the previous information to establish rational techniques for the operation of their extruders. At this point, a summary of the characteristics of each process and their effect on melt quality will help to ensure satisfactory
results in controlling the extrusion process.

4.33.1 Two-Stage Single-Screw Vented Extrusion

The overall high peak melt temperatures and wider temperature gradients across the melt have to be closely monitored and controlled not
only to avoid polymer degradation (which can be manifested by an incipient color formation, e.g., yellowness or the appearance of brownish streaks or a more serious polymer burning) but also to ensure development of physical properties like low-temperature impact resistance.
This is achieved through controlling the following process variables:
screw rpm, screw cooling, the crammer speed (which controls the rate
of feed), and also the barrel heat. As shown in Sec. 4.7, for a given
screw design, the screw speed is the primary operating variable con-

trolling the amount of work input and, hence, the shear and frictional heat.

Typically, to reach an equilibrium state in which the temperature input through zone heating is mainly used to maintain the melt temperature and as a secondary source of melt formation, the zone temperature profile from hopper to die should follow closely that of the melt in each zone; hence, the importance of proper instrumentation in this extrusion process.

A typical single-screw temperature profile follows:

Screw cooling, 130°C*
Barrel from hopper to delivery tube, 150 to 180°C
Delivery tube, 175 to 180°C
Die spider, 180°C
Die, 190 to 200°C

High melt temperature and wider radial temperature gradients result in melts leaving the die at high temperature and consequently at lower melt viscosity and melt strength. This can cause sizing problems as well as high and/or uneven extrudate gloss. When these problems occur at moderate extrusion rpm, they tend to limit the possibility of increasing output. Furthermore, if these problems are severe enough, the processor is forced to lower the screw rpm (and consequently output) as the most expedient way of controlling those problems.

Pressure fluctuations of the melt in single-screw extrusion can cause an overweight factor. Although this overweight factor varies from extruder to extruder and from plant to plant, a factor that has been frequently reported is 8 to 10%. As shown in Sec. 4.21, good barrier screw design, however, can reduce this number significantly. Also, the longer 30/1 and 34/1 L/D extruders, which permit lower stock temperatures due to longer residence times, can help reduce overweight, perhaps by reducing the surging tendency of a hotter melt.

Since the conveying mechanism is dictated by the frictional forces of the polymer against the barrel wall of the extruder, the crammer hopper, by force feeding the dry blend into the channels of the feed zone, influences the melting rate R_1 of the first stage of the screw and, consequently, extrusion output Q. Crammer speed, however, should be treated as an operating variable that should be controlled. Very high crammer speeds, for example, can cause values of R_1 higher than the pumping rate of the second metering zone R_2, and hence cause flow through vent. Perhaps the major use of a hopper crammer is in improving first-stage fluxing and thus its ability to increase the first-stage pressure profile.

*Higher screw temperature must be watched closely since they can cause feed problems.

The vent section of the extruder is designed to receive, for the most part, a fused material from the first metering section of the screw and to be partially filled to only ~25% of total capacity. This situation favors quick removal of volatiles.

4.33.2 Conical Twin-Screw Extrusion

As discussed before, the dry blend for this application is formulated for later fusion than single-screw formulations. Also discussed was the fact that the melting rate in these extruders is controlled mainly by barrel zone heat and screw temperature rather than by shear, which contributes to significantly lower amperage and higher power economy (pounds per horsepower-hour) at the higher rpm.

To minimize torque and screw and barrel wear, complete melting of the polymer is moved toward the metering zone of the screw, where back pressure buildup takes place. As expected, back pressure is directly related to the cross-sectional area of the metering zone.

In addition, since the ratio of external and internal lubricants are adjusted to a higher concentration of the external lubricants, to achieve later fusion, process engineers must reverse the extruder barrel temperature profile to achieve necessary early heat input to begin the compaction and fusion process, since there is very little friction-generated heat.

Contrary to single-screw vented extrusion, the polymer at the vent is only partially fused. Like the two-stage single-screw vented extrusion, however, the polymer is also decompressed at the vent, which is achieved through flight geometry. The effect of screw geometry on partially fused powder seems to favor the volatiles removal. If the polymer at the vent is too powdery, vacuum can pull a lot of the dry powder. If the polymer is too fused, volatiles remain entrapped in the melt.

A typical "reverse" barrel temperature profile for these extruders (i.e., CM 111) is also taken from Thacker's work at Cincinnati Milacron Chemicals, Inc.

Extruder zone								Rheostat Mandrel Heater
1	2	3	4	Screw	Adaptor	Spider	Die	
190-180	180-175	175-170	165-150	150-165	165	175	195°C	20%

Partially filled channels in the feed zone of single-screw extruders can cause serious output and quality problems. A starve-feed mode, however, is the safest way of running all twin-screw extruders, not only during the start-up operation but also during production. Competitive production economics however, dictate that each extruder must be run at its maximum output consistent with maintaining quality production. If higher output values are necessary, e.g., to manufacture schedule 80 large-diameter pipes, the screws must be run at a flooded mode and at a maximum rpm. This, however, has to be done progressively, starting with a starve-feed mode and increasing the screw rpm before increasing the feed. This is done while watching the extruder amperage very closely.

Once the desired output rate is reached, the temperatures of the extruder barrel at the metering zone as well as the screw oil are then adjusted to achieve the required pipe appearance, especially the inside surface (i.d.). Tuning these temperatures also produces the required degree of polymer gelation needed to achieve mechanical strength. Pipes with a wavy i.d. and high gloss, for example, suggest too high a melt temperature. Furthermore, pipes showing some disintegration when immersed in dry acetone suggest the need for higher screw oil and metering zone barrel temperatures.

4.33.3 Pipe Die Design: Matching Dies to Single- and Twin-Screw Extrusion

Since a twin-screw extruder barrel exit is not circularly symmetrical, the design engineers have a more difficult task streamlining the melt flow passages. As expected, the matching-blending operation of the barrel exit, adaptor, and die results in dies significantly larger and longer than those used in single-screw extrusion (circularly symmetrical). This larger size of head, die, and adaptor has, however, a beneficial effect on the cosmetics of the finished products: mainly, it tends to attenuate the screw marks typical of the twin-screw tips. In the extrusion of clear and translucent building panels, these marks can be quite objectionable.

Several factors seem to favor the use of a greater volume of melt at the die and adaptor passages without any detrimental effect on the process or product quality. Some of them follow:

The melt delivered by the screw is not only at lower temperature (150 to 165°C) but also exhibits greater thermohomogeneity.
The melt temperature delivered to the die is closely controlled by the screw oil and metering zone barrel temperatures. Process and quality problems, such as unbalanced melt flow entering the die, material too hot entering the die, and variable degree of inner surface gloss, are for the most part corrected by adjusting these two temperatures.

These characteristics of the melt temperature permit a larger inventory or volume of melt surrounding the spider zone, which in turn permits a greater die compression (reduction in cross section), desirable for ironing out the spider lines. Greater die compression after the spider zone produces pressure drops and temperature increases, which are augmented by the frictional heat induced by the die compression. This localized temperature rise lowers the melt viscosity and thus increases the diffusivity of the molecules, which tends to minimize spider lines. In other words, the final "work input," or generation of physical properties, actually occurs in the head and die.

Another beneficial side effect of a lower melt temperature around the torpedo and spider legs is that some of the incompatible volatile additives, which may plate-out, remain in solution while the melt travels around the spider zone, thus minimizing spider line problems induced by plate-out.

When designing dies, either for single-screw or for twin-screw extrusion of seamless pipes and tubes, there are certain common rules that, if followed, assure a more smooth extrusion operation. Already discussed is the importance of streamlining the flow to avoid any potential polymer stagnation. The rule of great importance is that of the flow path cross section, which should be gradually reduced at the exit side, after the spider legs and never before. In this manner, the pressure delivered by the screw drops while the temperature rises at the time that it can be of more value to help healing of the spider lines. Furthermore, an early pressure drop at the die reduces the chance of good forming in the exit slot.

A problem that remains a nightmare for die designers is that of the die mandrel temperature control. It is known that the die mandrel, with its overall bulkiness, has a decisive influence on extrudate quality, mainly at the inner surface of the pipe. It is also known that most of the waviness and orange peel i.d. surface problems are related to mandrel temperature and, consequently, can be reduced by adjusting this temperature. Mandrel cooling, however, is practically impossible since the cooling medium would have to flow through spider legs, which could create severe weld line problems.

Although close die mandrel temperature control is desirable for both die designs, it is perhaps more critical for dies for twin-screw extruders. Since the melt delivered by the screws is not only at a lower temperature but also exhibits a narrower temperature gradient, this colder melt finds higher resistance to flow on the mandrel surface, thus causing the common waviness. Usually, a rheostat heater on the mandrel, set at 20 to 25% of capacity, is sufficient to minimize this problem without overheating.

Figure 4.29 shows typical die characteristics and compression ratios for twin-screw dies [23]. A^1 is the spider leg length, and A^2 is the die gap. The land length L_1 of these dies is typically twice the die gap (wall thickness) (see Fig. 4.29c).

The overall characteristics of the PVC melt delivered by conventional two-stage single-screw vented extruders, mainly the wider radial temperature gradient and the higher peak melt temperature, dictate the

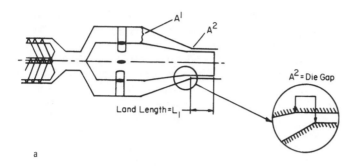

a

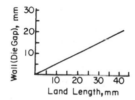

Pipe Diameter, in.	Spider Legs Length A^1, mm	Die Gap A^2 mm	Die Compression Ratios
Up to 8	30	10	3:1
6 to 10	15	8	1.875:1
12 & Up	24	15	1.6:1

b

c

Figure 4.29 (a) Twin-screw extrusion die characteristics. (b) Twin-screw extrusion die compression ratios. (c) Twin-screw die gap A^2 versus land length L.

need for smaller sized dies with narrow passages and shorter die lands so that the melt flows faster, thus minimizing polymer residence time. This is geometrically feasible with single-screw extruders because the extruder exit is circularly symmetrical.

As expected, the overall narrower melt passages along the entire die create pressure drops and temperature rises. Also, the narrower passages reduce the possibility of using higher die compression rates after the spider area, which is the most effective way of welding spider lines.

The higher melt temperature at the spider legs contributes to one or more of the following problems.

Incompatible volatile additives tend to plate out at the spider legs. The lower viscosity melt flows around the spider legs at faster rates, reducing the possibility of melt healing and thus producing weaker pipes.

As expected from a hotter melt, the inner surface of the pipes and tubes manufactured in single-screw extrusion systems tend to be more glossy than those manufactured in twin-screw extruders. On the other hand, a properly balanced heat profile (particularly screw oil and metering zone temperatures) can produce very smooth, semiglossy i.d. pipe from a twin-screw extruder.

Based on the preceding discussions of single and twin-screw extruders, it is not hard to visualize why, in extrusion plants having both systems, two-stage, single-screw vented extruders are preferably used to extrude the small-diameters (schedule 40) pipes or the thin-walled large-diameter pipes and conduits. As mentioned before, through a combination of screw and die designs, and calibration and cooling train know-how, some of the oldest and largest pipe manufacturers in this country extrude quite successfully commercial quality, large-diameter rigid PVC pipes (up to 12 in.) in two-stage single-screw vented extrusion systems. There are several reasons behind the amount of effort that these manufacturers have put into this development. First, due to their higher screw torque, single-screw extruder systems allow the use of a high concentration of reworked material in the extruder feed. (In the extrusion of rigid PVC siding, for example, quite frequently the siding regrind is used in single-screw extruders to extrude siding accessories.) Second, due to the lower cost involved in barrel and/or screw replacement, single-screw extruders are more economically desirable to extrude the lower cost, highly filled rigid PVC formulations, i.e., formulations containing high loads of ultrafine grade calcium carbonate. Third, with minor capital investment, these extruders can be used to extrude other thermoplastics resins in powder form, e.g., polypropylene and LLDPE (linear low-density polyethylene). Formulation costs savings via increased concentration of filler or rework should be weighed versus the increasing cost of replacing screw and barrel.

Although obviously requiring a more sophisticated technology to iron out some of the limitations inherent to the process, the extrusion of large-diameter rigid PVC pipes in two-stage single-screw vented extruders is quite feasible.

The subsequent discussions on die design and cooling bath requirements provides some help to those processors who are concerned with this subject.

Whatever die is used with a two-stage screw, its pressure drop should fairly well match the full pressure-developing capability of the second stage. As shown before, this happens when the filling length of this screw section coincides with the actual length of the second metering zone. This pressure-matching operation is of great importance, since only limited second-stage filling control, or balancing, can be achieved by varying operating conditions or hopper-crammer rpm. Also, a hopper-crammer has limited usefulness as a rate-controlling device for balancing the stages. By serving as a screw extension, however, it does improve fluxing or energy transfer to the material by increasing the first-stage pressure profile.

Low shear dies for thick-walled pipes, when used with a two-stage screw, however, require an additional pressure drop downstream from the spider section to minimize thickness variations due to the inherent surge of each screw revolution, or interior pipe marks due to spider drag effects. As discussed before, this can be obtained via choke restrictions after the spider section. Furthermore, by adjusting the annular choke gap one can match the pressure-developing capability of the pumping section of the screw, thus obtaining the ideal channel fillage for the given screw design (for more information on this, see Sec. 4.20). The effect of choke gap restrictions on the temperature profile across the exit flow stream should be measured and followed closely, since local temperature peaks ranging from 460°F (237.8°C) to as high as 500°F (260°C), as well as average temperature level in excess of 450°F (232.2°C), have been measured in the exit flow stream. The choke gap, however, should not be too small, since this makes wall thickness adjustments difficult. Nor should the choke length be too short, since this can make the choke pressure drop quite sensitive to local variations in melt viscosity or melt temperature, always present to some extent. Neither should it be too long, since sufficient length should be left to permit the flow to adjust to the final discharge geometry, which is that of the die land. The choke should be placed at a fair distance downstream from the spiders in order to provide a "flow equalization" chamber, which would allow some time for the melt to flow back around the spiders. It seems reasonable to expect that a high hydrostatic pressure in this region would be helpful in equalizing the flow and eliminating the weld defects.

In practice, one can state that the two best ways of minimizing spider lines are (1) higher die compression after the spiders, and (2) longer traveling distance (time) after the spiders, which aids in knitting together.

These practical design criteria are of great importance in designing dies for large-diameter, heavy-walled pipes. In fact, one die manufacturer has developed the concept of a "spiderless" die, because it is hard to see the spider lines. The spiders supporting the mandrel are not radially placed, but are placed axially at the entrance end of the die. In this manner, greater residence time as well as higher compression can be achieved.

Although previous discussions have been directed at dies for two-stage single-screw vented extrusion, the overall discussions on spider drag effects and/or practical design criteria are also applicable to pipe dies for twin-screw extruders.

The basic equations

$$\Delta P = \frac{2L}{h} \tau$$

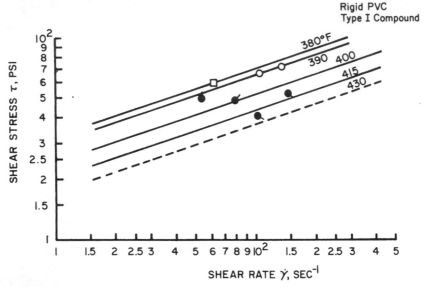

Figure 4.30 Viscosity flow curves.

and

$$\dot{\gamma} = \frac{6Q}{wh^2}$$

where

    L = choke length, which for the most part is selected based on die designers judgment
    h = choke clearance
    w = average annular circumference
    τ = shear stress at the wall
    $\dot{\gamma}$ = shear rate at the wall

are used to determine the dimensions of a restriction or choke on the mandrel for a given pressure drop and a known average temperature. This is done as follows. From the equation $\dot{\gamma} = 6Q/wh^2$, shear rate, (sec^{-1}) is determined. This value is then taken to the viscosity flow curves (shear stress versus shear rate) for the rigid PVC compound used at various processing temperatures (see Fig. 4.30). The corresponding τ value, the desired pressure drop ΔP, and the selected L value are then substituted in the equation $\Delta P = 2L\tau/h$. From this equation, the dimension h of the choke or restriction is then determined.

Obviously, the previous equations can also be used to determine the pressure drop at the die land, as well as the total pressure drop at the die after the spider section, which will be the addition of the pressure drop at the choke and at the die land sections.

4.33.4 Pipe and Conduit Takeoff (Calibration, Cooling Train, and Haul-off Units)

The main function of the various components of the cooling and takeoff equipment is to take over the extrudate at its exit melt temperature and to process it to the proper size and specifications. As discussed throughout this book, rigid PVC extrudates leaving the die swell in thickness and diameter. Also, this postextrusion swelling is modest and, for the most part, quantitatively predictable, depending on melt temperatures and pressures, among other factors.

Another characteristic of the rigid PVC extrudate leaving the die is its radial temperature gradient, which interferes with the cooling mechanism by partially reheating the layers cooled during the early stages of the calibration. The existence of the radial temperature gradient and all the potential cooling and even polymer degradation problems are, perhaps, the major drawbacks of the single-screw extruders in the extrusion of pipes.

The three basic components of the takeoff equipment of smaller diameter pipes, thin-walled hollow profiles that cannot be coiled and heavy-walled hollow profiles or conduit lines are a vacuum sizing unit, a flood or water spray cooling tank system, and a haul-off unit. These three areas need the most attention from process engineers, since it is necessary to use this equipment to freeze the dimensions of the extrudate while removing as much heat as possible from the polymer. The extruder output to take off puller linear speed is a critical relationship that governs the wall thickness as well as the i.d. of the pipe or conduit.

The interior of the hollow profile need not be sealed when using the vacuum sizing method. Atmospheric pressure inside the pipe or hollow profile inflates the molten polymer against the vacuum sizing sleeve.

Other components of the takeoff equipment, such as the traveling cutoff saw, inline printer, and dumping table, although important equipment features, do not require the attention given to the previous components.

4.33.5 Cooling Bath Requirements

Quite frequently, at high rates and summer water temperatures, the pipe or hollow profile cannot be cooled to a low enough average temperature to prevent sag after temperature relaxation has occurred or at times even to prevent distortion in the puller rolls. Actual polymer degradation has been observed in insufficiently cooled heavy-walled pipes. This degradation, sometimes manifested by longitudinal lines, can be easily detected under any UV lamp, before it can be seen by the eye in daylight. For this reason, most pipe plants must invest in water chillers to deliver refrigerated water to all cooling tanks.

Throughout this book, great emphasis has been placed on melt temperature, e.g., peak melt temperature, and temperature gradients

across the melt. We also discussed that, in two-stage single-screw
extrusion, temperature peaks ranging from 460°F (237.8°C) to as high
as 500°F (260°C) and average temperature in excess of 450°F (232.2°C)
are common. Even in twin-screw extrusion with its lower melt temperatures (385 to 395°F) (196 to 201.7°C) and narrower temperature gradients, an increase in output can produce a substantial increase in
melt temperature. As expected, this situation requires that the calibration unit (a vacuum sizing or pressure sizing sleeve), in combination with the cooling baths be efficient enough to cool the shaped pipe
conduit or hollow profile to an average temperature no higher than the
transition temperature of the rigid PVC compound used to extrude the
pipe, i.e., 74 to 80°C (165.2 to 176°F). This is achieved through a
combination of several equipment and process variables such as (1)
cooling water temperature, (2) bath length, (3) high water film heat-
transfer coefficient normally achieved through adequate agitation in
the flooded tank, and (4) in cooling train spray systems, increasing
the number of spray nozzles, volume, and/or velocity of the cooling
water. Most pipe plants, especially those located in the South and
Southwest, require *chilled* water and employ several water-chilling
units.

The average pipe temperature $\bar{T}$ or equilibrium temperature the pipe
or hollow profile reaches after leaving the cooling bath train, assuming
no further heat loss occurs during the temperature relaxation period,
is information of prime importance in the cooling-forming process of
rigid PVC pipes. Since rigid PVC starts softening or deforming at
about 80°C (176°F) (its transition temperature), it is obvious that
this temperature should be the absolute maximum average temperature
of the pipe leaving the cooling bath.

When using flood tanks, two major sets of data are required for
cooling bath calculations. The first set consists of the following.

1. The thermal properties of the rigid PVC compound used to extrude the pipe or conduit and its density

2. The initial average pipe temperature as it enters bath and overall pipe dimensions

3. The water bath temperature and water film heat-transfer coefficient and the pipe contact time in cooling bath

The information above is directly obtained from the polymer properties, the process, and the processed polymer. The other set of data
is, for the most part, derived from the previous one.

The thermal properties are thermal conductivity k = Btu/hr ft °F,
heat capacity C_p = Btu/lb °F, and its density is ρ = lb/ft³. Although
the thermal diffusivity, κ = k/ρCp is calculated from the above properties, if we use the following average values for rigid PVC, k = 0.11
Btu/hr ft °F; C_p = 0.36 Btu/lb °F, ρ = 85 lb/ft³, κ becomes 0.0040
ft²/hr. The pipe temperature and pipe dimensions used in these
cooling bath calculations are: T_i = initial pipe temperature as it enters

the bath, T_O = inside pipe surface temperature, and L = pipe wall thickness (in.). Relative to the water bath, we need the bulk water temperature T_w (°F), the water film heat-transfer coefficient h_w = Btu/hr ft^2 °F, and the pipe contact time in cooling bath θ. Also needed in cooling bath calculations are the following data.

1. The pipe relative conductivity m is given by the expression $m = k/Lh_w$. If we assume pipe wall thickness L = 0.20 in. and a rather low water heat-transfer coefficient, h_w = 50 Btu/hr ft^2 °F,

$$m = \frac{(0.11)(12)}{(0.20)(50)} = 0.132$$

2. A dimensionless term combining the thermal diffusivity κ of the polymer with the pipe wall thickness L and the pipe residence time in the cooling bath θ, as follows: $\kappa\theta/L^2$.

3. The expression for the average pipe temperature $\bar{T}$, or the equilibrium temperature the pipe would reach after leaving the cooling bath

$$\frac{\bar{T} - T_w}{T_i - T_w} = \bar{Y}$$

where $\bar{Y}$ is a dimensionless temperature ratio used for relaxation or heat transfer calculations. With this information available, we now proceed to use Heisler's well-known temperature charts for constant-temperature heating and cooling [27]. With the aid of these curves and a similar technique used by Latinen in Ref. 28, we organize the data and plot the cooling curves for rigid PVC pipes. From Heisler's charts, we select those for a semi-infinite plate to approximate the solution for rigid PVC pipes. Specifically, we use the charts summarized in Figs. 7 and 10 (of Ref. 27) as follows: Figure 7 represents a plot of $(T_O - T_w)/(T_i - T_w)$ versus $\kappa\theta/L^2$ for values of m ranging from 0 to 100. The expression

$$\frac{T_o - T_w}{T_i - T_w} = Y_o$$

gives the inside pipe surface temperature T_o. In practice, however, we are interested not in the inside pipe surface temperature, but in the average pipe temperature $\bar{T}$. The previous expression then becomes

$$\frac{\bar{T} - T_w}{T_i - T_w} = \bar{Y}$$

Now we use Fig. 10 of Ref. 27 to obtain the temperature profile through the pipe at fractional distances n, measured from the inside

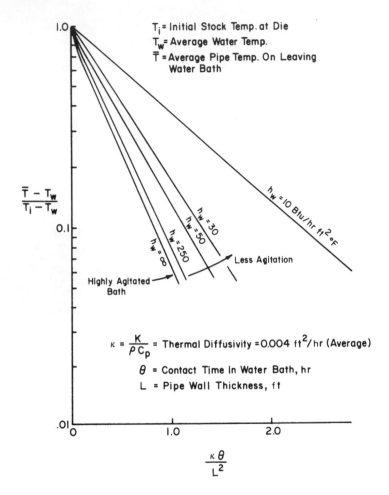

Figure 4.31 Cooling curves for PVC pipe in water bath with various water film heat-transfer coefficients h_w.

surface. From these data we then obtain the average $\overline{T}$ of the pipe leaving the cooling bath. This Fig. 10 represents a position correction factor for the dimensionless temperature ratio Y_n/Y_o versus m. Assuming m = 0.132, we obtain the following Y_n/Y_o from Fig. 10.

n   = 1.0   0.8   0.6   0.4   0.2   0

Y_n/Y_o = 0.175   0.43   0.67   0.85   0.96   1

By graphic integration, one gets the average pipe temperature Y_n/Y_o ≃ 0.70, or $\overline{Y} = (\overline{T} - T_w)/(T_i - T_w) = Y_o$ 0.70, $\overline{T}$ = average pipe tem-

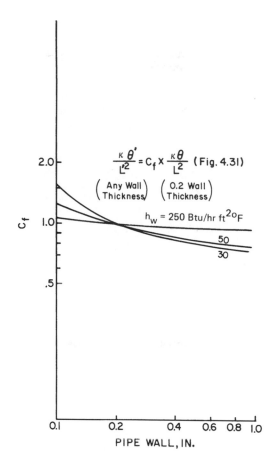

Figure 4.32 Wall thickness. Correction factor to be applied to $\kappa\theta/L^2$ from Fig. 4.31.

perature leaving the bath. From this simple relationship one calculates Y_0, which in turn can be used to obtain the dimensionless term $\kappa\theta/L^2$. The importance of this term is that it contains θ, the pipe residence time in the water bath, thus allowing the calculation of this process variable. For example, if $\bar{Y} = 0.30$, $Y_0 = 0.30/0.70 = 0.428$. This number, when taken to Heisler, Fig. 7, for $m = 0.132$, gives

$$\frac{\kappa\theta}{L^2} = 0.57$$

Figure 4.31 of this work, which is similar to Heisler's Fig. 7, shows a number of approximate cooling curves for rigid PVC pipes in a water bath with various water film heat-transfer coefficients h_w. It is important to note that all these curves were based on an arbitrarily selected pipe wall thickness of 0.20 in. and the same rigid PVC compound. The values in expression $\kappa\theta/L^2$ were obtained by varying only the contact time of pipe in water bath, in hours. Corrections have to be applied for other wall thicknesses, which are small when the water film heat transfer is high (most desirable case), but become appreciable when the water from the heat-transfer coefficient, or agitation, decreases. These corrections are given in Fig. 4.32. They were obtained by going through the same calculations for two other wall thickness, 0.10 and 0.40 in. The following is a sample calculation to illustrate the technique explained above.

Let us assume an extrusion output of 225 lb/hr, 1 in. schedule 40 pipe, leaving the die with an average temperature $\bar{T}$ of 440°F, and a water film heat-transfer coefficient $h_w = 50$ (i.e., rather inadequate agitation).

Problem: What bath length is required to cool the pipe to an average temperature of 150°F (i.e., just below the transition temperatures of about 160°F where softening starts). Bulk water temperature T_w is 80°F.

$$\frac{\bar{T} - T_w}{T_i - T_w} = \frac{150 - 80}{440 - 80} = \frac{70}{360} = 0.194$$

From Fig. 4.31 we find $\kappa\theta/L^2 = 0.79$, which has to be corrected for wall thickness L (which for 1 in. SCH 40 pipe = 0.133 in. or 0.011 ft) and h_w 50 (Fig. 4.32) is 1.12). Hence,

$$\frac{\kappa\theta'}{L'^2} = (0.79)(1.12) = 0.883$$

and the required contact time (as shown before, $\kappa = 0.004$ ft^2/hr)

$$\theta = \frac{(0.883)(0.0111)^2}{0.004} = 0.0271 \text{ hr}$$

Linear pipe velocity = output rate per cross-sectional area times density (85 lb/ft^3). The cross-sectional area of 1 in. SCH 40 pipe = 0.494 in^2. Hence,

$$\text{Linear pipe velocity} = \frac{225}{(0.499/144)85} = 771 \text{ ft/hr}$$

Therefore, the required bath length = velocity X required contact time = 771 X 0.0271 = 20.9 ft.

These examples are presented here to illustrate (1) the usefulness of these cooling curves in the determination of the length of the cooling tank and (2) how to manipulate the thermal properties of rigid PVC for engineering calculations. There is little difference, however, between these charts and the other cooling charts familiar to an engineer. Once all the thermal properties of the rigid PVC compound and their relationships, as well as the pertinent charts, are established, the pipe or hollow profile temperature history during the cooling process

can be determined in a rather simple operation. The same procedure
can be adapted to other hollow articles, such as window and door lineals and conduits.

For large-diameter, heavy-walled, rigid PVC pipes, internal air
pressure and a system of one or more circular plugs or stoppers,
chained to an anchor ring screwed into the center of the die mandrel,
with a water-cooled sizing sleeve attached to the die, provide most of
the pipe calibration. The correct dimensioning of the sizing tools, i.e.,
plug outside diameter, sizing sleeve inside diameter, and length, is of
prime importance. Because the rigid PVC pipe leaving the sizing sleeve
still contains a considerable amount of heat, and the pipe has not yet
been frozen to its final dimension, some designers tend to make the
i.d. of the sizing sleeve slightly larger (1 to 1.5%) than the specific
o.d. of the cooled pipe. When doing so, however, it is advisable to
consider also the shrinkage of the compound being extruded. Fortunately, in the case of rigid PVC compounds, especially those containing
fillers, high-molecular-weight resins, and/or emulsion resins, the difference between dimensions of the pipe when hot and cold is rather
small (low puff-up and low linear thermal expansion coefficient) and
predictable. This is an area that the extruder manufacturers supplying complete pipe package system and turnkey systems (e.g., extruders, dies, sizing sleeves, calibrating plugs, and cooling tanks),
dominate, and they can provide the sizing units with the correct dimensions.

From the sizing unit, the pipe travels through spray chambers provided with a curtain of nozzles surrounding the pipe. Meanwhile, air
is used, first, to keep the pipe from collapsing, so the plug can drag
inside without jamming; then, once the plug has reached the end of its
chain, air pressure is gradually adjusted (4 to 10 psi) to keep the
pipe against the sizing sleeve. This surface contact with the sleeve
should give a nice o.d. glossy surface. All cooling zones behind the
plug should be on for full cooling.

Water spray cooling, contrary to simple water bath immersion, by
impacting the surface wall of the pipe, prevents the creation of a warm
boundary layer, thus contributing to a more intense cooling effect.
Furthermore, since the water does not remain as a film on the surface
of the pipe, there is no risk of heat exchange interference.

In contrast to the flooded water bath-full immersion cooling approach, the determination of the length of the spray cooling tank is
mainly based on empirical data. This determination, however, is simplified because rigid PVC is about 10.0% crystalline only and its T_m
> 210°C (crystalline melting point) is, at least in twin-screw extruders,
above typical melt-processing temperatures; consequently, the heat of
crystallization, for practical purposes, can be ignored. This explains
why, in pipe extrusion plants processing rigid PVC and HDPE, the
cooling train length for rigid PVC is significantly shorter than that
for HDPE.

A new move that could be adapted to large-diameter rigid pipe,
which seems to have been started by Battenfeld-EKK, West Germany,

is that of using a novel modern vacuum calibration method in combination with intensive spray cooling, thus replacing the drag-plug calibration approach. Vacuum calibration offers several advantages. First, there is more freedom in design of the calibrating sleeve, thus enabling a closer match to the compound being processed. Second, without a drag-plug, heat exchange can take place in the interior of the pipe via air or water, for example. Third, the problems caused by drag-plug wear, such as the difficulties in start-ups, disappear.

These modern vacuum calibrators differ from the conventional in the manner in which suction is produced in the vacuum chamber. Suction in the vacuum chamber is generated by means of a water pump independent of the circulatory cooling and of the cooling water exchange [29].

Successful extrusion of rigid PVC pipe, siding sheet, and profile requires a good understanding and proper application of the previous discussions on polymer preparation, melt rheology, and overall melt processing. It should be emphasized, however, that there are other mechanical, temperature control, and electrical factors that also require good understanding and are of prime importance to maintain optimum balance between output and quality and its consequent economics.

In long extrusion runs of formulations containing high levels of inorganic fillers, such as calcium carbonate, clay, and titanium dioxide, the screw and barrel should be inspected periodically for excessive wear (both screw flight and root diameters should be checked), which is first manifested by a gradual loss in output per rpm at constant extrusion conditions, excessive head-pressure fluctuation, and so on. Output loss of 10% per month has been reported. With every screw revolution and with every pound of finished product put in inventory, the efficiency of the extruder is reduced, which contributes to increased operation cost. Eventually, product quality suffers.

It has been stated [30] that, from the moment an extrusion screw is first rotated in a barrel, the performance begins to deteriorate. It deteriorates faster, however, with high-viscosity filled PVC than it does with low-viscosity unfilled olefins.

4.33.6 Processing and Quality Problems That Can Be Remedied with Temperature Adjustments

Assuming that the extruder, die and downstream equipment are in good condition and all heating and cooling zones are working properly, most of the processing problems in the extruder are influenced by the quality of the powder blend being fed to the extruder. Throughout this book we have discussed the role that dry blend preparation, dry blend properties, and overall blend consistency has in minimizing extrusion problems.

In this section, we discuss those processing and quality problems that, for the most part, can be remedied or attenuated by making systematic temperature adjustments. Specifically, we discuss some temperature adjustments in the dry blend extrusion of pipes or hollow profiles in conical twin-screw extruders.* (We assume a high-output, flood-feed, high-rpm situation, which is how most pipe producers try to utilize full extruder capacity for optimum production economics.)

I. High amperage and material too fused at the vent, which generally results in a glossy and wavy i.d. extrusion.
   A. Screw temperature may be too high. Screw temperature should be about the same as the metering zone of the barrel for twins. Rear barrel temperatures either too high or front barrel temperature may be too cold. In any event, adjust these temperatures accordingly.
   B. Incoming dry blend feeding the extruder may be too hot.
   C. Other causes and remedies. If temperature adjustments do not work, formulation lubricant adjustment should be tried next: specifically, a higher external-internal lubricant ratio. A temporary remedy would be to "starve" the screws, i.e., reduce screw feeder rpm so that screw flights in zone 1 are half-full.

II. High back pressure caused by resistance of material being pushed through adaptor, head and die.
   A. Adaptor, head, and die zones may be too cold. Gradually increase temperatures in these zones. Melt may be too "stiff" coming from the screws. Increase metering zone and screw temperature.
   B. Other causes and remedies. If temperature adjustments do not correct the high-pressure problem, the adaptor orifice may be too small or the formulation may be underlubricated, causing friction drag in the die. Obvious remedies to these problems are a larger diameter orifice and/or a higher concentration of external lubricant.

III. Low amperage, material too "powdery" at vent.
   A. Fusion occurs too far down the barrel. Increase barrel and screw temperatures, especially in zones 1 and 2, causing an earlier fusion of powder.
   B. Other causes and remedies. Formulation is overlubricated on the external side or lacks sufficient processing aids.

IV. Low back pressure.
   A. Barrel and screw temperatures are too high. Gradually reduce these temperatures.
   B. Other causes and remedies. Formulation has too much external lubricant, or orifice in adapter is too large, which could be remedied either by reducing external lubricant, PE wax, and/or using a smaller diameter orifice.

---

*Some of these discussions are taken from George A. Thacker, Jr., "Rigid PVC Extrusion: Problems and Remedies."

   V. Material lumpy and with very low gloss.
     A. Material is too cold, not fully fused, which suggests that barrel and screw temperatures should be gradually increased.
     B. Other causes and remedies. Formulation may be overlubricated or low in processing aid. In any event, the concentration of external wax should be reduced and the concentration of processing aid should be increased.
   VI. Uniform yellowing, or discoloration of extruded product.
     A. Overall temperatures are too high. Gradually reduce heat and check controllers for temperature override.
     B. Other causes and remedies. Discoloration combined with orangepeel roughness and/or high amperage indicates underlubrication. Also, double-check stabilizer level.
  VII. Visible spider lines on i.d. with little or no gloss.
     A. Spider and die heat zones are too cold. Increase these heats carefully to lower melt viscosity around the spider legs, thus promoting the welding of the spider lines.
     B. Also, material may be too cold entering the head and die; observe if amperage is on the high side. Increase barrel and screw heat.
 VIII. Erratic flow out of die—smooth to rough to smooth again, including some surging: Too much cooling on barrel cooling zones, which cycle on and off, resulting in "temperature shock." Gradually decrease cooling water flow to barrel.

Obviously, this list of troubleshooting remedies is not universal. The listed remedies work best, however, for vented conical and cylindrical counterrotating twin-screw extruders. For cyclindrical corotating twin-screw extruders and two-stage, vented single-screw extruders, the remedies described should be considered broad suggestions, with the need for possible adjustments for differences in extruder behavior. In two-stage single-screw vented extruders, provided with commercially available barrier screw designs, for example, no two designs deliver the same kind of pressure and temperature uniformity. Consequently, the magnitude of temperature adjustments will not be exactly the same for the various barrier screw designs.

4.34 RIGID PVC EXTRUSION OF SIDING AND ACCESSORIES

The first processors of rigid PVC siding in the United States early in the 1960s, invariably used pelletized compounds and single-screw extruders, 20/1 to 24/1 L/D provided with metering screws. Even if some twin-screw extruders had been available, then, these extruders were used either for pelletizing (Anger A2-120) or to extrude rigid PVC panels (RC Roberto Colombo twin-screw extruders). The Anger A2-120, which was unvented and markedly underpowered, produced pellets that, when extruded as siding, produced a very desirable dull finish. Quite frequently these pellets were blended with Banbury mill rolls processed pellets to control surface gloss. In fact, the tendency to produce a rather high surface gloss was one of the nightmares of these early processors. With the introduction of embossing dies, which produce a wood-grain or matte finish, however, this problem has almost disappeared.

The single-screw extrusion lines, however, were of great value during the early days because they allowed the safe use of the high concentration of regrind generated during start-ups, operator training, and the development of cooling and takeoff equipment.

With the advent of the new twin-screw extruders (vented, higher power, temperature-controlled screws, and so on) and especially the family of conical twins with the same above-mentioned characteristics, a gradual switch to dry blend extrusion began. This switch was aided by the introduction of new resins, more efficient thermostabilizers, light-stable impact modifiers in powder form and other innovations. Obviously, this move was influenced by economic reasons, mainly lower compound conversion cost and higher extruder power economy. There were also some processing reasons, such as less extrusion surge, which contributes to cutting down product overweight, and more precise control of the melt temperature in the metering zone, which contributes to a closer control of extrudate surface gloss and dimensions. Also, there is less tendency to thermal and shear abuse of the material because polymer-metal friction is not the conveying mechanism. Finally, lower exit melt temperatures contribute to a much easier "sizing" job.

This trend, which started in the early 1970s, became the leading production method for rigid PVC siding and accessories. The development of the various new barrier screw designs in combination with the vented, high L/D, i.e., 30/1 or higher, two-stage single-screw extruders, which bring operating temperature and pressure within the range of those of the best conical twin-screw extruders, have narrowed the gap between both extrusion trends. At present, the selection of new extruder lines is up to the processors, who seem to be highly influenced by their past experience.

A claim has been made that the ideal siding extrusion plant should have both extruder systems, single screw to extrude the regrind into accessories or siding containing high concentration of reworked material in the dry blend, and conical twin-screw extruders to extrude dry blends containing no or low concentrations of regrind.

Another factor that seems to favor single-screw extruders in the extrusion of accessories is that the single-screw extruders allow the use of smaller easy-to-handle low-inventory shape dies and adapters. Some accessory shapes, for example, are extruded through simple plate dies attached to rather small adaptors using air tables to cool and size the profiles.

Since the early days, another area that has been heavily influenced by the processors' choice is that of the shape of the profile coming out of the die, i.e., a flat profile (sheet) followed by a postforming die versus the die-formed profile. The advocates of the flat profile-postforming die approach claim several processing advantages, such as the following.

They can use simpler dies, which can be designed to be adjustable via die lips or choke bar, for example, and are easier to tool and set up.

The flat profile produces a more uniform embossing.

Plate-out accumulation at the die, if any, has less effect on the profile edges. The jamming up of the flat profile at the forming die for butt-edge and anchoring flange, however, remains as a potential threat to this manufacturing approach.

Higher output rates are possible. A good profile die, however, can match sheet-die output rates and permit longer trouble-free production runs, owing to superior streamlining.

Regardless of the forming approach used, air remains the most commonly used cooling element, at least during the early stages of the cooling process. There are many reasons behind this choice. Some of them follow.

Air, if properly used, produces a more gradual cooling of the profile. Air also allows selective differential cooling of the lock portions of the siding.

Air also has a less detrimental effect on the impact resistance of the siding. Quick cooling (via a water-cooling jacket) of the rather thin siding web (~0.045 in.) tends to impair the impact resistance and other postforming operations and contributes to buildup of residual stress.

By keeping the profile pliable for a longer period of time, air cooling is also ideal for the early stages of the postextrusion forming techniques, such as cedar-grain or matte finish embossing, butt-edge and anchoring flange forming, and ribbing forming, if any.

When cooling the siding web, however, special care should be taken to expose both surfaces to the same amount (CFM) of air distributed as evenly as possible. This reduces the tendency to surface distortions across the profile, "oilcan" shape.

Profile extrusion lines generally run the profile through an embossing stand consisting of one or two steel-etched rolls on top and an unetched rubber roll on the bottom. The PVC passes between these two rolls, which are adjusted so that they emboss the PVC with the correct pressure to ensure adequate imprint, and then onto a small vacuum-sizing calibrator, or over metal templates, or one large pad that is placed over the siding as it travels downstream. This pad has holes drilled into it to allow air or water to circulate through. Some processors use a series of small pads instead of a large one. The pad approach has two functions: to hold the shape of the siding as it travels downstream and to cool it. The calibrator is an aluminum block, which is fitted with vacuum. After sizing and initial air cooling, the profile may be cooled with water mist or high velocity air. In the post-forming lines the siding is embossed as a sheet and then enters a pre-former, which is located downstream from the embosser where the physical shape of the siding is formed. The material then travels downstream to the postforming die, which accepts the PVC, cools it, and vacuum forms it into its final shape. Cooling can be done with water or air sprays and can also be done by immersing the PVC in a water bath.

Once the siding is formed and cooled it passes through the puller. The puller speed can be used to adjust the thickness of the siding exiting the die. When the siding leaves the puller it travels downstream to the weep hole (a small hole on the butt of the siding to allow moisture venting if collected in the back of the siding) and nail hole punchers. Once these holes are punched into the siding, the panels are cut to length and the notches and slots are cut into the siding.

Throughout this work we have discussed some typical rigid PVC formulations for siding and accessories applications. To these discussions we can add that overall processability in the downstream equipment of these formulations and properties of the finished siding can be greatly improved if (1) they are based on high-molecular-weight resins (high tensile modulus, high T_g) exhibiting good extrudability, and (2) if the compounds also exhibit good tensile melt strength and high heat-distortion temperatures DTUL*, °F at 264 psi, $\geq$ 167 (low concentration of liquids and/or additives lowering the heat-distortion temperature help to keep this property high).

Compounds based on high-molecular-weight resins also exhibiting high DTUL values, in general, produce the following beneficial effects.

Less tendency to exhibit surface distortion "oilcanning" during exposure.

Overall smoother and duller surfaces. This is especially noticeable if some high-molecular-weight emulsion resins are used in the formulation.

Higher rigidity and strength, which make it easier and safer for the installer to handle the 12.5 ft lengths.

Overall lower postextrusion swelling (smaller puff-up values).

Compounds exhibiting high melt tensile strength show the following desirable processing advantages.

---

*Deflection temperature under load.

Higher resistance to tearing at the profile edges. This property
is of extreme importance when profile shaping takes place at
the extruder die.

Overall higher strength, which permits sawing of the cooled profile
into conventional length without chipping at the profile edges,
"J" and "F" lock sections.

Easier threading of the hot profile through the embossing and forming dies, air cooling tables, and so on.

Melt tensile strength, however, is a property that can be greatly improved if some of the well-known all-acrylic processing aids are
incorporated in the formulation. Some of the acrylic elastomers
used as impact modifiers also improve the melt strength. High
filler and/or titanium dioxide loads, however, tend to lower the
melt tensile strength.

The modern single- and twin-screw extruders described thus far
achieve the same goal of producing a continuous shape out of the siding
or accessories dies. Throughout this work, we already discussed the
important formulation and process differences between the two systems
and how the use of rational techniques for extruder operation ensures
optimum quality at commercially acceptable rates for both systems.

Before pursuing any further discussion on siding, let us emphasize
that this particular engineering application of rigid PVC is the one
that has to meet the most stringent requirements, some of which are
listed below.

Aesthetics: Siding and accessories (also known as vinyl siding
trims) shall be uniform in color and gloss (not applicable to embossed siding). Variations in the glossmeter readings shall not
be more than ±10.0%. Obviously, siding surface shall be free
from other visible defects.

Physical dimensions: Siding shall meet very tight specifications for
thickness, weight, length, and tolerance for warp, as stated in
the ASTM D3679 specifications.

Mechanical properties: The unembossed siding shall have a minimum impact failure value of 2.0 in. lbf/mil (8900 J/m).* Also,
the rigid PVC compound shall have a minimum Izod impact strength
value of 0.65 ft lbf/in. of notch (34.7 J/m of notch) at 0°C.

Weatherability: The siding shall maintain uniform color and be free
from any visual surface or structural changes such as peeling,
chipping, or cracking, when exposed facing south at a 45° angle
of elevation for a minimum of 1 year or at an angle of elevation

---

*The embossed siding shall have a minimum impact failure of 1.74 in.
lbf/mil (7750 J/m).

representative of the manufacturer's normal installation recommendations for the siding for at least 2 years. Also, the siding
shall show permanence of its mechanical and physical properties
upon weathering exposure. There are also specifications for
heat shrinkage and coefficient of linear thermal expansion (not
greater than 4.5×10^{-5} in./in. °F (8.1×10^{-5} mm/mm °C).
The above properties must be *long-lasting*: 20 to 30 year guarantee
for installed siding are offered.
Economics: Since this is a very cost-competitive area, all the specifications stated above must be met while keeping a good balance between cost and performance.

To the above we can add that, of all the engineering applications of
rigid PVC, siding is perhaps the profile for which the postextrusion
operation has the greatest effect on the quality of the siding. Not only
because of key cosmetic (embossing) and other important postforming
operations (lock edges and ribbing when applicable), which take place
after extrusion, but also as described before, the various nonuniform
cross sections dictate the need for sectional or differential cooling to
delay setup of the thinner sections (web) and speed up the cooling of
the heavier wall portions. There are all kinds of quality problems,
such as warp and surface distortion, which are for the most part caused
by stresses due to improper cooling, i.e., lack of indiscriminately
uniform amount of air on the profile surface.

Problems and Possible Remedies in Siding Extrusion Operations

Based on the list of product quality requirements and the description
of the cooling and postforming operations, it is not hard to visualize
that in the extrusion of siding most of the process engineers' and line
operators' effort is after extrusion. In other words, the majority of
production line problems are postextrusion-related problems.
    There are potential major extrusion problems that must be watched
closely regardless of the type of extruder used. Some of these problems are listed below.

Screw and barrel wear, which is aggravated by the titanium dioxide
    and/or inorganic pigments, like chromium oxide, and fillers,
    like precipitated calcium carbonates. Screw wear produces var-
    iable flight clearance, which in turn is manifested by loss in
    output, difference in temperature profile setting, and so on.
Plate-out buildup on extrusion dies, which is induced by sudden
    restrictions in the die cross section has been addressed through-
    out this book. It is commonly manifested by either longitudinal
    die marks or uneven profile gloss.
Lack of proper die zone temperature control. Independent die zone
    temperature controls are of great importance when the shaping

    of the profile takes place at the die, since for the most part
    profile thickness is controlled by temperature adjustments.
    Some other extrusion problems are common to other extrusion opera-
    tions, like pipe extrusion, and have already been discussed.

The corrections to the siding quality problem that are related to post
extrusion conditions (warp, surface distortions, and lock dimensions)
depend very heavily on the line operators' skill and experience. They
make decisions on the location and the amount of the cooling air necessary to correct those problems. In this area they are guided by simple quality control tests that can be performed at the line or near the
line and also by the product specifications. It is up to the line operators and/or process engineers, however, to systematically correct these
problems without creating new ones.

    A simple test, ASTM D2152, which is normally used to determine the
degree of fusion of rigid PVC pipes and molded fittings, can be of value for line operators in their troubleshooting efforts. Specifically, the
location of the acetone attack can guide the operator in setting the
proper die temperature profile. Obviously, this suggests poorly fused
compound; which could lead to some other potential failure, e.g., poor
impact resistance or brittle impact failure.

    If to the previous discussions we add the fact that, as with auto-
mobiles, no two siding lines of the same basic design producing the
same profile from the same formulation run exactly the same, we can
visualize how important is the role of the operator's dexterity in
achieving a good balance between high extrusion while furnishing a
top quality siding product.

    Chapter 30 of the Dekker *Encyclopedia of PVC*, "The Recommended
Voluntary Product Standard PS-55-72 (TS159b) for Rigid PVC Siding,"
published by the U.S. Department of Commerce, and the ASTM D3679
method, are suggested to guide line operators in their troubleshooting
efforts.

4.35 RIGID PVC EXTRUSION OF WINDOW AND DOOR PROFILES

Although hardly a newcomer among the rigid PVC profiles for residential building product applications (window lineals have been manufactured in this country since the early 1960s), the emphasis on remodel-
ing as well as the existence of improved equipment (complete window
lineal extrusion lines are now readily available), resins, and improved
fusion-welding technique for frame and sash, have brought a lot of
renewed interest in this area (see Fig. 4.33). For example, a great
many of the existing producers are expanding their present facilities,
and some pipe manufacturers are becoming new producers by marrying
their rigid vinyl expertise to advanced window fabrication technology,
which is mostly imported from Europe. The overall U.S. window and

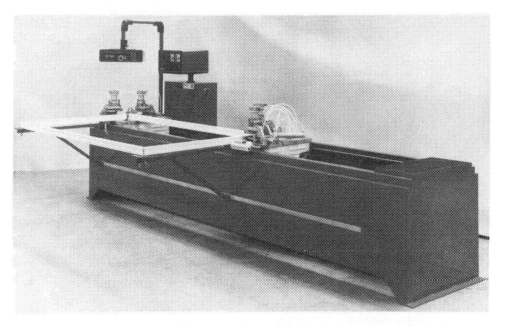

Figure 4.33 Fusion-welding equipment for window frames. (Courtesy
of ACTUAL Anlagen-, Maschinen-, und Werkzeugbau Ges. mbH, Haid,
Austria.)

door industry is currently using 55 to 60 million pounds of rigid PVC
[31].

As it was the case with siding manufacturers, the largest and oldest producers continue using single-screw extruders and the newcomers
tend to acquire complete lines based mostly on vented conical twin-
screw extruders. Assuming a well-balanced window profile turnkey
system, this is an area in which, contrary to manufacturing of siding,
the postextrusion operation does not need a lot of attention.* The
vacuum calibrator commonly used with hollow profiles, in conjunction
with the spray bath, freezes the profile shape and delivers the final
shape to the cutoff saw or saws. Any pipe manufacturer using vacuum
sizing can translate their experience to the extrusion of these hollow
profiles quite smoothly. This is even more true if the design and manufacturing of the hot die, corresponding vacuum sizing, cooling tank,

---

*At the 1981 Interplas, Birmingham, England, the author, for five
consecutive days, observed several extrusion lines (including Amut
and Ide) running very complex window profiles with minimum operator
assistance.

Figure 4.34 Calibrating table unit for window profiles, containing
multiple vacuum sizing dies. (Courtesy of ACTUAL Anlagen-,
Maschinen-, und Werkzeugbau Ges. mbH, Haid, Austria.)

puller pads, and saws are left to the well-known equipment suppliers
(Cincinnati Milacron, Gatto Machinery Development Corp., Durham
International Corp., Krauss-Maffei Corp., and Reinfenhauser-Nabco,
Inc.). Figure 4.34 is a calibrating table unit for a window profile of-
fered by Actual Kunststoffprofile, Austria.

The flow of rigid PVC melts in long-running dies for this application should be streamlined (free from dead spots) and balanced (extrudate exits through all parts of the die at the same speed). Special
attention, however, should be given to the streamlining of the transition adapter, which should be designed to allow for a continuous flow
of the PVC melt without any hang-up. Two types of dies are used
for rigid PVC profile extrusion: plate die for short runs and the
streamlined for long production runs. The plate dies are inexpensive,
since only the plate contains the profile shape and several plates can
be adapted to the same adapter and extrusion head.

Assuming that the die designer has been successful in selecting the
correct streamling angle of the die (usually 45° or less), thus eliminating potential stagnation points, the next step to follow immediately is
the initial die flow balance. In dies of complex profiles, this operation
is greatly simplified if the die land height h is treated as a constant,
and changes are made on the die land length only. This not an unreal-
istic proposition, since for a combination of a given compound-extruder

line-extrusion conditions-output and line speed extruding a given profile, the die land height h is mainly dictated by the wall thickness
specifications of the finished profile, and thus the h value can be
treated as a constant.*

Although it is not a common practice, some die makers tool the die
to the nominal dimensions of the profile, allowing the drawdown to compensate for the postextrusion swell.

As expected, long L values are desirable to better shape the melt
into the shape of the profile. Long land lengths, however, increase
the melt residence time in a high-shear region, contributing to increases in extrudate yellowness and to polymer degradation. They also
increase the back pressure, and in formulations prone to plate-out,
they seem to accelerate this phenomenon.

The initial die flow balance may be done by die designers. For the
final inline flow balance, however, die designers should work together
with process engineers, who, among other things, provide the rheological properties of the rigid PVC compound at typical processing temperatures (mainly, percentage memory, shear stress versus shear
rate curves at processing temperatures, and shear rate at which melt
fracture of the extrudate occurs). The determination of the correct
land length also depends on the shape of the individual profile as well
as on the overall melt quality (e.g., temperature, temperature gradient,
melt pressure, and pressure uniformity), which as shown before is a
function of the screw design, the feedstock composition, the nature of
the feedstock, and so on. These facts stress the importance of the
input of process engineers in die flow balance efforts.

The land length-land height ratios L/h normally vary between 10:1
and 30:1. Initial short land lengths, however, should be avoided,
since this handicaps die designers in their final die flow balance efforts. It is advisable to start with a high L/h ratio and then to shorten
the length by machining down the die rather than adding metal because
the land length is not long enough. Longer land lengths are also more
desirable when localized changes in the land length are necessary to
balance the flow in the die. This is normally achieved by locally varying the land height h at the entrance of the die land. It also contributes to localized changes in the land length.

The simple basic equations used in Sec. 4.33.3 to determine the
choke length of a pipe die:

$$\Delta P = \frac{2L}{h}\, \tau \qquad\qquad \dot\gamma = \frac{6Q}{wh^2}$$

---

*A typical window frame profile extrusion die, however, would only
be 1.5% to 2.5% over size of the final dimensions of the profile, allowing
for little drawdown of the extrudate prior to its entry into the calibration sleeve.

can be used to determine the desired land length of a profile die. Using the equation $\dot\gamma = 6Q/wh^2$, one determines the shear rate for the desired output Q. This value is then taken to Fig. 4.30 or a similar one, which is a viscosity flow curve for a rigid type I PVC at typical processing temperatures. From this plot we determine the shear stress τ at the processing temperature. For a given pressure drop through the die land, normally below 4000 psi, and land height h, we then determine the theoretically correct die land length L.

Another method of controlling flow in the die is achieved by selectively changing adapter and die temperatures or by selectively cooling the profile die. The die temperature balance is quite frequently conducted during start-ups when changing compounds or whenever there are signs of some flow imbalance. This die temperature balance is normally conducted by the line operator.

Selective die zone heating requires several heaters. This can be easily achieved with four-sided dies, where dies normally have one heater for each side with at least two temperature controls. More than one heater, however, is impractical for small cylindrical profile dies, which normally are provided with one band heater only.

Despite the various compensation approaches to balance flow in complex profile dies discussed thus far, a significant amount of tooling costs, production downtime due to processing problems, and so on, can be avoided if some basic know-how in balancing the profile walls is applied when designing the dies. Furthermore, knowledge of the cooling characteristics of the downstream equipment helps in selecting the right profile wall thickness; for example, inner walls, which are more difficult to cool, should be thinner than exterior walls.

The fully integrated manufacturers of turnkey profile extrusion line systems with their own extruder, die, and downstream equipment specialists are the places to go when in search of very complex profile dies with a balanced flow of melt through all parts of the profile.

Cooling and sizing systems today still remain the limiting factors on line speeds. As expected, they are also primarily responsible for keeping the profile dimensions within tight tolerances of about ±0.005 in. Just how fast and how efficiently you can cool and size will determine the line productivity. Since air cooling is a less efficient cooling medium than water, the trend is toward direct and/or indirect water cooling. Another trend is toward multiple vacuum sizing dies (see Fig. 4.34) with spaces or water baths between them. This design approach tends to reduce the drag. Aluminum, because of its lower cost, excellent machinability, and heat transfer properties, is the favored metal for the cold vacuum sizing dies. Stainless steel, however, remains the most desirable material because of its durability and surface smoothness.

For short runs and less complicated window profile shapes, the so-called air tables provided with flexible air fingers are still in use.

Mainly because of the similarity in weathering characteristics, window profile formulations are, for the most part, identical to those used in the extrusion of sidings. For the most complex window profile shapes, which require larger sizes and longer die lands, which in turn lengthen the melt residence time, higher levels of heat stabilizers as well as external lubricants and process aids (to increase melt strength mainly) are suggested. Due to the sharp corners of most window profiles, plate-out accumulation in the corresponding die areas should be watched very closely since it may produce rough corners or a variable degree of gloss. If the plate-out (mostly lubricants or titanium dioxide) is severe enough, it can be seen building up on vacuum sizing and/or takeoff equipment.

Because of the potential plate-out buildup problems and its consequences, selection and balance of external-internal lubricants is of prime importance in window profile formulations. This is more so than in the extrusion of sidings. In a long run, it is more profitable for processors to stay away from the lower cost paraffin waxes and to switch to the more expensive but less plate-out prone waxes, such as ethylene bisstearamide, montanic acid esters, or glyceryl monostearates.

Although the rigid PVC window lineals technology developed in the United States remains fundamentally based on single-screw extrusion, tin stabilizers, all-acrylic impact modifiers, and high concentrations of titanium dioxide pigment (10.0 to 15.0 phr), the European technology is based on twin-screw extrusion, barium-cadmium or lead stabilizers (dibasic lead phosphite mainly), EVA or CPE powders, and lower concentrations of titanium dioxide (5.0 to 10.0 phr). What is common to both window profile technologies (U.S. and European) is the inclusion of ultrafine (average particle size $\leqslant 0.1$ μm) surface-treated calcium carbonate to cut formulation costs. Although modestly, these ultrafine grades of calcium carbonate also contribute to other side benefits, mainly die plate-out reduction, improved powder flow properties, gains in impact strength, and lower postextrusion swelling. Most of these benefits have been discussed throughout this work.

Most recently of interest—both for siding and window lineal production—is the concept of "capstock" coextrusion. A high-quality, low heat buildup, weatherable compound (usually lightly plasticized PVC, but it can also be PVF or acrylic) is coextruded as a thin protective film (0.003 to 0.004 in. thick) onto the substrate rigid PVC profile at the most weather-exposed locations. Since UV degradation is strictly a surface phenomenon, such a thin layer of highly stable material provides added protection where it is most needed. In addition, many feel this technique offers greater opportunity to produce darker colored products having greater UV and heat stability. Special attention, however, should be given to the heat buildup tendency of the capstock compound, since it may produce excessive cumulative expansion of the rigid PVC substrate, which can in turn produce warp of the window frame assembly. It has been reported [32] that composite material containing approximately 10% capstock of Geon vinyl has been granulated and reincorporated into virgin PVC with good test results.

This concept, together with dual durometer extrusion (including semi-rigid PVC portions) provides very broad product design opportunities for both PVC siding and window lineal markets that must displace alternate materials in order to realize maximum growth potential. Furthermore, since one of the primary reasons favoring rigid PVC in window and door construction is because of its excellent insulating characteristics, consumers' demand for more energy-efficient windows and doors should benefit this thermoplastic in the displacement of alternate materials—aluminum and wood.

Small 3/4 to $1\frac{1}{4}$ in. single-screw "piggyback" extruders are used in both single- and twin-screw substrate extrusion lines. Die design is the key to success of the capstock extrusion and it is highly proprietary. In the specific case of semi-rigid capstock formulations, some die design and process conditions have been suggested [33] that, if followed, will help personnel involved to handle the significant differences in rheological and thermal properties between the substrate and capstock formulations:

1. In profiles of complex shapes the capstock should be introduced after the rigid PVC substrate has achieved its final shape. For simple dies, the capstock may be introduced farther back in the die.
2. Assuming that both the main and the capstock distribution dies have been designed following the rheological properties of both compounds, final balancing of the die should be done simultaneously in small steps by removing metal from behind the die land areas.
3. To avoid capstock layer distortion, the profile exiting the die should be annealed at about 60°C for about 15 sec prior to any rapid quenching in cold water.

REFERENCES

1. B F Goodrich Chemical Group, Extruding rigid Geon vinyls, Bulletin G-40, Cleveland, Ohio.
2. P. R. Schwaegerle, *SPE ANTEC 40th,* May 10-13, 1982, p. 516. (The 39th Annual Technical Conference and Exhibition of the Society of Plastics Engineers, Boston, MA.)
3. H. T. Kim and E. A. Collins, *Polymer Eng. and Sci. 11(2):* 1971, pp. 83-91.
4. J. M. McKelvey and N. C. Wheeler, *SPE Trans. April:*138, 1963.

5. R. G. Griskey, C. A. Gellner, and M. W. Din, Modern Plastics *July*:119, 1966.
6. R. G. Griskey and N. Waldman, Modern Plastics *March*:119, 1966.
7. J. M. McKelvey, Lopac extrusion studies at Bloomfield, Monsanto Company, October 1971 to May 1972, unpublished.
8. J. M. McKelvey, Lopac container process reports 2 through 10, Monsanto Company, February 1973 to October 1974, unpublished.
9. P. H. Squires, *SPE ANTEC 39th, May*:643, 1981. (The 40th Annual Technical Conference and Exhibition of the Society of Plastics Engineers, San Francisco, CA.)
10. R. F. Westover, *Processing of Thermoplastics Materials*, Section III, Reinhold, New York, 1959.
11. C. Maillefer, Swiss Patent 363,149.
12. G. Latinen, ACS Adv. Chem. Series *34*:235, 1962.
13. J. Stevenson, Principles of temperature control. Part 1, Plastics Design Processing *October/November*:1981, pp. 22-25.
14. J. Stevenson, Principles of temperature control. Part 2, Plastics Design Processing *December/January*:1982, pp. 23-24.
15. D. M. Burcham, Are your temperature controls in tune? Plastics Technol. *February*:1982, pp. 70-73.
16. Temperature control-1 and 2, Plastics World *December*:1981.
17. M. Hartung, Process controls, Plastics Technol. *August*:1981, pp. 51-54.
18. G. A. Kruder, and R. E. Nunn, *SPE ANTEC 39th*, May 1981, p. 648.
19. H. T. Kim, J. P. Darby, and G. F. Wilson, Polymer Eng. Sci. *13*(5):1973, pp. 372-381.
20. C. I. Chung, Trans. N.Y. Acad. Sci. *October*:1973, pp. 311-323.
21. R. F. Lippoldt, *SPE ANTEC 36th*, April 1978, pp. 737-739.
22. L. I. Nass, *Encyclopedia of PVC*, Vol. 1, Chap. 9, Marcel Dekker, Inc., New York, 1976, pp. 295-384.
23. G. A. Thacker, Jr., *Rigid PVC Extrusion*, Cincinnati Milacron Chemicals, 1976.
24. S. H. Collins, Screw and barrel wear, Plastics Compounding *July/August*:1982.
25. C. J. Rauwendaal, *SPE ANTEC 39th*, May 4-7, 1981, p. 618.
26. G. Schenkel, *Plastics Extrusion Technology and Theory*, American Elsevier, New York, 1966.
27. M. P. Heisler, Trans. ASME *April*:227-236, 1947.
28. G. A. Latinen, Rigid PVC dry blend pipe extrusion, Monsanto Company, September 1966, unpublished.
29. U. Neumann, *MM Maschinenmarkt* 86(151):1980.
30. D. C. Lounsbury, New studies on screw wear and its effects on output, stock temperature, uniformity and processor economics, *SPE ANTEC 40th*, 1982.

31. D. E. Stroud, PVC in window and door construction: A market
     overview, *SPE Retec Vinyl in Building and Construction*, Septem-
     ber 20-22, 1982.
32. B F Goodrich Chemical Group, Capstock compounds of Geon
     vinyl, Bulletin G-65.
33. L. G. Shaw and J. W. Summers, *SPE ANTEC 42*, Apr. 30-
     May 3, 1984.

BIBLIOGRAPHY

For those wishing additional information on the topics in this chapter,
the author suggests the following bibliographic sources:

E. C. Bernhardt, *Processing of Thermoplastic Materials*, Reinhold,
     New York, 1959.
E. G. Fisher, *Extrusion of Plastics*, Interscience, New York, 1964.
L. P. B. M. Janssen, *Twin Screw Extrusion*, Elsevier Scientific, New
     York 1978.
J. McKelvey, *Polymer Processing*, John Wiley & Sons, New York, 1962.
L. I. Nass, *Encyclopedia of PVC*, Marcel Dekker, Inc., New York,
     1976.
Z. Tadmor and I. Klein, *Engineering Principles of Plasticating Extru-
     sion*, Van Nostrand Reinhold, New York, 1970.
G. A. Thacker, Jr., *Rigid PVC Extrusion*, Cincinnati Milacron Chemi-
     cals, 1976.
G. A. Thacker, Jr., *Rigid PVC Extrusion: Problems and Remedies*,
     A Troubleshooting Guide, Cincinnati Milacron Chemicals, Inc., 1978.
J. L. Throne, *Plastics Process Engineering*, Marcel Dekker, Inc., New
     York, 1978.
G. Schenkel, *Plastics Extrusion Technology and Theory*, American
     Elsevier, New York, 1966.

5

Extrusion Blow-Molding of Rigid PVC Containers

W. J. FUDAKOWSKI/Metal Box p.l.c., Wantage, England

5.1   Introduction 246

5.2   Twin-Screw and Single-Screw Extruders: PVC Dry Blend
       Powder Versus Granules 246

5.3   Two-Stage Extrusion System 251

5.4   Single-Screw Extrusion Systems   257

5.5   Extruder Dies: Parison Control   260

5.6   Bottle-Blowing Machines 264

5.7   Platen-Type Machines 264

5.8   Rotary Machines 271
       5.8.1 Vertical Wheels 271
       5.8.2 Horizontal Table: Rotary Machines 277

5.9   Biaxially Oriented PVC Bottle Equipment 278
       5.9.1 General 278
       5.9.2 Solvay-Sidel MSF-BO Equipment 280
       5.9.3 Bekum BMO-4D 285
       5.9.4 Kautex 285
       5.9.5 Johann Fischer 287
       5.9.6 Automa 287
       5.9.7 ADS 288

5.10   PVC Compounds for Bottle Blowing 289
       5.10.1 General 289
       5.10.2 Compounding: Fusion 290
       5.10.3 Food Grades 291
       5.10.4 Impact Modifiers 292
       5.10.5 Stabilizers and Lubricants 293
       5.10.6 Processing Aids 295

5.10.7 PVC Compounds for Oriented Bottles   296
5.10.8 Formulations   296

References 300

5.1 INTRODUCTION

The extrusion blow-molding of rigid PVC bottles began in Europe in
the 1950s. It was an outgrowth of the development of the extrusion
process for rigid PVC pipes, which itself was a product of the German
wartime industry [1]. This source of rigid PVC extrusion, and the
later emergence of the blow-molding equipment industry in Europe,[2]
examplified by LMP-Colombo machines, Blow-o-Matic in Denmark, and
and later Johann Fischer, Kautex, and Bekum in Germany, and the short-
lived Marrick process and equipment in England, provided the required
technology for the rigid PVC bottle-blowing industry. From this beginning, France being the leader with the Sidel, Pont-a-Mousson, and
HMS technologies, developed a myriad processes and machines. Major
incentives for rigid PVC bottle blowing of drinking water, oil, and
wine bottles, were created throughout the 1960s. This development
matured fully with the biaxial orientation option in the 1970s.

In connection with the pipe and cable extrusion of PVC, the twin-
screw classic extruder of Colombo must be mentioned. With this, a
very early patented blow-molding machine, a rotary or mold conveyor
basis, was used, and this extrusion-blown system was eventually marketed by the LMP Company in Turin. In the United Kingdom, R. H.
Windsor, Ltd., produced the Colombo twin-screw extruders under license, and this author remembers well the early 1957 experiments when
we combined a twin-screw R.C. 100 extruder with a Blow-o-Matic single-cavity bottle blower in an effort to blow rigid PVC bottles from the
best stabilized rigid PVC compound one could get. After we managed
to extrude and blow-mold a few experimental bottles, bad discoloration
was experienced. Suddenly, the extrusion rate decreased, and within
minutes, to everyone's amazement, the operator witnessed the extrusion head dropping down in a cloud of fumes after the extruder flange
gave in. With appropriate conclusions for equipment and materials
development, such lessons had to be learned by everyone converting
rigid PVC to bottles.

5.2 TWIN-SCREW AND SINGLE-SCREW EXTRUDERS: PVC DRY
BLEND POWDER VERSUS GRANULES

Historically, twin-screw extruders have been used for PVC dry blend
powders, starting with the Colombo machines. This was mainly because of two factors; the more natural conveying system for powder

with degasing possibility at both the hopper end and at the vacuum vent and the controllable heating balance between shear development heat and electric barrel heater input. This approach was continued on the multimold machines, requiring high outputs, such as the Frohn-Alfing AFRB-6 rotary bottle blower or its Bell platen machine successor, both utilizing modified twin-conical screw extruders of the Anger-Cincinnati type.

The better known Solvay process PVC bottle machines were built originally by the HMS Company of Belgium and, more recently, by SMTP (Société des Machines pour Transformation des Matières Plastiques). Sidel in France utilized twin-screw extruders made by German manufactures like Kestermann, Weber, and Krauss-Maffei.

For practical reasons, i.e., better extrusion uniformity of PVC powders, a change was made more recently in favor of the latest "combination" extruders. These have a twin-screw extruder for PVC dry blend conveying and melting and a single-screw extruder for metering and delivery of the melt to the die in an "L" configuration, developed by Andouart and now supplied by EMS Company of France. The twin-screw extruders, a natural degasing-conveying system, require most of the heat input from the outside for melting, and this enabled the temperature control, separate from the screw speed-dependent shear heat generation, to give an equilibrium balance between output and melt temperature. Precise heat input control is achieved by (1) electric heating power control and (2) jacketing of the extruder barrel for heat transfer fluid coil use with an appropriate heat-exchanger system.

The disadvantage of twin-screw thrust bearing proximity and drive construction has been identified for a long time. These disadvantages have been remedied by the conical screw design, as in the Anger extruders, or by the opposite-end drives, as in the EMS extruder. Although not yet utilized in any bottle-blowing application, the twin-screw extruder by Maplan also successfully resolved the twin-screw extruder drive-end problem. Twin-screw design also needs optimization for a specific PVC dry blend, or vice-versa, and the dry blend formulation needs optimization for a given extruder screw design and for the required output. But generally, twin-screw designs use lower cost formulations than those for simple-screw extruders, mainly due to the lower concentration of heat stabilizers.

One of the claims of the twin-screw extruder-equipped bottle-blowing machines, such as Bell, is that a higher molecular weight PVC resin can be used with resulting improvement in the mechanical properties of the product, leading to potential bottle weight saving, and, hence, to cost reduction.

The application of twin-screw extruders to bottle blowing on the Frohn-Alfing, Bell, and Solvay equipment has always been for a specifically optimized combination of PVC dry blend powder formulation, screw design, and output requirements of long-running, large-quantity bottles, such as those made for liquid detergents, drinking water, oil, wine, and even beer.

Typical output rates achieved with twin-screw Anger-type extruders have been in the range of 150 to 350 Kg/hr. In the Solvay-type machines, equipped with Kestermann extruders producing 10,000 to 11,000 of 1 1/2 liter bottles per hour, the range has been 500 to 600 kg/hr.

Those readers interested in an up-to-date, theoretical approach to twin-screw extruder design with the merits of counterrotation versus corotation (Fig. 5.1), conical versus parallel screw arrangement, specific screw design, and mathematical models, should consult these Plastics and Rubber Institute (London) publications: "Polymer Extrusion 1" June 1979; "Polymer Extrusion 11," May 1982; and "PVC Processing," April 1978. Similarly, the *Annual Technical Papers of the U.S. Society of Plastics Engineers 1973* (and the following years) present a good source of information on twin-screw extruders, which gained some popularity on the American Continent over the traditional single-screw extruder design. Reference should also be made to Chap. 4 of this book.

The twin-screw extruders, although they are the classic workhorses for rigid PVC extrusion, mainly in the pipe and profile field, have a serious handicap, particularly in parallel screw arrangement: the screw tips are far apart, creating a pool of slow-moving melt, causing problems in the converging of flow to the small-diameter die entry

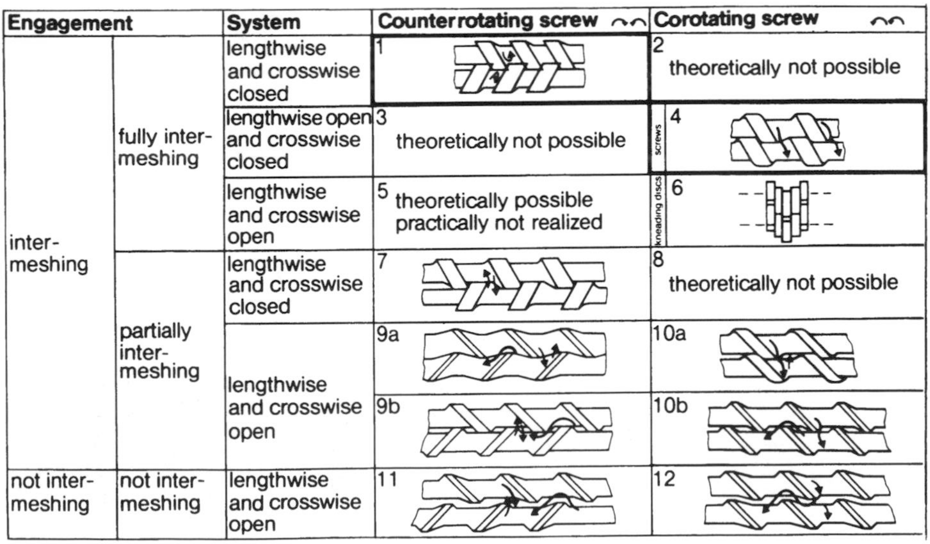

Engagement		System	Counterrotating screw ∩∩	Corotating screw ∩∩
inter-meshing	fully inter-meshing	lengthwise and crosswise closed	1	2 theoretically not possible
		lengthwise open and crosswise closed	3 theoretically not possible	4
		lengthwise and crosswise open	5 theoretically possible practically not realized	6
	partially inter-meshing	lengthwise and crosswise closed	7	8 theoretically not possible
		lengthwise and crosswise open	9a	10a
			9b	10b
not inter-meshing	not inter-meshing	lengthwise and crosswise open	11	12

Figure 5.1 Types of twin-screw mechanisms. (Courtesy of Werner & Pfleiderer Ltd., England.)

normally required for blow-molding applications. The Anger-type
conical twin-screw arrangement (Fig. 5.2a,b) offers some improvement
in this respect.

The typical parameters of the Anger-type conical twin-screw extruders, offered by Cincinnati Milacron, follows.

Parameter	CM65	CM80
Diameter, mm	65/120	80/130
Effective length, mm	130	165
Arrangement	Conical, meshing	Conical, meshing
Rotation	Counterrotating	Counterrotating
Speed	1-35	1-36.4
Total torque, mN	9000	14,100
Screw temperature control, °C	50-200	50-200
Drive power, kW	32	52
Heating, kW	43	52
Cooling	Heat-transfer oil	Heat-transfer oil
Rigid PVC output, kg /hr	250	400

These extruders, together with the CM55, have been adapted in the
range of the Bell bottle-blowing machines, offered until 1981.

Cincinnati's series starts with model CM45, and the whole range output was recently improved by 20 to 30% after incorporating a new
type of plasticizing screw called the superconical. This provided
greater volume and surface area, accounting for the output increase.
A new additional model, CM90 with a rigid PVC output of up to 750
kg/hr in pipe extrusion, was recently introduced.

Krauss-Maffei's new version of the original Anger conical screw
design is the "double conical screw system" (Fig. 5.2b), with a variable
channel depth distinct from that of the conical screw system, combining advantages of the parallel and the conical twin-screw extruders [3],
and claiming benefits similar to the superconical design of Cincinnati.

An advantage claimed for the double conical screw systems is that,
with high back pressures occurring in profile and small-diameter tube
extrusion for blow-molding, venting of the screws is better designed
without the problem of vent blockage.

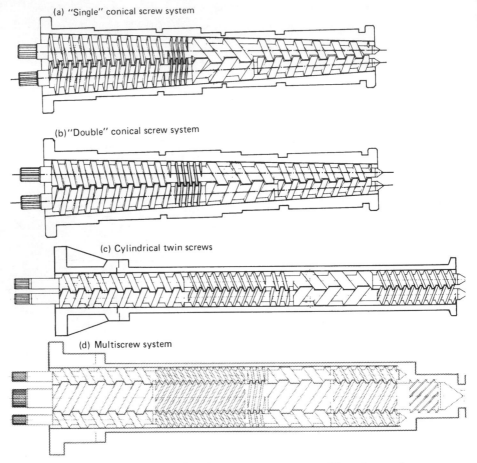

(a) "Single" conical screw system

(b) "Double" conical screw system

(c) Cylindrical twin screws

(d) Multiscrew system

Figure 5.2 Various types of twin-screw systems. (Courtesy of Krauss-Maffei AG, Munich, West Germany.)

Double conical counterrotating extruders exist, namely, the KDM 50KK and KMD 50K. The first gives an output of up to 150 kg/hr and the other up to 250 kg/hr in rigid PVC at about 37 rpm. This extruder design is incorporated in the Frohn-Alfing rotary bottle-blowing machine.

The Krauss-Maffei multiscrew extruders (Fig. 5.2d) and cylindrical (Fig. 5.2c) and conical systems can have their output boosted by a preheating hopper with a rotor, adding another 20 to 30% rigid PVC output at the same screw speed and drive motor load.

The "single" conical type has a metering depth equal to the feed depth; in the "double" type, feed depth is greater than that of the metering end.

Conical twin-screw extruders are also produced by Weber in Germany and are manufactured under license in the United States and in Japan. Two models are offered, CE7 and CE9.

It must be stressed here, however, that the monolithic twin-screw extruder systems have only limited practical application, so far, in PVC bottle-blowing equipment and are restricted to the specialized, high-quantity blow-molding machines mentioned, running on one specific PVC formulation and at one optimized output.

5.3 TWO-STAGE EXTRUSION SYSTEM

For various reasons, twin-screw extruders (as the only source of melt) are losing ground in PVC bottle-blowing applications. Instead, L-shaped combinations of single-screw or of twin-screw plus single-screw extruders are favored.

The L-configuration separately driven "cascade" extruders, started by arranging two separate, single-screw extruders, one feeding the other, was probably initiated by Bandera of Italy but developed into the one machine form by Andouart of France for blow-molding applications.

The original Andouart cascade extruder combination was, and still is, utilized in combination with the 14/10 station rotary blow-molding machines called "Mills Wheels," built in the 1960s in France under license from Continental Can Company (CCC) by St. Gobain Company. It is used, to date, by the major French drinking-water PVC bottle manufacturers, like SAEME and Vittel. It is also used for commercial bottle blowing by other French companies, like Seprosy and Carnaud, as well as by some other companies outside France.

The original Andouart L-configuration extruders had a vertical extruder, driven from its lower delivery end, that delivered a coarse PVC melt through a decompression-vented transfer pot to a horizontal extruder driven separately by its feeding end. Its main function was to take a devolatized, coarse melt from its first-stage extruder, homogenize it, and meter it to the parison extrusion head. This extruder configuration has very significant advantages for PVC dry blend extrusion. It is self-feeding, with the vertical screw protruding into the feed hopper and finishing there in the form of an agitator and feeding auger, which could be changed to a different configuration, best suited for a given PVC dry blend formulation and bulk density. Another important advantage of this extruder system was that the multiple extruder functions of solid conveying, compression melting, homogenizing, devolatilizing and melt metering are separated and optimized in two independently driven screws.

The obvious advantage of two separate drives is that the two screw functions can be balanced to give perfect venting, homogenizing, and delivery with independent dwell times in the two screws, thus providing

an equivalent of the variable single-screw design for PVC. Such a
system is also better designed to handle lightly stabilized (food-
grade) PVC formulations. Some of the disadvantages of the original
design were the weight of such an extruder combination, its complicated and heavy construction, with two separate motors and gearboxes,
and the need for a rather skilled operation to obtain balanced conditions.

In the 1970s this design was improved to offer a horizontal L-shaped
configuration, intended to match better the Solvay-designed wheels,
produced originally by the Belgian/British company HMS, and since
1980 by Sidel in France.

The vertical L-shaped Andouart extruders for bottle blowing had
the following basic characteristics.

	Vertical extruder	Horizontal extruder	Output kg/hr
BLV 60	40 mm dia. 10/1 L/D	60 mm dia. 15/1 L/D	40
BLV 90	90 mm dia. 15/1 L/D	90 mm dia. 15/1 L/D	80
BLV 120	120 mm dia. 10/1 L/D	120 mm dia. 15/1 L/D	230-280

The horizontal, L-shaped Andouart extruders, now produced by
the EMS Company, have the following basic characteristics.

	Feeding extruder	Delivery extruder	KW	Output kg/hr
BLH 120	120 mm dia. 10/1 L/D	120 mm dia. 15/1 L/D	122	300
	120 mm dia. 15/1 L/D	120 mm dia. 15/1 L/D	175	400-500

A typical extruder layout and construction principle for right-hand feeding is shown in Fig. 5.3 and Fig. 5.4 shows the drive-end side of a left-
hand feeding version. The heating elements are formed as U-shaped
rods, fitting tightly in barrel periphery as shown by M. Cooling is by
two finned shells of high heat-exchange rate capacity with water circulation over the feed and compression zones and with air cooling of the other
zones by high-capacity fans (Y).

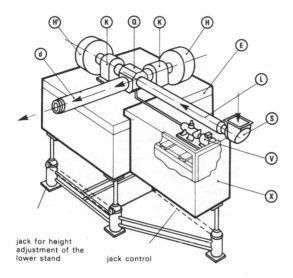

jack for height
adjustment of the
lower stand   jack control

Figure 5.3 Construction principle, right-hand feeding. (Courtesy of
EMS Industrie, France.)

A further improvement in the Andouart extruders for bottle blowing
came about when the first single-screw extruder was replaced by a
twin-screw extruder feeding, as in the other models, a single-screw
extruder connected to a parison die head. This third generation of
Andouart extruders in the BDV series of the EMS Company has an interesting solution to typical twin-screw extruder construction: the
two screws are driven by two opposing gear boxes connected by a
transmission shaft (C), as seen in Fig. 5.5. The transfer of the melt
to the single-screw delivery extruder is via the venting connector
(Fig. 5.5 g) and, as in the previous series, a separate variable-speed
motor driven through a separate gear box (d).
     There are two BDV extruder systems, BDV 90 and BDV 120. BVD
90 is exemplified in Fig. 5.6 . These have the following basic characteristics. The twin-screw extruder is heated by means of ordinary

	Twin-screw feeding extruder	Single-screw delivery extruder	KW	Output, kg/hr
90	90 mm dia. 13/1 L/D	120 mm dia. 10/1 L/D	110	300-400
120	120 mm dia. 13/1 L/D	150 mm dia. 10/1 L/D	182	500-700

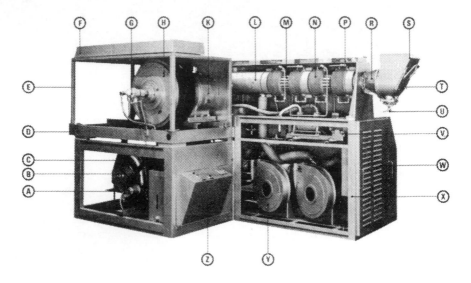

A Tachometric magneto
B Outlet motor pulley
C Outlet screw motor
D Slinging points
E Upper stand
F Protection cover
G Outlet screw cooling
H Outlet barrel pulley
H' Inlet barrel pulley
K Thrust box
L Inlet barrel
M Armored elements

N Finned Shells
P Airing
Q Breech
R Forcing Screw
S Feed Hopper
T Inspection plate
U Emptying trapdoor
V Vacuum pump
W Airing removable panel
X Chassis
Y Fans
Z RTV 22 type set

Figure 5.4 Extruder without cover, BLH 120 RETH 120, 15 + 15D
model. (Courtesy of EMS Industrie, France.)

band heaters (C, Fig. 5.6), but the single-screw extruder is heated
by tubular elements, as before, and is cooled by finned-heat exchange
shells. This extruder arrangement provides a very steady PVC powder blend extrusion with a typical water bottle weight variability,
guaranteed to be within 1 g at outputs not yet achieved on other bottle-blowing extruders. Its suitability for rigid PVC bottle-blowing
formulations can be illustrated by the fact that, although the mono-

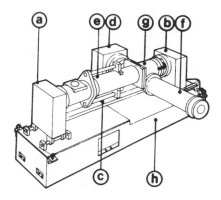

a-b Opposite reducing-gears
c   Transmission shaft   **f** Second stage barrel
d   Reducing gears   **g** Junction breech
e   First stage barrel   **h** Stand

Figure 5.5 Construction principle. (Courtesy of EMS Industrie, France.)

A. Driving motor   **D.** Shells
B. Screw temperature regulation **E.** Electrical elements
C. Electrical elements   **F.** Moto-pump unit

Figure 5.6 A BDV 90 extruder without cover. (Courtesy of EMS Industrie, France.)

lithic twin-screw extruder system had to be stopped and cleaned every 2 days, the BDV extruders ran for 8 days nonstop before preventive cleaning needed to be done at one of the major PVC water-bottle installations. These extruders are now regarded as the workhorses of the large-capacity bottle-blowing machines, operated in France and other countries, for 1.5 liter PVC water bottles and recently for biaxially oriented equivalents. For this reason they have been described here in greater detail.

The advantages of the cascade-type single-screw extruder arrangement for rigid PVC and for other polymer extrusion were also recognized by other manufacturers. Rheotec in France offers a range of such vertical L configurations, including up to three single-screw extruders in the large-output models. Their range of models is as below.

	Vertical extruder	Horizontal extruder	Outputs, kg/hr
VT 50	75 mm dia.	50 mm dia.	100
VT 75	100 mm dia.	75 mm dia.	200
VT 120	Two 75 mm dia.	120 mm dia.	350
VT 150	Two 100 mm dia.	150 mm dia.	700

The special feature of the horizontal extruder is that it has a specially designed mixing screw for a high degree of melt homogeneity. Two vents provide efficient evacuation of volatiles from low bulk density materials with the help of vacuum pumps. Independent extruder controls provide the means for the desired balancing of the two extruders. Vertical extruders have exchangeable feed augers to provide an appropriate compression ratio for bulk density and a cooled grooved feed zone for positive conveying.

In the United States, a different arrangement of the vertical L-configuration extruders is available from Wilmington Plastics Machinery. The vertical extruder is driven from above the feed hopper by a hydraulic motor, and the two extruders are arranged in the same vertical plane with melt transfer from the screw tip to the fourth flight of the horizontal extruder and with the vacuum vent behind it. Here again, the advantages of high output rates with various bulk densities, low melt temperatures, and good homogeneity of melt are claimed.

Another form of the single-screw cascade extruder for blow-molding applications, including rigid PVC, is offered by some Japanese companies, such as Mitsubishi Heavy Industries and Ikegai. Both these companies also manufacture (in cooperation with European principals) twin-screw extruders for rigid PVC extrusion.

5.4 SINGLE-SCREW EXTRUSION SYSTEMS

The vast majority of producers of PVC bottles, however, rely on single-screw extruders, either designed and manufactured by the blow-molding equipment maker or subcontracted from a specialized extruder manufacturer. Thus, the best known French PVC bottle-making machines produced by SMTP-Sidel or ADS are equipped with single-screw extruders with a maximum diameter of 90 mm. This is also true of the German-Bekum, Kautex, Battenfeld-Fischer, and Hesta machines or the United Kingdom-based Hayssen Company blow-molders. Likewise, the Italian PVC bottle-blowing machines have single-screw extruders, designed and built in-house. It is only when higher rigid PVC outputs are required, as for larger bottle size and for wheel combinations, that blow-molding machines need to be designed for and linked to a special PVC extruder, such as the L-shaped, twin-screw, or combination single-screw configurations. Some blow-molding machines are designed for universal material use, and the same extruder can be used for polyethylene (PE) and for PVC duties. In such a case, different design screws would be used.

Generally, PVC single-screw designs are within the limits of the 16 to 24 L/D ratios, with a compression ratio of about 2.5, but individual makers have their own preferences, often using mixing "torpedoes" cut to a special pattern. In clear PVC formulations such screw extensions are intended to provide an additional homogenizing function to the screw and to enable the polymer melt to recover from its screw flow plastoelastic memories before being shaped in the extrusion die.

Single-screw extruders of relatively small diameters can be used satisfactorily for PVC dry blend powders, but above a critical screw speed, air entrapment often occurs. Also, with the use of regranulated PVC bottle regrind of different bulk density, output stability is difficult to maintain. The advantages of small-diameter screw extruders for PVC powder processing can be seen from a number of single-parison bottle blowers, such as those offered by Hesta, Bekum, Battenfeld-Fischer, Kautex, Sidel, or ADS. The majority of such single-screw extruders are of horizontal configuration. Hesta, as well as Hayssen, uses vertical extruders with a hydraulic motor drive on top. uses vertical extruders with a hydraulic motor drive on top.

Each manufacturer likes to retain this PVC screw design as proprietary information, discussing specific applications with prospective customers only. However, a general guide on screw dimensions and their outputs can be seen in the digest classification in (Table 5.1).

Table 5.1 Extruders Available with Blow-Molding Machines: Screw
Characteristics and Outputs with PVC

Manufacturer		Screw diameter, mm	L/D	Claimed output, kg/hr
Kautex	KEB 1-10	40	20	16
		50	20	32
		60	20	50
		80	20	90
Bekum	BAE 1-3	50	20	40
		60	20	50
	BAE 11	80	20	80
		90	20	90
Bekum	HBD	60	24	65
	BMO	80	24	100
		90	24	110
Sidel	DSL 1	65		70
	DSL 3	90		150
ADS	3 Station	65	33 (2 stage)	110
	11 G1	40	20	30
Hayssen	2150	50	22	45
	2125	45	21	35
Hesta	B 33	33	17	16
	B 40	45	18	35

There are, however, other independent blow-molding machines,
such as the U.S. Graham Wheel, Mills Wheel, the Canadian Comatic,
and the U.K.-U.S. Rotablow, which require separate extruders, usually supplied by specializing manufacturers. These are available from
Davis Standard, NRM, HPM, or other manufacturers in the United
States and elsewhere. The most up-to-date PVC extruder design improvements can be incorporated by these extruder manufacturers.

In the United States there has been rapid development of single-
screw extruders, competing in the PVC field with the European twin-
screw, and those interested in greater detail would be advised to
refer to an article entitled "Battling for PVC Extrusion: Twin vs.
Single Screw," which appeared in *Plastics World* (October 1981).

The barrier screw design for PVC started with the Maillefer extruder. As already mentioned in Chap. 4, it is now apparent that the

barrier screw design is favored for PVC in dry blend extrusion in single-screw systems. Davis Standard barrier screw design, the Sterling barrier, and the HPM double-wave screw design are very popular barrier designs.

The typical output of a 90 mm double-barrier screw, working with PVC dry blend powder, is in the range of 160 to 200 kg/hr at 40 to 50 rpm with improved homogeneity and melt temperature.

Among others, Fairex of France offers special design extruders and screws for PVC extrusion, claiming success in blow-molding. More recently, the Uniloy Division of Hoover Ball Bearing Company has added to their range a suitable extruder modification on their bottle blowers for PVC conversion.

The majority of European PVC bottle-blowing extruders are designed to work with PVC dry blends for reasons of economy. Also, this allows the convenience of tailoring the PVC dry blend to suit a particular extruder design and bottle requirement. Dry blend, PVC powders can be bought from a number of PVC suppliers at prices below the granulated compound level. Another reason the dry blend form is preferred is that the powder form has no previous heat history apart from the compounding stage and is thus inherently more thermally stable than the pelletized material.

In North America, PVC suppliers can offer pelletized PVC at the same price level as powder. Quite clearly, the powder form requires the special handling equipment, causes housekeeping problems, and may result in bubbles appearing in the extrudate if either moisture is absorbed on this high surface-volume ratio form or if air is entrapped in the melting process. These disadvantages caused a preference in North America for the pelletized form. In Europe, however, another reason for powder use is the scale of blow-molding operations. Big savings can be made on huge quantities of PVC being transformed to water, oil, or wine bottles. On the other hand, for the smaller operator, the "do-it-yourself" approach gives the desired flexibility. A medium-sized convertor will have a preference for pelletized granular form if the cost differences are not very significant.

A number of European PVC suppliers, like those in North America, can supply PVC bottle-blowing grades with a choice of formulations to suit both the final application and the extruder requirements. Granular compounds are used with single-screw extruders exclusively, most of which can be adapted to cope with the powder form. Twin-screw extruders are ideally suited to low bulk density powder and are used in blow-molding for that reason. They can also convert the higher molecular weight materials more effectively, due to controlled shear heating, and may enable stabilizer reduction with further cost savings.

The vast majority of the clear PVC bottle-blowing compounds are in the K value range 55 to 60, and with the growing rivalry of the brilliantly sparkling polyethylene terephthalate (PET) bottles, is likely to remain so.

Whatever the original form of PVC, the reworked material, amounting
to some 20 to 30%, is always added and this is another consideration.
Ideally, the reworked material should be in the same form and with
the same bulk density as the virgin material, but this is seldom achievable. In any case, it should be as clean as possible, avoiding any degraded particles, and should be mixed efficiently with the virgin component in a consistent way. As indicated before, some extruders with
the vertical configuration have the advantage of a changeable feed
helix in the hopper to suit a particular bulk density and mix.

5.5 EXTRUDER DIES: PARISON CONTROL

The process of tube extrusion led naturally to blow-molding and remained for a long time an experimental art, especially so in the design
of extrusion dies. This position has changed, particularly in the 1970s,
with numerous theoretical approaches. This is discussed in January
1971 *Plastics International:* "The Scientific Design of Fabrication
Process—Blow Molding" and in *Designing Machines and Dies for Polymer Processing with Computer Programmes—Fortran and Basic,* by
Natti and Rao, published in 1981.

In terms of extrusion die design for PVC processing, swan-neck
"spider-torpedo head" has been the standard approach for most blow-
molding machines, with parison wall thickness control being made
available only on some more sophisticated equipment. Again, in the
1970s progress was made here as well in utilizing microprocessor techniques in parison wall thickness control. This was summarized in
an article that appeared in January 1980 in *Modern Plastics International.*

For rigid PVC extrusion dies, the most successful and widely accepted approach has been a fixed spider and mandrel with moving
extrusion nozzle by means of vertically moving yoke, operated normally by a hydraulic servovalve and controlled by a 10 to 50 valving
step electronic system. Such systems were evolved by major PVC
blow-molding equipment manufacturers such as Bekum, Voith-Fischer,
Bell, Sidel, and others.

There have been, however, specializing companies in the design
and manufacture of parison wall thickness control equipment, such as
Moog and Hunkar in the United States.

An early spider head design for PVC with parison wall thickness
control by a moving nozzle principle controlled by 32 points, Hunkar
controller, was available in Germany from Fuchslocher.

A more advanced rigid PVC extrusion head with parison wall thickness control was developed by Solvay, involving a closed-loop computer
control of both vertical and annular wall thickness, utilizing a wall
thickness scanner, and displacing the extrusion nozzles with three
hydraulic cylinder actuators to provide annular wall thickness control.
Naturally, such an approach could be warranted only by a high-volume

production approach, such as over 10,000 water bottles per hour, so
typical of French water-bottling installations.

The most common approach, however, for rigid PVC bottle blowing
has been a simple spider-head design die without parison wall thickness
control, typified by the Sidel DSL range of bottle-blowing machines
or, as used with Andouart extruders, by the St. Gobain Pont-à-Mous-
son version of the Mills Wheel machines. A valved head is exemplified
by a typical design in Fig. 5.7.

There have also been cases of adapted cross-head design for rigid
PVC bottle blowing with and without parison wall thickness control,
but generally, such an approach had limitations to specific duty requirements [4].

In both types of die design for rigid PVC, the so-called weld lines,
resulting from divided flow by the spider legs or by the flow around
the cross-head mandrel, was a major problem to overcome. This normally necessitated careful PVC material selection and appropriate die
profiling, compatible with the no dead spot requirement, which could
lead to stagnant, or slow-moving PVC degradation centers. Major converting companies in PVC blow-molding have often developed their own
dies with wall thickness control that matched their requirements and
were not available on the open market. It has to be said that such
sophisticated developments required technical skills for operation and
maintenance that were not normally available to a smaller PVC bottle-
blowing converter and could lead to undesirable complications. For
satisfactory utilization of electronically controlled parison dies, particularly for PVC, a good backup of electronic and hydraulic expertise
is considered essential.

Because of the unstable nature of PVC, materials of construction
for moving parts of the die in particular require careful selection, as
do the dimensional tolerancing and/or selection of sealing means. Normally, working surfaces in contact with PVC melt should be stainless
steel or hard chrome, but sliding contact metals should be of different
and suitable characteristics, an example being Sulphonisation, or Tef-
lon impregnation of hard chrome or nickel coating.

A careful start-up and shutdown procedure is mandatory to avoid
decomposition lines, which can often develop into autocatalytic degradation with long corrective efforts required. The use of polyethylene or
other polyolefin and even special start-up and shutdown compounds is
often desirable, particularly if food contact formulations are being used.

Performance of the parison wall thickness control system depends
on the swell ratio characteristic of PVC melts as much as on the temperature dependency of melt viscosity, so a careful selection of rigid
PVC formulations for the duty envisaged is recommended. This aspect
has been dealt with in previous chapters as well as in other publications
on PVC technology [5].

In more recent years higher molecular weight PVC resins with K values in excess of 60 have been introduced on some equipment with twin-

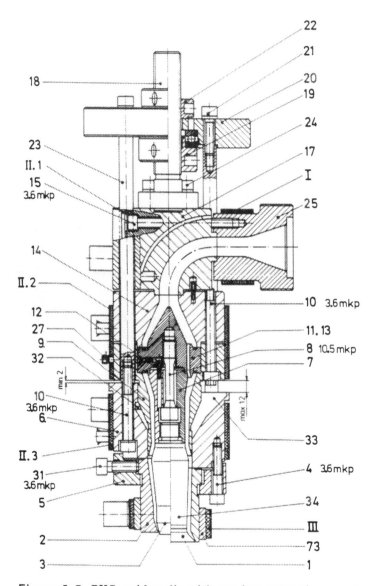

Figure 5.7 PVC spider die with moving extrusion nozzle. (Courtesy of Bekum AG, West Germany.)

screw extruders, and these resins require particular care in parison control. The Bell microprocessor-controlled die heads have been particularly studied in this respect and are now part and parcel of the Uniloy technology. Frohn-Alfing equipment, from which the Swiss Bell originated, has developed into a separate entity with similar improvements, which are currently being used by some German in-plant installations. Both these machines are typical of the multidie type, where a number of dies, usually two to four, are fed from one or more extruders.

Where a single extruder, as is the case with Bell and Frohn-Alfing systems, is used, a manifold distributor of melt flow has to be used to divide the flow in a uniform manner. The design of such manifolds with satisfactory performance is an art in itself, whether the dies are placed on the same axis as the extruder or at right angles to it. As expected, it requires careful heat input to balance the flow in the flow-corrective channels. These design considerations led some manufacturers, including Bekum and ADS, to adopt an approach where each die is fed by a separate extruder of suitable output characteristics. The control of parison wall thickness in multidie systems is also more complicated, and ideally, each die should have its own independent valving and control unit (Fig. 5.8).

The majority of commercially operating PVC bottle-blowing machines available use two die heads per extruder, mostly without parison control. In the large-output bottle-blowing machines, ranging from 2000 to 12,000 bottles per hour, single die extrusion is the standard approach. Here the main problem in die design is one of rheological nature, to obtain a high flow rate without the disturbances that would adversely affect bottle appearance and quality. It can be generalized that, for low cost, high bottle contents of, say, 1 to 1.5 liters, this approach now prevails. For smaller bottles, below 1 liter in size and intended for better appearance high-gloss applications, particularly in the cosmetics packaging field, slow extrusion single- or multidie machines are used.

Figure 5.8 Double PVC spider head with parison control viewed from the front and from inlet side. (Courtesy of Bekum AG, West Germany)

There are specializing die manufacturers outside the main bottle-blowing machine companies in Germany, Italy, and Japan able to offer their own design interpretations.

A trial evaluation with the intended extruder and material would, however, be appropriate before final commitment and selection.

5.6 BOTTLE-BLOWING MACHINES

Bottle-blowing machines available commercially can be classified in a number of ways, but the prime consideration normally is the productive capacity in terms of bottle use and hourly output. In technical terms, the type of neck finishing, bottle deflashing, and handling to downstream operations must be considered. Extrusion from powder or granular PVC compound leads to appropriate extruder choice, as outlined at the beginning of this chapter.

The most common type of commercially available machine is a single-screw extruder, single- or twin-die platen-type machine with neck-forming capability by a calibrating blow mandrel unit to which the mold-mounting platen moves after collecting the parison tubes from extrusion dies. A high-speed moving blade is normally used for cutting off parisons at the die nozzle face.

The second most common type of PVC bottle blower is a vertical wheel carrying a plurality of molds and collecting the parison tube from a single die. Parison cutoff can be made by a number of different means, such as scissors, as in the case of the Sidel DSL-type wheels, to in-mold cutoff by mold edges or separately operated blades.

Most of the wheel-type machines are designed for a blow-molded neck, requiring posttrimming, usually by a slicing device. Blow-molding air is normally introduced into the preform parison by a side-moving perforating needle. Classic examples of such machines are the Mills Wheel and Solvay Wheels.

Another form of a wheel-type machine is a horizontal rotary table, either indexing under extrusion die or continuously rotating, where the parison tube, extending in a vertical plane, is cut off at the die nozzle and collected by molds with their individual blow-molding and/or neck-calibrating heads. Examples of such blow-molding machines are the ADS wheel in France, Frohn-Alfing in Germany, the Rotoblow wheels in the United States, and some proprietary equipment described in the patents of the Plax Corporation, operated currently by the licensees of the U.S. Monsanto Corporation and known as BDS equipment. Biaxial orientation variants of PVC bottle-blowing machines are to be treated separately in this chapter.

5.7 PLATEN-TYPE MACHINES

Most platen-type machines in this category are built in Europe, mainly in Germany, Italy, France, and in the United Kingdom, but more recently, PVC equipment of this type has become available from Japanese and other Far East suppliers. Although it is beyond the scope of this chapter to describe each and every machine, some of the better known or more interesting variants will be described or referred to.

Since Bekum has secured a significant share in PVC bottle-blowing machines throughout the world, it would be appropriate to consider some of their equipment as typical in this class.

For bottles below nominal 1 liter capacity, the well-established BAE-1 machines and the new designs BM-08 and BM-08D represent the proven and the more up-to-date design concepts in mold moving and clamping (Fig. 5.9). Typically, a separate hydraulic pack provides the driving force for platen movements, mold closing, and bottle-finishing operations. A choice of three extruders with either 38, 50, or 60 mm diameter and 20D length provides BAE-1 machine with maximum rigid PVC corresponding output levels of 20, 40, and 50 kg/hr. Such a choice provides a possibility of fine-tuning against the envisaged production application. It is worth pointing out again that, on account of higher bulk density, granular PVC compounds give about 15% higher output for the same screw than powder. Actual output depends on the PVC material, regranulated material contents and form, extrusion die back pressure, melt temperature requirement, and/or possible screw speed.

The newer design BO-08 with reduced dry cycle and D twin station has an even better choice of extruder with either 20 or 24D lengths, namely, 38, 50, or 60 mm diameter. Its higher capacity electric drive motors provide up to 45 and 70 kg/hr for 50 and 60 mm diameter 24D extruders (Fig. 5.10).

Standard Bekum extrusion dies with or without parison wall thickness control are available for either single- or double-parison extrusion. For the single-extrusion die, the maximum die diameter would be 70 mm, and for a double-parison die, the maximum die diameter would be 35 mm (in the case of BM-08 PVC specificaton). This would enable a single 1 liter production on BM equipment and maximum single 1.5 liter production on BAE-1 machine or a double 0.5 liter on BM equipment and a double 0.7 liter on BAE-1. The improvement in dry cycling from BAE-1 to 1.4 sec in the case of BM is due to the improved platen movements, the use of special light aluminum alloy for mold platens, the clamping system as well as the proportional hydraulics. This means that while the BAE-1 machine can work up to 15 cycles per minute per platen system, the BM would be able to produce 19 cycles per minute. The actual production rate naturally depends on bottle design and wall thickness, but for the same parameters, the new generation platen machine produces more per hour than its predecessor.

This sort of improvement is now typical for the other competing equipment and impinges on the economics of production. Another aspect worth noting is that, in the age of energy conservation, the use of electrical power is being carefully studied. In the case of the 50 mm

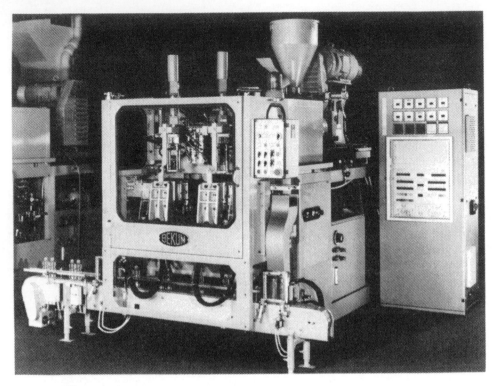

Figure 5.9 BM-08D. (Courtesy of Bekum AG, West Germany.)

extruder BAE-1, the average energy consumption was quoted at 20.3
kWh, but the BM-08D equivalent is being quoted at minimum of 14.2
and a maximum of 19 kWh with a possibility of increased hourly production by a significant factor.

If one takes as an example the 0.7 liter PVC bottle, weighing 33
to 35 g, this could be produced on the BAE-1 single-die machine at a
rate of 780 per hour, but the equivalent rate on the BM-08 would not
be much higher owing to the cooling rate limitation. However, for a
0.5 liter bottle, without the assumed heat-transfer limit, there could
be a factor difference of 1.8 between the twin-die single-platen BAE-1
rate and the BM-08D twin-die double-platen rate with these two machines equipped with appropriate rate extruders. If it were not for the
machine design and performance improvement, the difference between
single-die single-platen and single-die double-platen would be of the
order of 1300/780 = 1.7, and a similar ratio would be applicable to
twin dies, in terms of respective bottles per hour outputs.

The ratio of productivity between a single and a twin die has normally been taken as being below 2, owing to the loss of efficiency with

Figure 5.10 BM-08D. Details of dies and mold platens. (Courtesy of Bekum AG, West Germany.)

twin-die operations. Machine suppliers normally tend to quote optimized nominal production rates, which should not be used in any down-to-earth production or costing estimates, as no allowance is made for start-ups and shutdowns, PVC cleaning time, breakdowns of all kinds, and many easily unforeseen obstacles. As a yardstick for a good production run of PVC bottles, an efficiency correction factor of not more than 0.8 should be used.

The introduction of 24 L/D screws for PVC in place of the 20 L/D equivalents between these above-cited models, was based on the experience that the 24 L/D screw with improved mixing characteristics and better design enabled a lower melt temperature by about 15°C at maximum output and a far better homogeneity of melt. This enabled a much better performance for twin dies with improved radial temperature distribution and resulting wall thickness consistency, which could mean lower bottle weights. The use of the 24 L/D mixing screw is now considered mandatory by Bekum for any twin-die PVC machines, particularly of the larger kind, such as in the HBD and the BMO range.

There are five types of double-platen Bekum machines of the H series. The old established PVC machine HBD 51, followed by HBD–111, 121, 151, and 201 types with increasing sizes of platens, can be used for single- or double-extrusion die duty on increasing capacity containers in that series, including the handleware type. A selection of extruders of 50, 60, 70, 80, 90, 100, and 120 mm diameter allows at least a choice of two for each platen model. A typical line diagram and specification for HBD–111 shows the possibilities (Fig. 5.11). Kautex, Voith-Fischer, and other machine builders can offer comparable choice, although normally PVC is not converted on extruders larger than 100 mm in diameter.

The old established Italian PVC bottle-blowing machine builder Co-Mec Spa can offer a comparable choice with extruders up to 90 mm diameter in its CS series of blow-molding machines for container capacities up to 5 liters. Here again, consideration of PVC bottle handling and finishing as well as the monoblock principle for smaller PVC bottle production offered by majority of machine builders led that company to the development of the MS-1000 PVC with a choice of single or twin-die operation. In the Speed 3000 DM range, another Italian machine producer, Automa, and its North American associate in Ontario, offer a range of comparable single- and twin-die machines up to 5 liter capacity equipped with hydraulic motor-driven extruders up to 90 mm diameter. Other Italian PVC bottle-blowing machine builders, including Moretti and Automazione Minghetti, compete in similar range.

Not many French machines are of the platen type. Among them are the ADS monoblock machines and Serta ESH 10 variant for PVC with a twin die in the extruder axis. The extruder in this case is pivoted with about 2 in. vertical head movement to facilitate mold indexing in the extruder axis. Both extruder and platen motion are hydraulically powered.

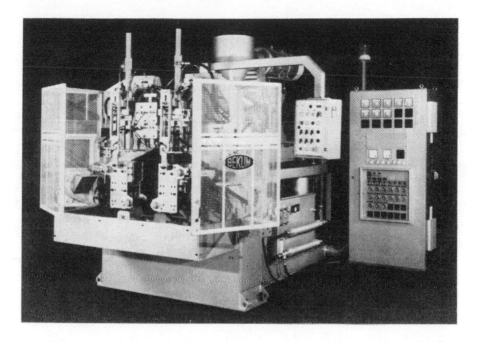

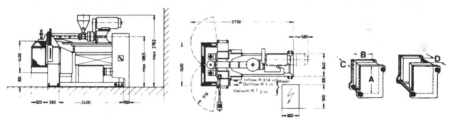

Figure 5.11 HBD-111. (Courtesy of Bekum AG, West Germany.)

The Swiss Bell machine of single- or double-platen type, mentioned before, fulfills all the performance parameters of other machines with high precision and with microprocessor control. Its TB-1 and TB-2 models were characterized by the adaptation of the Cincinnati Milacron twin-screw PVC extruders, as discussed in this chapter, and were aimed at the use of high K value PVC powder blends.

The U.S. Hayssen Manufacturing Company offers its Duablow 1, controlled by a Maco VI microprocessor, for larger handleware bottles, including PVC.

The machines described thus far in the platen category have horizontally mounted extruders, requiring cross-head die mounting.

The other group of platen machines have exploited vertical extruders, giving PVC processing dies the advantage of the same axis as the screw. Another advantage of single-screw extruder vertical configuration is that it is autofeeding in case of powder blends, eliminating the usual "bridging" problem. The leading exponents of this approach are the United Kingdom based Hayssen Europa Company, offering a broad range of PVC bottle blowers, and the German Hesta machines, built by Staehle Maschinenbau GmbH, offering machines for containers up to 1500 cm^3 capacity.

A new arrival in the same group is the Italian Plastimac Spa with its Mini-Gamma and Maxi-Gamma machine concepts. The special feature of this equipment is that it has three vertical extruders mounted on the same frame in a star configuration with three corresponding platen systems operated separately but from the same material hopper. Any of the extruders can be shut while the others can keep working. Mini-Gamma, equipped with three 50 mm extruders, can produce up to 1800 single liter PVC bottles per hour. ADS in France also had a small vertical extruder-platen machine.

The Hayssen Europa range of Monablow machines, which are also manufactured by its US parent company, are based on a fundamental concept, established in the 1960s, of a vertically mounted extruder with a hydraulically actuated vertical motion that enables continuous extrusion with a horizontally swinging mold platen. A single or double die with or without parison control can be used for molds that can be equipped with top blow and neck calibration, bottom blow spigots, or simultaneous top and bottom blow and side blow with needle or hydraulically operated blow pin assembly. All the normal techniques, such as "preblow," deflashing in the mold, off-center blow, oriented bottle takeoff, and scrap collection, are available. Parison cutoff is by a high-speed reciprocating knife or a hot wire mechanism. Extruders are hydraulically driven by low-speed, high-torque motors, particularly suitable for PVC processing. Monablow 2125 and 2200 machines have the same 45 mm diameter, 21/1 L/D extruders and platen sizes. The 2200 machines has, however, bigger maximum daylight, enabling containers up to 5 liter size to be made with a single mold, or two 0.75 liter containers on twin-mold operation. The 2125 machine can produce single mold containers up to 2.5 liter and twin 1 liter.

These machines, of fairly simple construction and operation, found application in many countries, including the developing nations. Another variant of this design, the Monablow MB 2150, is equipped with a 50 mm diameter extruder and with larger platens suitable for a 5 liter container, and up to twin 1.5 liter bottle production.

Developed at about the same time is the Hesta range of machines, which initially served the mainly small-capacity bottle market but now extends up to the 1500 ml size. A hydraulically driven, vertical extruder is stationary, which can be tolerated for small-sized, fast-moving molds. Three basic models are available, the B 33, B 40, and B 50 with

numbers corresponding to extruder screw diameter, and suitable for
up to 500 ml, 1 liter, and 1.5 liters, respectively. A fourth one, which
is a variant of B50, called B51, has become available with larger clamping force and daylight. All these models can be equipped with single-
or double-parison dies, with or without parison control by a 20-point
electronic or three-step timer-controlled movement.

Container deflashing, separation from "moils," and bottle alignment
on discharge from the B51 machine are available. Wide-necked containers can be produced automatically in specially equipped, flying
knife molds.

Like the majority of other bottle machine suppliers, a leak tester is
also offered by Hesta.

A particular field of application of Hesta machines has been in the
PVC shampoo and other cosmetic bottles, requiring good appearance.

The above description of some typical platen machines is by no
means exhaustive and is dictated by the available space of this chapter.
Although the author has been conscious that some machines, even
major ones, have not been described in the same detail as others, it is
hoped that our fair cross section of equipment is of help as an introduction to the more detailed survey that may be required by some
readers.

5.8 ROTARY MACHINES

Although the earliest rotary machines originated just before or during
the World War II in the form of the Plax and the LMP-Colombo machines,
the real growth occurred in the late 1950s and 1960s, with the advent
of high-density polyethylene (HDPE) use in blow molding. This, in
turn, stimulated the application of such machines for PVC.

5.8.1 Vertical Wheels

Mills Wheels

The best example of vertical wheels was the adaptation of the U.S.-de-
signed and developed *Mills Wheels* for PVC blow-molding by the French
St. Gobain-SEVA in Europe by way of manufacturing license obtained
from the Continental Can Company group.

Although many variants of the Mills Wheel were developed over the
years for specific applications with various numbers of mold stations,
such as 5, 6, 10, 14, 18, or more, they were all based on a relatively
simple continuous rotation mechanism with stationary cams in a rigid
frame, controlling the mold closing, opening, and blowing as well as
bottle ejection.

The molds, radially mounted and moving around a horizontal axis,
can be fixed at various distances from the hub, depending on the re-

quired mold length and thickness. The lower mold half is fixed on the hub face or its extension, a mold block; the upper mold half is fixed in a guide turret and is capable of moving up and down on the guide tie-bars or rails by the rotation of the wheel against the stationary cams of the main frame. The upper mold half, closed position, is adjustable in relation to the position of the lower mold half by means of spacer blocks and the springbox, which contained some form of spring, such as a bellow spring, to provide the mold locking force with some safety. The position of the springbox can, of course, be changed, as it sometimes done, to the lower mold half in order to provide for the maximum mold length in the periphery of the outer most mold closing.

Each mold has its own needle blow unit, which penetrates the continuously extruded and clamped parison after mold closing. The upper mold halves have ejector pins, also operated by a fixed cam. The extruder die is located between the open mold halves at the top of the wheel and extrudes parison tube, tangentially to the lower mold halves.

The blow chamber, where the needle penetrates the parison, is at the top of the bottle moil, which is joined to next bottle base tail, so that each bottle is ejected with its own top moil and with the next bottle's tail. Depending on the bottle design and, particularly, on its neck specification, a separate deflashing and trimming unit is employed as a downstream operation.

The usual method of neck finishing is by means of a slicing blade, either as a straight-line, angled blade in line with the conveyor or as a semicircular blade on the reversing leg of the bottle conveyer

The most common extruders used on PVC-operating Mills Wheels were the L-configuration Andouart with either vertical or horizontal configuration.

A 14-station Mills Wheel was commonly used for PVC bottles up to 1.5 liter capacity. In some cases, where it was required to make 2 liter bottles, the wheel was modified to 12 stations in order to provide the necessary space. Typically, a 14-station Mills Wheel produces up to 6000 single-liter bottles/per hour, 5800 half-liter bottles/per hour; a 12-station modified one, 4800 two-liter bottles/per hour with a high overall efficiency of about 90%. Mills Wheels are used for PVC bottle blowing mainly in Europe; in France, Spain, and Italy, where they continue to produce large quantities of PVC water bottles, as well as bottles for edible oil, vinegar, wine, powdered foodstuffs, liquid soaps, and so on. They were also used as proprietary equipment in the Americas and in Japan for selected PVC bottle blowing. They are predominantly suitable for long-running orders with only periodic shutdowns for purging, cleaning, and maintenance.

Extrusion heads used with these wheels are mainly single-nozzle ones without parison control, although this could be obtained. Twin-extrusion head system could also be employed if the run length and bottle design would warrant it. By 1982, Mills Wheels scored some 15 years of satisfactory operation with PVC.

Solvay Wheels

Another variant of horizontal axis, vertical wheel was developed by Solvay. This equipment was initially produced in Belgium by the Heuze, Malevez & Simon Company, and, subsequently, the manufacturing license was taken by SMTP-Sidel Company of France.

Continuous rotation wheels have mechanical operation, but molds open and close on a book hingelike basis with a positive, side-locking mechanism, all movements again controlled by cams. All models have the designation MSF. The MSF machine thus has a fixed mold split-line diameter, and ideally, it should be correctly designed for a given bottle size.

In fact, three different wheels were constructed: the MSF 300 model was suitable for bottle heights of 75 to 120 mm with a maximum diameter of 60 mm, a maximum nominal rotation speed of up to 10,000 bottles per hour (bph), and a PVC output of 100 kg/hr. MSF 500-10 was intended for bottles up to 320 mm high and mainly for 1.5 liter size with a nominal running rate of 5000 bph and extruder output of up to 280 kg/hr. MSF 500-20, which was designed for similar bottles but had 20 mold stations instead of 10, was intended for a PVC output of the order of 500 kg/hr.

PVC extruders of adequate output could have been used from among those described in the earliest part of this chapter. Twin-screw extruders and L-shaped extruders of Andouart design were normally used in European installations. Several machines of this type were installed in France, Belgium, and in other countries, including Japan, but mainly for in plant operation in manufacturing PVC bottles for drinking water and for oil. One MSF 300 machine was reported to be used with a twin-parison, twin-cavity, 10 mold configuration to produce about 10,000 bph 30 ml PVC bottles for hair dye in France.

Single or double dies with or without parison wall control were available, but for the majority of installations working on mineral-water applications, simple spider-type dies without parison control were used. In one such installation MSF 500-20 was running regularly with 1.5 liter bottle molds at a rate of about 11,000 bph. All the MSF machines have positive, mechanical bottle takeouts with neck grippers, enabling regimented bottle handling through neck finishing by fixed blade trimmers to downstream equipment.

Other Vertical Wheels

Machines of similar vertical wheel construction with fixed-mold opening diameters but with straight mold-opening mechanisms and up-the-wheel parison extrusion are being offered in the United States by the Graham Engineering Corporation, and some of these machines are reportedly being used for PVC bottle production both in the United States and in Europe. Here again, a given machine design is best suited for a given bottle height and type.

Extrusion dies with or without parison control were available, and
suitable output extruders, as mentioned before, are used provided that
they can be of the low enough type for upward parison extrusion.
Other vertical "wheels" in existence and in PVC use at one stage or another were constructed as proprietary equipment and are not generally
available on the open market. Such equipment was constructed in
Japan and in the United States.

Other vertical wheels, originally designed for polyethylene use,
were also adapted for PVC. Perhaps best, however, was the Canadian
Comatic design, with one to six-mold holding stations, eventually built
under license by the US Belloit Corporation and more recently by NRM
Corporation.

ADS Company of France was also offering similar machines with
three molds and with a claimed output of up to 2400 single-liter bottles
per hour, which would represent excellent mold utilization, if achievable. A similar concept for a three-mold vertical rotary machine, described as BB-3, was prototyped in 1977 by the Austrian branch of
Cincinnati Milacron. As described before, it was intended that these
machines operate with their twin conical screw extruders. The prototype machine was connected to a CM 55 extruder with a capacity of up
to 120 kg PVC per hour and had a claimed output of 1800-2000 of 1.5
liter bottles per hour. All these Comatic concept-connected machines
had molds mounted on radially spaced platens with an axial mold closing and opening by means of either pneumatic or hydraulic cylinders.
The advantage of such a configuration was that blow-molding mandrels
with neck forming could be utilized to produce finished neck bottles.
In-mold bottle deflashing was possible and employed in some machines,
such as in the B.B.-3.

Spider heads with or without parison wall thickness control were
available from machine suppliers, and from these, parison extrusion
was straight down in the vertical axis. The molds closed "on the
move" to catch the correct length while passing in tangential position
or, as may be the case with Comatic, in a radial one. Parison cutoff
from the die was normally by a high-speed knife blade, which was
sometimes heated to ensure a clean, open tube end, into which the
blow-molding mandrel, coupled to the mold station, could be easily in-
serted.

Another advantage of Comatic-type machines was that large-sized
molds could be utilized and changed early, enabling the production of
a relatively small series of handleware PVC bottles. The handle deflashing is done either by the operator or by one of the handleware bottle
trimmers commercially available in the United States or in Europe.

Sidel DSL-Type Wheels

Perhaps the best known among the continuously rotating vertical
wheels with radially mounted and axially opening and closing molds is

the range of the Sidel DSL-type machines. Originally, these were specifically developed for PVC bottles, commonly utilized throughout Europe, and scattered around the world. Designed in France by and for the Lesieur Company's edible oil packing, these machines were enlarged in range and types of the Sidel company and marketed throughout the world largely for in-plant operation alongside filling equipment for mineral water, oil, wine, vinegar, and for a score of other applications. The current series of Sidel DSL machines designed for PVC bottle production consist of the following designations: DSL 1000, DSL 1CS, DSL 2CS, and DSL 3CS (see Figs. 5.12 and 5.13). Their designed performance parameters are shown in Table 5.2.

All these machines consist of a horizontally mounted extruder with a spider cross head, typically not equipped for parison programming, extruding continuously a parison tube, downward. Molds rotate continuously on rotary platens with a common horizontal axis. One of the platens, closest to the extrusion head, has mold halves mounted on

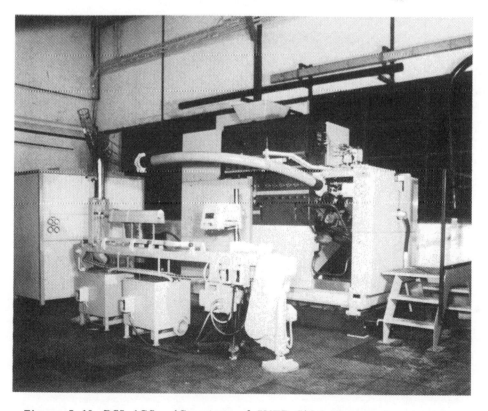

Figure 5.12 DSL 1CS. (Courtesy of SMTP-Sidel, France.)

Table 5.2 Performance Parameters of Sidel DSL Machines

	DSL 1000	DSL ICS	DSL 2CS	DSL 3CS
No. molds	1	2	4	6
Bottle size, cm^3				
min	200.	200	200	200
max	2000	1500	1000	2000
Output nominal max no./hr	900	1600	2700	3200
PVC output, kg/hr max.	50	50	65	150
Extruder dia., mm; L/D	65 mm; 20	65 mm; 20	65 mm; 20	65 mm; 26

Figure 5.13 DSL 3CS. (Courtesy of SMTP-Sidel, France.)

two tie bars with two strong springs behind them for best mold closing.
Mold opening and closing is operated by cams behind the platen. With
the parison tube extruding continuously downward and with the radially
mounted molds moving in and out of line with it, the molds must move
very fast to catch the parison, which is cut off with fast-acting scissors.
This is the reason for placing the fixed mold halves close to one side
of that parison while the moving mold halves on the opposite side move
rapidly on release of springs. The DSL 3CS has precisely that system,
but the DSL 2C has a fast-operating cam movement only. The DSL
1000 machine has both mold halves operated by cams and closing on the
position in the middle. This allows a faster ended movement and,
hence, a better cycle time. Bottle tails are removed in the molds by
the moving tail pinchers while the bottle tops are sliced off on a rotary
fixed-blade slicer placed on a downstream conveyor.

Emphasis is laid on simplicity of the design and the operation with
mechanical movements only. Filling plant operators and Third World
users can operate the machine with a minimum of maintenance and skill.
Single-screw extruders are designed to operate on PVC dry-blend
powders but granular compound can be used as well. High-power air
blowers are used to cool extruder barrels in case of overheat or during
the shutdown procedure.

5.8.2 Horizontal Table: Rotary Machines

In another group of rotary machines for PVC blow-molding, the rotary
movement is in the horizontal plane around a vertical axis with molds
spaced on the periphery of the rotary table, with indexing under extrusion head to pick up the parisons. Here again there is a problem
of the parison extrusion not being interfered with by the movement of
the molds. This is often resolved by moving vertically either the extrusion head or moving the molds down under the extrusion head.

The first solution was adopted by the classic Plax machine design,
well described in the patent literature, some variants of which were
used in America as well as in Europe and Japan for PVC bottle manufacture. This proprietary equipment is, however, not available commercially outside the licensed operating companies. As can be seen
in the patent literature, either the eight or four molds with book-type
mold closing and opening index under the extruder head is lifted immediately after the index and parison cutoff to enable continuous parison extrusion [6].

Generally, this equipment was being used for bottles up to a 1.5
liter capacity. Bottle neck finishing could be done by any conventional
method, if required, including facing and limited drilling, on special
bottle trimmers. For PVC bottles, in-mold neck finishing, with limited
trimming only, was generally preferred.

The Frohn-Alfing rotary table machine represented the other solution in which indexing molds "dipped" under extrusion heads after picking up the parisons. The first Frohn-Alfing machine, designated AFRB-6 and demonstrated in 1973, had a six-station rotary horizontal mold arrangement, each mold unit with either two or three cavities opening and closing on radially mounted platens operated by hydraulically actuated toggle levers.

The extruder was a twin-screw conical one, CT 111, with a claimed output up to 300 kg/hr and intended to produce 1 liter PVC mineral water bottles at up to 5400 per hour with triple-activity molds per station. With the hydraulic actuation of the mold stations, the vertical table movement, and the neck-forming mandrels, this machine reflects in a way the opposite concept of the mechanically operated wheel-type equipment described before. In fact, this design did not reach a wide application, and in 1981-1982, a new, simplified design was produced. This machine had the capability of producing large-capacity PVC containers up to 4 liters in size in a single-parison operation. Another six-station rotary table machine, but with a "bobbing" extruder, was produced in the United Kingdom by Femco and distributed in the United States by Strong Plastics Corporation. This was known as Rotoblow. Very limited application for PVC blow-molding was reported using single-screw extruders.

Chemica of Aarau (Switzerland) also produced for some time in the 1960s a six-station rotary table machine for PVC bottles, which could be operated with a selected extruder, but it was really intended to run with a twin-screw extruder, such as Colombo LMP and PVC dry blend to be produced alongside the extruder.

ADS of France also included in its range of machines a nine-mold rotary horizontal table machine claimed to be capable of producing 1 liter PVC bottles, up to 5000 per hour.

5.9 BIAXIALLY ORIENTED PVC BOTTLE EQUIPMENT

5.9.1 General

The benefit of orienting PVC in the blow-molding process was recognized in the early 1960s mainly through the investigations carried out at the 4P Company in Germany, which led to a series of patents on the injection blow-type process. The equipment was eventually produced under license by Johann Fischer Company.

In the extrusion blow-molding process, the main developments leading to biaxial orientation were carried out at the Solvay Company in Belgium, and they, in turn, obtained a series of patents under which equipment builders, first HMS Company of Belgium and then Sidel of France, were licensed to make and sell the MSF series of

machines. Sidel, Bekum, Kautex, and Fischer came up in the late
1970s with biaxially orienting, platen-type machines, and later some offered conversion kits for standard platen machines. In the 1980s
biaxially orienting equipment also became available from the ADS
Company of France and Automa of Italy.

Limited orientation was also achieved in the Gildemeister-Corpoplast
process, in which extruded PVC tube was first formed into test-tube-
like preforms, which were then reheated and blow molded. A similar
limited orientation process was operated (in the 1960s) from precut
tube lengths by the U.K. Marrick process in cooperation with ICI.
Even before that, the Danish Blow-o-Matic Company manufactured tube-
reheating process equipment, which was used in limited application by
a French water bottle manufacturer. In 1981-1982, Nissei Company of
Japan developed a modified injection blow molding process in which
limited biaxial orientation could be achieved.

In the development of the biaxial orientation of PVC, the main driving force was a combination of cheaper PVC formulations with a reduced bottle weight but with a comparable drop strength, or better,
as in extrusion blow-molded PVC bottles.

In addition, biaxially oriented PVC bottles can have a slightly enhanced barrier property to gas permeation, mainly due to formulation
differences and also excellent transparency with surface brilliance.
The disadvantages to be mentioned here are (1) the surface improvements lead to high friction in bottle-to-bottle contact, which may affect bottle handling and filling on high-speed equipment, and (2) some
biaxially oriented bottles may display stress-cracking phenomenon.

The improvements in biaxially oriented PVC bottles stimulated further
advances in the standard blow-molding process and equipment with
the result that, in the 1980s, the margin of difference in bottle weights
became narrower against the original estimates, thus making the justi-
fication for extra equipment cost more critical. In 1982-1983 the volume
of biaxially oriented bottles produced in Europe, mainly in France,
was still a low percentage of the total PVC bottle market. Even bigger
differences existed in America and in Asia, where biaxial orientation
of PVC has not met with the success known in Europe.

In the injection-blow process, biaxial orientation is derived from a
combination of melt-flow orientation in the injection stage, giving a
form of "onionskin" structure. The hoop orientation is derived from
the final blow process after appropriate thermal conditioning.

In the extrusion blow, biaxial orientation is introduced through
conditioning of blow-molded preform and by subsequent stretch blow.
Injection blow is suitable for both thin- and thick-walled containers,
but the extrusion blow is essentially suitable for thin-walled bottles.
This chapter deals with the latter process only as the predominant
one at the time of writing.

5.9.2 Solvay-Sidel MSF-BO Equipment

In terms of numbers of bottles produced in 1982, in biaxially oriented PVC, the MSF equipment currently built under Solvay license by the SMTP-Sidel Company of France was at the top of the league.

The predominant model in this group of equipment was the MSF BO-5000 model (Fig. 5.14), the prototype form of which was constructed by the Belgian HMS Company for the French BAP Company-Solvay subsidary engaged in blow molding of PVC and also of Solvay high nitrile polymers for beer bottles.

A simplified version of that equipment was subsequently built by SMTP-Sidel together with the platen-type machine model MSF BO-3000. A typical MSF BO-5000 line, installed at one of the French water bottle manufacturers, was capable of producing about 11,000 PVC bottles of 1.5 liter capacity at 38.5 g under continuous running conditions, achieving up to 95% efficiency. At that bottle weight, a saving of some 3 g on the conventional bottle equivalent was being achieved. The MSF BO-5000 machine can work with any high-output PVC extruder, and both twin-screw extruders and L-configuration twin-and single-screw extruders have been used. A preference for the Andouart-EMS-BDV 120 extruder was clearly evident in the latest installations.

Although the whole line, including the extruder, was computer controlled, no parison programming was deemed necessary, but Solvay-designed spider dies were available with Hunkar parison programmers and could be justified for heavier, gaseous water bottles running at lower speeds.

Parison extrusion was downward into a 20-station vertical wheel, of similar construction to the MSF 500 wheel, with book-type mold opening, and a positive, mechanical neck gripper take out attachment. For better heat transfer, Be-Cu molds are normally used for the heavy wall, preform making. Needle blow was used for blow-molding preforms.

Preforms with a solid, chilled outer surface were delivered by the rotary takeout to a transfer conveyor in which preforms were suspended by the neck. Partial reheat of the preforms took place by balancing heat transfer the thick wall to the chilled surfaces.

A 64 neck support conveyor carried the preforms to a temperature conditioning oven where they rotated in front of short-wave infrared (IR) heaters. Prior to entry in the oven, preforms had the top blow moil removed by a slicer trimmer. In the first section of the temperature conditioning oven, quartz lamps, controlled in eight rows, were used to raise the preform temperature. Air blowers in the oven improved the convection heating. In the second section of the oven of about the same length (~10 m), preforms were heat stabilized to balance surface and bulk temperatures.

Preform temperatures were checked by IR scanners, and the heating operation was controlled in a zoned fashion by the line computer.

Preform temperature distribution was more important than preform wall thickness control. The regranulated preform moils and any other scrap were normally fed back in a closed loop to the extruder hopper.

Heat-conditioned preforms were transferred by an indexing, chain-type, 64-neck collets conveyor into a horizontal table, rotary stretch blow machine with 24 stations.

Air pressure of about 280 psi was used in the final preform blow to a finished bottle. Mechanical mold locking was provided. Blown bottles were given a final neck trim on a slicing unit on the bottle conveyor to storage.

The bottle base is the least oriented apart from the bottle neck, and in drop tests this is usually the failure area. The base design can compensate for the lack of orientation and is normally researched for best results. With a good base design, a 1.5 liter water bottle produced on MSF BO-5000 equipment should withstand, on average, a 1.4 m drop height. A typical formulation for biaxially oriented bottles contains only 1% impact modifier against a normal blow-molding formulation for water bottles, which contains about 9%.

As can be expected, a major installation of the MSF BO-5000 type involves a heavy capital investment of the order of $2 million in 1982. This is obviously in excess of the capital required to blow ordinary PVC bottles at an equivalent rate, and the difference must be justified by both lower bottle weight and cheaper raw materials formulations. It can be expected that at least a 10% bottle weight reduction would be achieved, and raw material costs saving would be of the order of 5%.

The Sidel MSF-BO-3000 platen-type machine has an improved version in the 1982 model with a Weber 85 mm twin-screw extruder feeding twin parison dies (Fig. 5.15). It was designed with 1 liter PVC bottle market as the main application, and nominally it can produce up to 2800-single-liter PVC bottles per hour.

This machine comprises two symmetrical assemblies left and right of the extrusion heads. Each assembly consits of a preform mold with a twin cavity, oscillating between the extrusion dies and the first neck-calibrating blow station, a base tail deflasher, and a chain-mounted preform neck gripper transport to the stretch blow station. It also has two heat-conditioning stations with IR elements and IR temperature probes, controlling the conditioning process, and a stretch blow unit with telescopic movement stretch and blow mandrel. In addition, a slicing trimmer is required to remove the neck top blow moil. A microprocessor controls the whole cycle: preform conditioning, extruder speed, melt temperature, water temperature, and parison wall thickness controller, if one is used.

The latest model was demonstrated running on a relatively heavy, low-carbonation water bottle at about 1400 bottles per hour and on a 1 liter oil bottle weighing 42 g at 2400 bph.

For comparison with the MSF-5000, the BO-3000 cost around $550,000 in 1982.

Figure 5.14 MSF-BO-5000. (Courtesy of SMTP-Sidel, France.)

Figure 5.15 MSF-BO-3000. (Courtesy of SMTP-Sidel, France.)

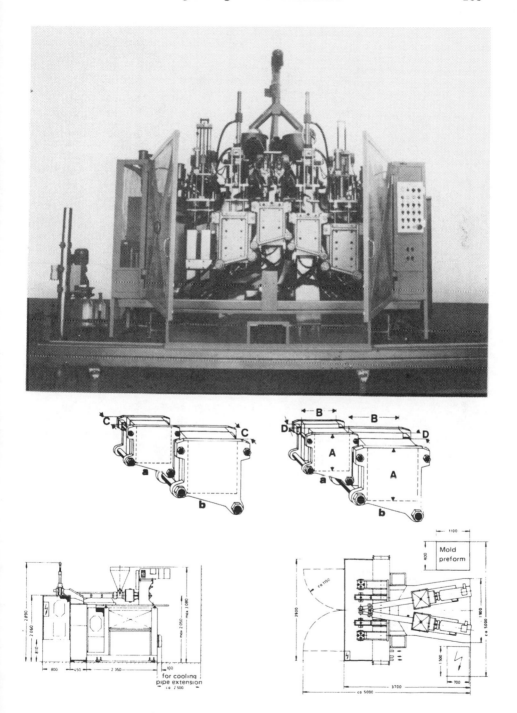

Figure 5.16 BMO 4D. (Courtesy of Bekum AG, West Germany.)

Figure 5.17 BMO-4D. Details of dies and mold platens. (Courtesy of Bekum AG, West Germany.)

5.9.3 Bekum BMO-4D

Bekum constructed in 1979/1980 a stretch blow, platen-type machine
model, BMO-4D, with a typical left- and right-hand configuration and
with two extruders connected to separate extrusion heads   (Fig. 5.16).
BMO-4D was fully developed by 1982 and became their advanced technology unit for PVC and high nitrile polymers.   In a way, this equipment operates on a similar basis as the Bekum HBV machines, described
before, but the 15° incline sliding carriages consist in case of the BMO
machine for these clamping mechanisms opening and closing in unison
the preform molds, the stretch blow molds, and the discharge clamp
for placing the bottles on the conveyor (Fig. 5.17).

As was the case with Sidel equipment, final bottle wall thickness
distribution is more dependent on thermal conditioning of bottle preforms than on parison wall control. It is, therefore, understandable
that the Bekum parison wall thickness control system is not required
with this equipment. For some applications a single extruder with
twin-extrusion heads can be used, but normally either two 60 mm
diameter or two 70 mm diameter extruders with a 24 L/D ratio, as described in the earlier part of this chapter, are being offered here.
The smaller of the two extruders is offered for bottles up to 1.5 liters
capacity and the larger for bottles up to 2 liters in size.

With two extrusion heads, two preforms are molded in sequence on
both sides. Preform molds are zoned for controlled cooling down to
about 90°C from the recommended melt temperature of about 200°C.
Bottle necks are calibrated at the first preform molding station in the
normal Bekum way, and the deflashed preforms are then picked up
by the second stretch blow station for the final bottle blow-molding
step and then discharged on to a collecting conveyor (Figs. 5.18, 5.19).
The total molding cycle depends, as always, on final bottle design and
wall thickness, but typically it is of the order of 8 to 9 sec and so BMO-
4D can produce a maximum of 1700 to 1800 lightweight PVC bottles per
hour. For experimental applications, only half of that machine can be
constructed, and it operates to one side only. Such a variant is known
as BMO-4.

BMO equipment, with its relatively low output and with good bottle
handling on the discharge conveyor, found its application mainly for
in-plant installations for such products as drinking water, edible oil,
or soft drink bottles. According to Bekum's calculations it can produce PVC bottles competitive with PET alternatives up to a 1.5 liter
size.

5.9.4 Kautex

In 1982, Kautex manufactured a double- and a single-cavity machine
for stretch blow-molding of PVC. The two models were KEB 2-2 and

Figure 5.18 Picture of the Bekum BMO-4D machine showing two blown preforms and two blown bottles on the blowpins after molds open. (Courtesy of Bekum Plastics Machinery, Inc., Williamston, MI.)

Figure 5.19 Overall picture of the Bekum BMO-4D machine with the collecting conveyor on the side of the mold platens. As shown, bottles come out of the machine in an upright position. (Courtesy of Bekum Plastics Machinery, Inc., Williamston, MI.)

KEB 2/14, respectively. They have made a thorough analysis of bottle properties achieved, together with comparative costings for a 1 liter bottle, showing significant benefits achievable against conventional PVC platen blow-molding equipment. The principle of operation is similar to the other platen machines described above, and the two-mold machine has a nominal capacity of up to 1200 single-liter bottles per hour at a nominal bottle weight of 29 to 32 g. A single mold variant produces up to 727 bph. A positive bottle takeout to a conveyor or to a leak tester is a feature of this machine design. These machines are intended to operate as multiple unit batteries, providing flexibility of production loading, especially for a medium-sized custom molder.

5.9.5 Johann Fischer

In 1981-1982, Johann Fischer, now part of the Battenfeld Group in its Voith Fischer Division, introduced a modified version of their VK 1-2 blow molder, called VK 1-2 ESB, capable of stretch blow-molding PVC bottles. Initially this was limited to a single preform, single stretch blow-mold version, to enable field evaluation of stretch blow-molding. In fact, a conversion kit for standard VK 1-2 machines was offered with a view to battery-type operation of production lines. Here again, a positive bottle handling, in line with the extruder and mold axis, was a feature.

Bottles up to a 1.5 liter capacity could be made on VK 1-2 ESB with either 50 mm/20D extruder or optional alternatives of 60 mm/20D and 60 mm/24D variants. With a standard extruder, up to 35 kg PVC can be converted per hour and the 60 mm/24D extruder extends this to 65 kg/hr. It would thus be capable of producing normal lightweight bottles as well as the heavy-walled variants for carbonated soft drinks, although for that category, the IBS equipment previously offered by this company is preferred.

5.9.6 Automa

The Automa Company of Italy has designed a single-sided platen machine, similar to others described above. Known as Speed 1.5-SB, it can be equipped with either a 50 or a 60 mm extruder for outputs of 30 or 40 kg/hr, respectively. Single- or double-extrusion heads are available, with the double one having a center distance of 125 mm. This, and the extruder output, define whether a given bottle can be made one-up or two-up. Typical molding cycles given for a 42 g 1.5 liter PVC bottle was 8.5 sec, for a 32 g 1 liter bottle, 8 sec, and for a 26 g 330 cm^3 PVC bottle, 10.5 sec. This machine is a derivation of

their Speed 1500 model, and likewise, it is intended for a side-by-side
battery-type operation. For example, for a production volume of 5000
bottles per hour of 1.5 liter capacity, a battery of 15 machines, operating with an efficiency of 80%, would be indicated.

5.9.7 ADS

ADS of France developed a biaxial orientation variant of its ADS H
modular, single-station machine with a designation ADS "bio-modular-
type" machine. They offered a BIO kit for conversion of standard
machines, if required. Machines of this type were intended for a
PVC output below 30 kg/hr. A characteristic of this equipment is a
"tailless" bottle process, enabling a higher extruder output utilization
a significant factor in PVC conversion.

Preforms were made on a standard ADS H unit and transferred by
a system of scoop and tongs to the adjoining hydraulically operated
platen machine for final blow-molding. The temperature of the preform
was adjusted by the first blow-molding stage with equalization and
correction in the transfer operation. It was claimed that, with PVC
melt temperatures achievable on this equipment at below 200°C and with
tailless operation, high material and production efficiency was attainable. To illustrate that point it was claimed that, for production of up
to 5000 single-liter bottles, a battery of eight ADS-11 machines would
be adequate. For heavier wall bottles, such as the 1.25 liter (carbonated water), the output per machine would be of the order of 400 bottles per hour, nominal rate, at 80% efficiency, indicating an installation
of a 15-unit battery.

In 1981, a further improvement of the BIO system was introduced,
involving a conditioning conveyor between the ADS-11 preform machine
and the bottle blow-molding station arranged in the same axis as the
extruder. This new machine arrangement was designated ADS-11 H2
BIO, and the BIO kit was available separately as well. In this arrangement, preforms are blown "neck down" from a sealed-end tube,
taken "up-and-down" on the ADS-11 mold system with needle blow and
the neck moil deflashing in the mold. On mold opening, the blow-
molded preforms are discharged downward onto a vertical pin-chain
conveyor.

Preforms on a pin-chain conveyor are then indexed into a thermal
conditioning station, where vertically moving sleeve-type heating
elements can be used to correct the heat content of the preform in a
zoned manner. This, and the reheat time available on the transfer
conveyor, ensures a better thermal conditioning of the preform to just
above the second glass transition point of the PVC compound used.
The bottle blow-molding station is equipped with a stretch-rod mechanism on the preform locating pin, thus providing a proper stretch-blow
orientation. From the blow-molding station, bottles are conveyed on
the pin-chain indexing conveyor to the bottle takeout station, where a

vertically moving suction hood cooperates with a mechanical off-the-pin ejector to take the bottles onto a pneumatic conveyor to silos.

With such improvements in biaxial orientation and material-related weight savings achievable, it is very important not to lose sight of one issue: the final bottle performance in the hands of the consumer. While it is possible to produce oriented PVC bottles with much reduced weight and with satisfactory drop strength and adequate top load resistance for filling, transport, and distribution, such bottles may not have adequate squeeze resistance in the hands of the consumer while a full bottle is being opened, causing more or less disastrous spillover. This was the experience of one European oil bottle company that got carried away with claims of achievable cost reduction.

5.10 PVC COMPOUNDS FOR BOTTLE BLOWING

5.10.1 General

It is seen in previous chapters that PVC offers a wide range of compound formulations aimed at specific applications. In terms of bottle blowing, this requires a fine balance of final bottle properties together with processing characteristics for a given equipment design and process involved. As it is well recognized, PVC is an unstable polymer on exposure to heat, shear, and catalytic effects. This is normally linked with evolution of hydrochloric acid due to the breakdown of molecular structure, starting with the less stable groups of tertiary carbon links or with allyl groups in end groups or in side branches. This leads to almost autocatalytic unzipping of molecules with the well-known carbonized appearance of degraded PVC, so often worrying a PVC bottle-blowing operator when things do not go as they should. It is in situations like these that one appreciates best why working surfaces should not expose the catalytic iron and why they should be of a high surface finish stainless steel or high-quality nonporous chrome.

The bulk of PVC bottle blowing in Europe is from PVC dry blend powder compounds; in America granular compounds are preferred because the cost margin between the two forms is not as significant.

Dry blending is normally done on the same site as bottle blowing with possibilities of easily adjusting the formulation to suit best the equipment and product requirements. More recently, however, there has been a noticeable trend for the major PVC resin producers in Europe to supply the bottle-blowing dry blends at no significant disadvantage to the converter. This is normally done after a detailed and often prolonged trial-and-correction period of cooperation between the supplier and the user. Quite often the user has a preferred formulation the supplier agrees to provide.

The standard starting point is a PVC homopolymer of medium molecular weight, which in Europe is typically between the K values of 57 and 62.

Most PVC bottles are in clear or transparent tinted form. As a
majority of the bottles are for liquid products they require adequate
drop-strength and top-load resistance,w hich call for toughness and
rigidity in the PVC compound. Typically, then, a good PVC for bottle
blowing represents a compromise of processing requirements and optimum clarity drop-strength and top-load resistance.

With this in mind, it will be understood that although in the past
lower molecular weight polymers were favored at the level of 55 to 57 K
values, the more recent improvements in lubricants, stabilizers, and
impact modifiers enabled an increase in K values to 62, or above, yielding some of the better mechanical properties associated with higher
molecular weight resins. When mass polymerization of PVC became a
reality in the 1960s, it was anticipated that this would be the best
route to high-clarity PVC bottle-blowing compounds. It was, however,
a useful stimulus to suspension PVC producers to improve their process, and so most of PVC bottle-blowing compounds are still based on
suspension-polymerized resins.

The main reason for this is that the suspension process, in general,
produces a polymer with better initial color and less gel (overall less
batch-to-batch carryover contamination), and better control on molecular weight in the manufacture of low-molecular-weight resins. Also,
through the years the producers of suspension resins have been upgrading their own processes, and for the most part, there is no royalty
payment involved. Another, but minor reason is that, from an energy
point of view the suspension polymer has typically a heat of fusion of
3.1 J/g PVC, but the mass polymer one of 4.1 J/g PVC [7].

The energy standpoint has, however, become an important consideration and is one of the main reasons PVC bottles have been successfully replacing glass 70 TOE (tonnes of oil), for 1 million 1 liter PVC
bottles versus 230 TOE for 1 million 1 liter glass bottles [8].

Because of the brittle nature of rigid PVC, it has been customary
to counteract this in compound formulations by the addition of impact
modifers, which create a shock-absorbing phase in the PVC bottle
wall. These have been typically of the acrylonitrile-butadiene-styrene (ABS) or methylmethacrylate-butadiene-styrene (MBS) type, but
because of clarity requirements and impact improvements, a preference
has developed in Europe for the "acrylates," such as the elastomers derived from the butyl or octyl acrylates with a glass transition
temperature below $-50°C$.

5.10.2 Compounding: Fusion

Although PVC is essentially a glassy polymer, with only about 10%
crystallinity in a typical homopolymer, the crystallites, which are
small and imperfect, with melting points near 210°C, have an influence
on processing and final bottle performance. As most of them disappear
only at about 200°C, they have some importance in fusing and in the

melt flow properties of PVC compounds.

The fusion process of PVC polymer particles, or "grains," has been subject to detailed studies as it affects the compounding and the processing stage as well as product properties such as impact strength, clarity, and surface finish. Fusion (or gelation) is defined as complete when PVC polymer grains disappear in the molten phase. As expected, the end temperature of fusion depends on the K value, and a K 58 resin, for instance, has a fusion temperature of 187°C, but a K 64 resin, 198°C [9].

In PVC compound preparation, the first step is the dry blending process, when PVC resin is mixed with the required additives in a high-intensity speed mixer where the work put in by the impeller blades generates internal friction heat, raising the mix temperature to about 130°C, the so-called pregelling stage. At this stage the PVC grains should be coated with additives and the liquids absorbed by the porous grain structures to yield a free-flowing powder after discharge from the high-speed mixer to a jacketed, slow-speed cooler, where the temperature is brought close to room temperature prior to moving the dry blend to a storage silo or a hopper. During the extrusion process, higher temperatures are attained by internal friction and external heat input. During this stage, PVC grains are densified and deformed without significant comminution of the grains, such as normally occurs in the high-shear operations like the Banbury mixer [10].

External lubricants added to PVC blends reduce the frictional heat development and delay the fusion, particularly below 200°C. When fusion occurs, the original PVC grain boundaries begin to disappear, resulting in a homogeneous appearance.

Recent studies at different laboratories confirm that the best impact resistance of PVC compounds is attained at the fusion level of 60% to 70%. This can be correlated to the melt flow index (MFI) measurements with the ungelled blend giving the minimum values and the 100% fused state reached when maximum MFI values are reached and level off. The fusion process is best observed by microscopic techniques related to MFI readings [7,11].

Any given combination of PVC compound and bottle-blowing extruder has specific fusion characteristics, and a gelation diagram can be derived as a series of plots of percentage fusion for screw speed and temperature variants [11]. For a high degree of bottle transparency and for best surface finish, the fusion should be more complete, and normally a balance of mechanical properties with the required appearance of finished bottle is agreed between supplier and customer. Those requirements goven the choice of PVC bottle-blowing compounds.

5.10.3 Food Grades

Although PVC bottles for domestic cleaning products or shampoos require a formulation with the resulting compromise of processing, mechanical properties, appearance, and cost, food-containing bottles have additional requirements of adequate taint and odor characteristics, coupled to the necessity of compound ingredients being suitable for food contact. After the vinyl chloride monomer (VCM) problem, in the United States in particular, the number of food-contact PVC applications were restricted and this was echoed in Japan and to a lesser extent in Europe. At the beginning of the 1980s, the VCM content of PVC bottle walls was restricted by the European Economic Community (EEC) regulation to not more than 1 ppm, and this was reflected throughout the world, although some U.S.companies offered PVC resin with "nondetectable" VCM.

The major issue in PVC bottles for foods is the choice of adequate PVC stabilizer systems approved by such authorities as the U.S. FDA, the German Bundesgesundheitsamt, and other national bodies. In Europe and in France, in particular, vast numbers of PVC bottles were used since the early 1960s for oil, water, and wine packaging. These required specific formulations for the various bottle-blowing systems involved, and the finely "tuned" compounds were normally of a proprietary, well-guarded nature. More recently, such an approach was changed with the enormous improvement in the extrusion equipment design for PVC bottle-blowing, to the extent that major PVC resin manufacturers now offer specific compound formulations either in powder or in pelletized versions and with such designations as, for instance, "for drinking water," "for orange concentrate," "for oil," or "for shampoos and toiletries."

5.10.4 Impact Modifiers

Traditionally, impact modifier levels have been set at "low," "medium," or "high" levels of about 8, 10, and 14 phr (parts per hundred of resin) PVC resin, so that impact requirements of small and large bottles could be satisfied economically.

The selected impact modifier must fulfill several requirements in both processing and final bottle properties. The predominant requirement is obviously adequate impact resistance to enable to the bottle to be handled, filled, and used without damage. Low-temperature behavior is obviously important in this context. Modifier addition may affect processing characteristics, and its effect on melt viscosity, and, above all, on the "swell" characteristics of the parison, need to be matched to the equipment used. Other melt quality-related problems, such as weld and die line, can also be affected. In the finished bottle, stress whitening, particularly in contact with some pine oils, in a phenomenon to be watched. Some vegetable oils can produce stress cracking in the modifier phase, and a suitable selection is necessary.

Quite clearly, for food-contact application the selected modifier should not cause any tainting or odor transfer. In terms of formulation interaction, care should be taken in matching the impact modifier performance with the stabilizer system used; for example, Ca-Zn formulations require a different grade of modifiers than the tin-based formulations. Effects such as color reversal and fluorescence, should be discussed with suppliers before a final selection is made. Experienced suppliers of impact modifiers are able to provide specific detailed recommendations, as can be seen from examples of formulations quoted in Sec. 5.10.8.

Methylmethacrylate-butadiene-styrene are the most commonly used modifiers in clear bottles. Typically, commercial examples are in the range of Paraloid modifiers produced by Rhom and Haas, in the Blendex BTA 111 grades produced by Borg Warner Chemicals, Kane Ace B from Kaneka Company, and others. The ABS modifiers are used less for bottle applications than for other products requiring impact resistance. They are normally available from major ABS suppliers, and some, like Borg Warner, offer the choice of MBS, ABS, or the MABS types. The Japanese companies, such as Kureha, Kaneka, Mitsubishi Rayon Company or Ryo-Nichi Company, offer at least as wide a choice.

5.10.5 Stabilizers and Lubricants

The final product application and the type of processing equipment to be used determine the nature and detailed formulation of stabilizers and lubricants. In terms of bottles, the main division is between food and non-food contact applications.

In Europe, calcium-zinc stabilizers have been used for food-contact applications and particularly for the enormous volume of drinking water bottles. These were used with synergistic admixtures of epoxidized soybean oil, aminocrotonates, and as used for oil formulations in the past with α-phenylindole. More recently, organophosphite compounds have been used to replace the phenol ring compounds (antioxidants) in order to eliminate the often occurring free phenol odor problem.

In both the United States and in Europe, organotin stabilizers have become the preferred stabilization system on account of their efficiency and ability to give crystal-clear bottles for special requirements, such as toiletries. They are also the most expensive category of stabilizers, with the ever-increasing cost of tin.

Octylthiotin stabilizers have been adapted in Europe for both food and non-food PVC bottles, including oil packaging with its own specific problems of color and taint.

In the United States, methyltin stabilizer systems became quite popular on account of their stabilizing efficiency and adaptability in various non-food grade PVC formulations. Tin stabilizers are subject to some regulatory restrictions for food-contact applications, and both their nature and dosage need checking with appropriate national requirements.

In the last decade more concentrated forms of liquid stabilizers were developed, requiring lower dosage levels and reducing fumes during

processing. Normally, octylthiotins are not used above 1.5 phr of PVC
resin. A. U.S. supplier of such stabilizers, M&T Chemicals, Inc.,
has approached the FDA for a wider approval [12].

For non-food contact applications, more efficient tin-based stabilizers are normally used. For more detailed recommendations the reader
is advised to study the extensive manuals available from the main stabilizer compound suppliers, such as Interstab-AKZO, Lankro-Diamond
Shamrock, Hoechst, Barlocher, M&T Chemicals, Advance Chemicals
Corporations Ciba-Geigy, or Associated Lead Manufacturers Limited.

Food-contact formulations often contain organic stabilizers, particularly in combination with Ca-Zn systems. One such example is Rhône-
Poulenc's Rhodiastab 50, which is a mixture of diketones.

All food-contact formulations should comply with the national regulations controlled by such bodies as the U.S. FDA, the British Plastic
Federation-British Industrial Biological Research Association, the BGA
in Germany, the Ministry of Agriculture in France, the Ministry of
Health in Italy, and so on.

Some of the liquid forms of stabilizers, such as metal soaps, also
provide a measure of "external" lubrication, required for compounding
and processing. Lubricants are classified as either "internal" or "external." In theory, the internal ones are more soluble in the polymer
matrix, increasing the distance between molecules and thus reducing
internal friction. The external lubricants are required to form a continuous boundary layer between processing surfaces and PVC compounds. These are normally less polar in nature than are the internal
ones are and partially absorbed on metal, thus providing greater ease
of processing and smoother surface to the product. Irrespective of
their nature, lubricants reduce the rate of fusion of PVC [13]. There
are, however, exceptions, as with the unoxidized low-molecular-weight
polyethylene waxes, which, for the most part, do not affect the rate
of fusion of PVC. On the other hand, the oxidized polyethylene combined with low-molecular-weight paraffin provides controlled external
lubrication while calcium stearate increases the melt strength at higher
concentration and acts more as a processing aid [14].

Normally, lubricants are categorized by observation of their pro-
cesing characteristics and by their ability to delay gelation in PVC.
With this approach they can be divided in accordance with the criteria
listed in Table 5.3.

Many compounds have been proposed as lubricants for rigid PVC
bottle compounds, several of which offer little or no advantage over
other products available. The products suggested usually fall within
one of the following chemical types.

Fatty acids
Fatty acid soaps
Fatty acid esters
Fatty acid amides
Fatty alcohols
Hydrocarbons

Table 5.3 Properties of Lubricants

Properties	Internal lubricants	External lubricants
Compatibility with PVC	High	Poor
Chemical interaction with PVC	None	None
Affinity with PVC	Some	None
Affinity with metal surfaces	None	Some
Influence on melt flow properties	Strong	Slight
Influence on gelation rate	None	Strong
Polarity	High	None

It is not essential that a compound fall strictly within the external or internal lubricant category. In fact, most products exhibit some properties related to each. It is essential, however, to determine the optimum use level of a lubricant or lubricant combination for a given end product, as an incorrect choice can lead to undesirable side effects in the finished article, such as surface delamination and reduced impact strength and softening point [15]. Normally, external lubricants are chosen from a group of waxes or soaps for example, and internal lubricants are selected from a group of fatty acids, esters, and alcohols.

As a result of many years of experience, suppliers of stabilizers and lubricants or stabilizer-lubricant combinations already formulated for the various types of equipment and product requirements [15]. This aspect requires particular attention in clear bottles and in bottles intended for food contact. In Ca-Zn-stabilized formulations a yellowish color occurs and must be corrected by a blue tint where this is not acceptable.

5.10.6 Processing Aids

The role of the processing aid additives is to help in the compounding of various additives with PVC resin and act as a "compatibilizer" in the fusion and in extrusion blowing stage. Unmodified PVC melts are often considered as having a tendency to melt fracture and to break while the parison is blown (poor melt strength). On the other hand, this lack of melt strength can lead at higher melt temperature to undue parison extension due to gravity action on the extruding tube. This last phenomenon can affect adversely the parison thickness control. Processing aid additives are supposed to ameliorate all these PVC resin

shortcomings as well as improve the surface finish and hence clarity.
Processing aids are often derived from acrylic polymers, and hence
the addition of acrylate impact modifier can be a substitute for the
normal "aid," as is sometimes the case with compounds for biaxially
oriented PVC bottles.

Processing aids do have a role in the fusion process and can accelerate or retard it according to their nature and formulation [16].
They must be selected for specific properties and their effect on clarity and bottle appearance, and processing effects such as plate-out
should be investigated. Some processing aids include in their function
a degree of lubrication, and this is normally taken into account when
formulation of lubricants is considered. Such an aspect is particularly
important for some blow-molding dies [17], and specialized processing
aids are available [18].

Some products packed eventually in PVC bottles can affect bottle
performance and result in stress cracking or whitening. Both impact
modifiers and the processing aids must be selected to avoid such problems, particularly with oils and with some disinfectants [19]. Processing aids suppliers are aware of such problems and provide specialized guidance in arriving at a correct balance of properties [20].
Processing aids are either free-flowing powders or more like particles
and can be added as are modifiers at the dry-blending stage.

5.10.7 PVC Compounds for Oriented Bottles

With the growth of biaxial orientation in PVC bottle blowing, the need
for impact modifiers was decreased so much so that, for injection
stretch-blown bottles, a nominal amount, of say 1 phr, is currently
used to act more as a "processing aid." Such compound variants can
now be also obtained from the major PVC resin suppliers or from specialized compound manufacturers, of which the French Dolryl and the
British Vinatex Companies are typical examples in Europe and the
Ethyl Corporaton, Occidental, Tenneco, and others in the United
States.

5.10.8 Formulations

Normally, formulations are derived from theoretical considerations to
meet all possible requirements envisaged but are fine tuned by trial
and error on production equipment. Attempts were made at using basic
data in computerizing such an approach, and an early presentation on
this topic was given at the ANTEC conference in 1973 [21] for general
purpose and PVC hose applications.

As an example of starting formulations for bottle blowing, the following examples can be used.

General purpose:

Ingredient	Formulation
S-PVC[a], K 57	100 parts
Acrylate impact modifier	10.0 phr
Process aid	2.0 phr
Octylthiotin stabilizer[b]	1.5 phr
Internal lubricant	0.6 phr
External lubricants	0.7 phr

[a]Suspension PVC
[b]Di-n-octyltin-S,S'-bis(isooctylmercaptoacetate)

Crystal-clear food-contact bottles [15]:

Ingredient	Formulation
M-PVC[a], K 57	100 parts
MBS, impact modifier	8-14 phr
Di-n-octylthiotin stabilizer[b]	1.5 phr
Lubricant based on a long-chain fatty acid ester	0.8 phr
Lubricant based on a fatty alcohol	0.2 phr
Amide wax (Acrawax C, Advawax 280)	0.2 phr

[a]Mass (bulk) PVC resin
[b]Di-n-octyltin-S,S'-bis(isooctylmercaptoacetate)

Clear bottles good on taint and odor [22]:

Ingredient	Formulation
S-PVC K 58	100 parts
Paraloid KM522 (impact modifier)	10 phr
Paraloid K 120N (process aid)	1.0 phr
Butyltin stabilizer[a]	1.5 phr
Glyceryl monostearate	0.7 phr
Partially modified montan ester wax	0.3 phr

[a]Di-n-butyltin-S,S'-bis(isooctylmercaptoacetate)

Ca-Zn formulation (mineral water) [23]:

Ingredient	Formulation
S-PVC, K 57	100 parts
Kane Ace B modifier	10.0 phr
Kane Ace PA 20 process aid	0.5 phr
Epoxidized soybean oil	5.0 phr
Saturated partial ester of glycerine	1.2 phr
Ester of montanic acid	0.4 phr
Calcium stearate	0.16 phr
Stearoyl benzoyl methane	0.15 phr
Zinc stearate	0.14 phr

Ca-Zn epoxy recipe for 1 liter bottle made on Bekum BA 3-S50 [24]:

Ingredient	Formulation
S-PVC, K 58	100 parts
MBS-impact modifier	8 phr
Epoxidized soybean oil	5.5 phr
Acrylic processing aid	0.5 phr
Glyceryl ester of montanic acid	0.50 phr
Di-stearyl pentaerythritol disphosphite (antioxidant)	0.30 phr
Calcium stearate	0.25
Zinc octoate	0.17
Stearoyl benzoyl methane	0.12

Ca-Zn epoxy recipe with special lubricant combination [25]:

Ingredient	Formulation
S-PVC K 57-60	100 parts
Impact modifier	10-20 phr
Processing aid	1-3 phr
Ca-Zn stabilizer	1.2-1.5 phr
Epoxy plasticizer D.81[a]	1-2 phr
Loxiol GH 4	1-1.5 phr
Ester of montanic acid	0.1-0.4 phr
PE wax	0.01-0.05 phr

[a]Epoxidized soybean oil. (Manufactured by Henkel.)

Water Bottle formulations for small machines, such as DSL I Bekum and Fischer [26]:

Ingredient	Formulation
S-PVC, K 57 to 60	100 parts
BTA 111, Borg Warner impact modifier	10 parts
K120 N processing aid	0.5-1 phr
Epoxidized soybean oil	5 parts
D 507[a]	0.15-0.2 phr
Calcium stearate	0.25-0.7 phr
Zinc octoate at 10% (SL2016)	0.15-0.20 phr
Lubricant 6203	1.7-2 phr
Polyethylene wax (AC-316)	0.2 phr
BHT Topanol OC (ICI) (Antioxidant)	0.05 phr

[a]Dihydro-1,4 dimethyl-2,6-dicarbododecyl-oxy-3,5-pyridine. (Manufactured by Sapchim, Paris, France.)

Water bottle formulations for big machines, such as DSL-3 or ADS-3 [26]:

Ingredient	Formulation
PVC	100 parts
Impact modifier	10 parts
Processing aid	1.5-2 phr
Epoxidized soybean oil	5 parts
D 507	0.15-0.2 phr
Calcium stearate	0.27 phr
Zinc octoate at 10%	0.15 phr
Lubricant 6203[a]	2 phr
Polyethylene wax	0.25 phr
Lubricant Loxiol G 78	0.5 phr
BHT-Topanol OC (ICI) (Antioxidant)	0.05 phr

[a]Glyceryl triester of montanic and hydroxystearic acids. (Manufactured by Sapchim, Paris, France.)

REFERENCES

1. *German Plastics Practice,* DeBell and Richardson Springfield, Massachusetts, 1946.
2. Machine developments for high speed blow moulding, Europlastics, *August:*1972.
3. Krauss-Maffer Aktiengesellschaft, Bulletin WB 12803.5, Munich, Federal Republic of Germany
4. Parison Programming, *Modern Plastics Encyclopedia,* McGraw-Hill, New York, 1976-1977.
5. *Principles of Polymer Processing,* Z. Tadmor and C. G. Gogos, Wiley-Interscience, New York, 1979, Chapter 13, 521-583.
6. British Patent 818 904.
7. M. Gilbert, D. A. Hemsley and A. Miadonyé, Assessment of fusion in PVC compounds, *PVC Processing II,* Plastics and Rubber Institute Conference, April 1983.
8. W. J. Prinselaar, The commercial future of PVC, *PVC Processing II,* Plastics and Rubber Institute Conference, April 1983.
9. G. Schoukens, Fusion characteristics of PVC, *PVC Processing II,* Plastics and Rubber Institute Conference, April 1983.
10. M. W. Allsop, Mechanism of gelation of u-PVC during processing, *PVC Processing II,* Plastics and Rubber Institute Conference, April 1983.
11. E. Kruger and G. Menges, Extrusion of PVC on single and twin screw extruders, *PVC Processing II,* Plastics and Rubber Institute Conference, April 1983.
12. Plastics Technol., *March:*78, 1983.
13. T. E. Fahey, Classification of lubricants for PVC by compaction, *40th ANTEC,* San Francisco, 1982, pp. 542-544.
14. D. W. Riley, Lubrication for rigid PVC: A slightly different perspective, *ANTEC 1983.*
15. Interstab Limited, General information on lubricants for PVC, Hoechst Plastics "Hostalit" manual.
16. Mitsubishi Rayon Co., Plastics Division, Metablen.
17. Rohm and Haas Co., Plastic Intermediates, *Paraloid K175.*
18. Ryo-Nichi Co., Ravender.
19. ICI Petrochemicals & Plastics Division, Welvic PVC Compositions.
20. Rohm and Haas Co., Paraloid Modifiers for PVC and Other Plastics.
21. G. Bareich, Formulating PVC compounds with a computer, *31st ANTEC,* Montreal, 1973.
22. Rohm and Haas Co., Paraloid KM522.
23. Kaneka Belgium NV, Kane Ace B: Impact Modifier for PVC.
24. Borg Warner Chemicals, Blendex modifier BTA 111, Amsterdam, September 1980.
25. Henkel & Cie, GmbH, Loxiol GH4.
26. Centre Pharmaceutique Europeen, Division Sapchim, Paris, Technical notes 14.11.80.

6

Rigid PVC Injection Molding

JOHN A. BACLAWSKI/BF Goodrich Chemical Group, Cleveland, Ohio

JERRY L. MURREY*/BF Goodrich Chemical Group, Avon Lake, Ohio

6.1 Background and History 303

6.2 Market Trends and Opportunities 306
    6.2.1 Business Machines 308
    6.2.2 Appliances 309
    6.2.3 Electrical and Communications 309
    6.2.4 Consumer Home Entertainment 310
    6.2.5 Medicine and Health 314
    6.2.6 Specialty Plumbing 314
    6.2.7 Furniture 314
    6.2.8 Summary 315

6.3 Economics 316

6.4 Properties and Characteristics of Current Products 320

6.5 Formulations and Their Effect on Material Properties 324
    6.5.1 Testing 325
    6.5.2 PVC Resins: The Starting Component 332
    6.5.3 Heat Stabilizers 334
    6.5.4 Lubricants 337
    6.5.5 Processing Aids 338
    6.5.6 Impact Modifiers 339
    6.5.7 Fillers 343
    6.5.8 Colorants 343

6.6 Product Design and Mold Construction 344
    6.6.1 Product Design 345
    6.6.2 Mold Construction 347
    6.6.3 CAD/CAM 348

---

*Current affiliation: Mobay Corporation, Pittsburgh, Pennsylvania

6.7 Processing Recommendations 349
     6.7.1   Equipment 349
     6.7.2   Machine Size 349
     6.7.3   Screw Design 350
     6.7.4   Nozzle 352
     6.7.5   Sprue Bushing 353
     6.7.6   Runners 353
     6.7.7   Cold Slug Wells 354
     6.7.8   Gates 354

6.8 Processing Techniques 356
     6.8.1   Drying 356
     6.8.2   Mold Temperature 356
     6.8.3   Stock Temperature 356
     6.8.4   Heater Band Settings 357
     6.8.5   Screw Back Pressure 357
     6.8.6   Screw rpm 357
     6.8.7   Injection Speed 358
     6.8.8   Injection and Holding Pressures 358
     6.8.9   Processing Summary 359
     6.8.10 Machine Start-up Techniques 359
     6.8.11 Purging and Shutdown Techniques 359

References 360

PVC is well known as a versatile thermoplastic material. Alterations in PVC polymer morphology, molecular weight, and/or the compound additive system can yield dramatic transformations in the processing characteristics and physical properties of PVC compounds. Over the last 40 years this versatility has enabled PVC to participate in a number of diverse extrusion market areas, such as pipe, house siding, windows, and wire and cable. During the late 1970s, a new family of low-molecular-weight, higher melt flow rigid vinyl compounds designed specifically for injection molding emerged. These new materials have successfully overcome many of the processing problems formerly associated with injection molding of rigid vinyl. Combined with the excellent processability are a gamut of other specialized property potentials: toughness, transparency, weatherability, and higher modulus, among others. Of course, these unique features are always combined with the inherent chemical resistance, flame retardance, and favored economics of vinyl. This chapter describes the evolution of the new higher melt flow rigid vinyl injection-molding compounds, as well as discussions of the product characteristics, market trends, typical formulations, part design, and processing of these materials.

6.1 BACKGROUND AND HISTORY

For decades, rigid vinyl compounds were classified as "unmoldable." Early attempts at injection molding the high-molecular-weight, heat-sensitive compounds on standard plunger machines were rarely successful. Because of limitations in both the early compound formulations and processing equipment, rigid PVC rapidly earned a reputation of poor processability. Early molded products were restricted to heavy-wall, short flow length, single-cavity sprue-gated pipe fittings that were often characterized by an objectionable appearance blush emanating from the gate and covering a side of the part. While the development of the first in-line reciprocating screw machines during the mid-1950s improved the processability of heat-sensitive materials, rigid vinyl molding was still considered a specialized operation limited to pipe and conduit fittings.

Early PVC injection-molding compounds were formulated using the high-molecular-weight resins available at that time. The first major PVC markets (gaskets, waterproof fabric, and wire insulation) grew out of a 1933 patent issued to Waldo Semon of the B. F. Goodrich Company. These flexible PVC materials were produced utilizing a unique characteristic of vinyl resin; i.e., although it is insoluble in many solvents and oils at room temperature, PVC becomes soluble in certain selected solvents at elevated temperatures. While this discovery was the origin of early flexible vinyl markets, it was found that higher strength, greater toughness (tear resistance), better chemical resistance, and higher service temperature capabilities could be achieved by utilizing high-molecular-weight resin, i.e., resin with over 900 repeating units in the polymer chain (K value 66 to 70).* In addition, early PVC resin provided the high porosity needed to absorb plasticizer.

During the 1950s, PVC found application in other major rigid markets. Pipe and building product markets developed and later pushed PVC to its current commodity status. These early rigid extrusions did not require absorption of plasticizer; thus, there was no need for high porosity. It was later found that a lower porosity, higher bulk density resin was desirable in promoting higher extrusion output rates from powder formulations. Since the ultimate in PVC physical properties and extrusion dimensional stability were also required in rigid building products and pipe applications, a high-molecular-weight, intermediate porosity resin was the order of the day.

Piping systems also required a family of assorted fittings for assembly. The compounds based on high-molecular-weight PVC resins had a tendency to shear burn at the higher shear rates associated with injection molding. They simply did not possess the processability, melt flow, and thermal stability required for injection molding into even the

---

*Estimate based upon number average molecular weight.

heavy-walled pipe fitting accessories. So, at a minimal sacrifice in
physical properties (particularly hoop stress), an intermediate molecular weight (700 repeating units in the polymer chain, K value 58 to
63) was developed. Through this initial moderate reduction in molecular weight, coupled with optimum compound formulations and the introduction of reciprocating screw injection-molding equipment, rigid
PVC heavy-walled pipe fittings have been successfully produced since
the mid-1950s. Melt flow and thermal stability improvements over-
shadowed a modest sacrifice in physical properties. Injection molding
of rigid PVC had progressed to the state of producing single- and
double-cavity, simple geometry, heavy wall, short flow length functional parts. Appearance aesthetics were still marginal.

During the late 1960s and early 1970s, designers and fabricators
in various custom injection-molded applications were attracted by the
excellent combination of physical properties, inherent flame retardance,
chemical resistance, and cost performance of rigid vinyl. Many inex-
perienced, first-time PVC molders attempted to injection mold specialty
parts from the rigid PVC pipe-fitting compounds, which were based
upon the intermediate-molecular-weight resins. The results were disastrous. PVC, an amorphous polymer with a broad melting point range,
was evaluated using inappropriate machinery often designed for high
melt flow, crystalline (sharp melting point) polymers like polyethylene
or polypropylene. Inappropriate screw design contributed to poor
melt temperature control. Forcing the viscous, shear-sensitive PVC
melt through restricted nozzles, runners, and gates into thin-walled,
multicavity molds taxed the material beyond its inherent thermal stability limitations. Processing a relatively cool melt at normal injection
speeds resulted in either short shots, or brittle parts due to high
levels of molded-in stress. Higher injection speeds caused localized
frictional and shear burning.

Unaware of the potential complications, unsuspecting molders (many
using rigid PVC for the first time) attempted to fill thin-walled parts
by raising stock temperature to reduce melt viscosity. As with earlier
attempts at molding higher molecular weight PVC compounds on plunger
machines, thermal degradation, noxious gassing, and disillusionment
resulted. Stories of these PVC burn-ups and equipment corrosion pre-
vailed in the injection-molding industry. A gradually subsiding stigma of poor processability remains attached to injection molding PVC
today.

Figure 6.1 demonstrates the improvements in melt flow that have
resulted from the recent developments in low-molecular-weight, high
melt flow, rigid PVC formulations. Note the significant flow length
difference, 24 in. (61 cm) versus 42.5 in. (108 cm), between an intermediate-molecular-weight PVC and a low-molecular-weight PVC, on a
standard spiral flow mold test.

During the mid-1970s, several PVC manufacturers recognized the
market potentials for a better processing, rigid PVC injection-molding

Figure 6.1 Spiral flow molding test measures the relative melt flow of materials. The low-molecular-weight, higher melt flow PVC resin demonstrates a dramatic improvement over a traditional intermediate-molecular-weight rigid PVC pipe-fitting formulation.

material: improved economics, flame retardance, chemical resistance, weatherability, and transparency. A new generation of low-molecular-weight (500 repeating units in the polymer chain, K value = 50) resulted.

Improved melt flow compounds based upon these low-molecular-weight PVC resins are available today and are targeted for more demanding application areas, where an excellent balance of rigid PVC properties is attractive.

A low-molecular-weight PVC resin may be combined with an apropriate additive package designed to enhance its properties even further. Dramatic improvements in processability have resulted. Higher melt flow rigid PVC compounds perform much more like other thermoplastics familiar to designers and custom molders. Rigid vinyl is now capable of being processed into the thinner walled, larger surface area, deep draw and intricately surface-detailed parts with lower internal molded-in stress. Surface appearance defects typical of earlier generations of PVC molding compounds have been largely eliminated. Rigid PVC parts can now compare favorably with general-purpose acrylonitrile-butadiene-styrene (ABS) in gloss and overall appearance.

Of greatest importance to the injection molder, the horrendous processability problems formerly associated with intermediate-molecular-weight injection molding rigid vinyl are stories of the past. Over the last several years, a growing number of custom molders have developed a confidence in the novel, less shear-sensitive, high-flow rigid PVC injection-molding compounds. A wider processing latitude allows the new generation of compounds to be molded comfortably and reliably in multicavity molds and molding machines designed to produce parts from ABS and other commercial injection-moldable thermoplastics. Combining a few simple hardware modifications (such as appropriate nozzle, sprue bushing, and gate sizes), with several appropriate, good molding practices for heat-sensitive materials (melt homogeneity and a low ratio of barrel capacity to shot volume), has reduced the apprehensions formerly associated with molding PVC. Equipment advances in screw design, as well as today's sophisticated solid-state temperature and closed-loop process controllers, contribute to making the job of molding PVC much easier.

In summary, a combination of marginal compounds, poorly defined processing techniques, and improper equipment resulted in the notorious processing problems of the early generation of rigid injection-molding vinyl. Recent improvements in formulations based upon low-molecular-weight PVC resin and use of molding techniques appropriate to heat-sensitive materials provide new confidence and reliability to custom injection molders. The fears of the past can be completely eliminated with the new generation of rigid PVC molding compounds. Designers can now consider the benefits of rigid PVC for demanding injection-molded part geometries that were unthinkable just a few years ago. Rigid PVC will become the polymer of choice in an ever-increasing number of demanding specialty injection-molding markets.

6.2 MARKET TRENDS AND OPPORTUNITIES

Vinyl has long been recognized as a very versatile material. Its capability of radical physical property transformations via a variety of formulation adjustments is well known. In spite of this great versatility, prior to the mid-1970s, rigid PVC injection-molding compounds were almost exclusively limited to two functional markets: general-purpose pipe and conduit fittings, and electrical outlet boxes (see Fig. 6.2).

Designers and manufacturers in both these markets were inclined to accept the marginal characteristics of the early rigid injection-molded vinyls. Poor processability, part or mold design constraints, molding productivity, and aesthetics were offset by a unique combination of attractive economics, excellent chemical resistance, inherent flame retardance, and good electrical properties. Manufacturers preferred the economic advantage and adjusted to the inadequacies. Rigid injection-molded PVC had found a small niche in the functional (non-appearance) plumbing and electrical construction markets.

Figure 6.2 General-purpose rigid PVC pipe fittings and electrical outlet boxes represent two functional construction markets of the early rigid PVC injection-molding compounds.

Through the last four decades, the versatility of PVC has enabled it to develop from an academic curiosity to the second largest thermoplastic in sales volume. Vinyl has penetrated a number of major market segments, including construction, communications, packaging, upholstery, and entertainment. However, on a volume basis, PVC sales have become largely dependent on traditional construction markets, which include pressure pipe, drain waste and vent pipe, sewer systems, electrical conduit, house siding, window frames, gutters, interior moldings, trim, flooring, wire and cable, and wall covering. Major PVC materials suppliers now recognize that a broad diversified market base is essential to building a strong stable business foundation. Market diversification of PVC away from dependence on construction-related markets toward new specialty areas has emerged as a popular strategy.

Largely due to its early processability problems, rigid PVC injection molding has, until recently, remained a very small segment of the
overall PVC business. And here again, even in the more specialized
injection-molding area, the influence of construction markets have been
apparent with pipe fittings and electrical outlet boxes.

This situation is rapidly changing. A new generation of versatile,
more thermally stable injection-moldable rigid PVC compounds, capable
of an excellent balance of physical, chemical, and electrical properties,
has evolved. Compounds based on specialty low-molecular-weight PVC
resins are enabling rigid PVC to penetrate vast new market areas.

6.2.1 Business Machines

As of late there has been a considerable amount of publicity related
to the blossoming of business machine, electronics, and small computer
markets. "Today's miniature micro-processors containing a tiny $3-$5
silicon chip enable a hand-held calculator to perform more tasks than
could an air-conditioned room full of hardware 25 years ago [11]. This
has put an enormous amount of computing power within the financial
reach of many users. Let it suffice to say that the business machine
markets—small computers, word-processing equipment, point of sale
equipment, information systems, copiers, and printers—are expected
to expand exponentially during the balance of the 1980s.

Most of this equipment must be housed in an aesthetically appealing,
tough, color-stable, chemically resistant, EMI-shielded, flame retar-
dant cabinet. Thermoplastics have dominated thermosets and metals
in these applications. Although the business machine market is growing
very rapidly, it is also very competitive. With participation of over
140 companies, costs are a primary consideration. No longer can one
afford the luxury of overengineering, i.e., specifying a high-priced
engineering thermoplastic when a more cost efficient material meets requirements. Although materials must satisfy strict performance criteria, cost performance is essential.

Rigid PVC injection-molding compounds currently enjoy a very significant cost advantage over most competitive materials in this market.
Of course, the attractive cost of rigid PVC would be of little interest
if it were unable to satisfy other physical criteria. Ergonomics (engineering to interface with humans) is a growing concern in these applications. General material requirements for this market are listed below.

Flammability: Underwriters Laboratory (UL) 94V-0 at 0.0625 in.
                (1.59 mm); UL 94-5V at 0.125 in. (3.17 mm).
Moldability: High melt flow required to fill larger, intricately de-
                tailed parts. *Note:* There is a trend toward down-
                sizing business machine equipment.

Shrinkage: Low and consistent shrinkage is needed to satisfy
        strict dimensional tolerances.
Appearance: Excellent.
Chemical resistance: Excellent, to resist staining and/or attack of
        common cleaning agents, for example.
Color stability: Current high area of interest. Well-defined speci-
        fications on tests simulating 3 years' indoor ultra-
        violet (UV) light exposure are anticipated.
Heat   A number of business machine manufacturers are
distortion   currently reevaluating the heat requirement for each
temperature: application. The advent of solid-state electronic
        components has reduced heat buildup considerably.
        Many components, even printers, have a maximum
        continuous use temperature of 140°F.
Toughness: High-impact materials with 6 ft lb/in. (3.2 J/cm) room
        temperature Izod impact.
Tensile strength: 6000 psi (41.3 MPa) minimum.
Flexural modulus: 350,000 psi (2411 MPa) minimum.

For the most part, the new rigid PVC is capable of satisfying these
criteria. Figure 6.3 demonstrates the use of a specialty formulated,
high-modulus, rigid PVC in a typewriter base application.

6.2.2 Appliances

A recent survey revealed that 14 different types of plastics are used
extensively in appliances [2]. Materials are selected by many combinations of requirements: strength, impact, heat resistance, appearance,
and feel. The same survey indicates that, of the products currently
being designed, 80% of the design changes will involve replacement of
a metal with a plastic.

Rigid PVC injection-molded parts are finding use in applications
requiring improved chemical resistance over metals, an example being
a garbage disposal housing, where die-cast aluminum has been attacked
and corroded. Since the new high melt flow rigid PVC is capable of
obtaining an 90°C, UL Recognized Component listing, it is also being
used in clotheswasher and drier and dishwasher control housing applications. Again, the rigid PVC must also have high impact resistance,
appearance gloss, UL 94V-0 flame retardance, chemical resistance,
and cost efficiency (see Fig. 6.4).

6.2.3 Electrical and Communications

Rigid PVC is an excellent electrical insulator. It does not absorb
significant amounts of moisture and retains good insulating properties
even under water. Transparency, inherent flame retardance, chemical

Figure 6.3 The typewriter base application demonstrates the use of
a new higher flow rigid PVC compound formulated to develop a combination of high modulus and toughness. (Courtesy of IBM, Lexington,
Kentucky.)

resistance, and toughness have enabled rigid PVC to be specified in
a demanding lead-acid battery cell that furnishes a backup, direct
current power supply to telephone systems in the event of power
failure (Fig. 6.5). A disposable transparent rigid PVC communications cable splice is filled with a urethane foam when used for temporary splices during military field manuevers (Fig. 6.5). A high-impact,
flame-retardant rigid injection-molded PVC qualifies for a communication terminal connector application (Fig. 6.6).

6.2.4 Consumer Home Entertainment

A new ultraclean, high melt flow, low-molecular-weight PVC homopolymer resin, similar to that used for injection-molding compounds, is the
base resin used in the injection-compression molded videodisc. The

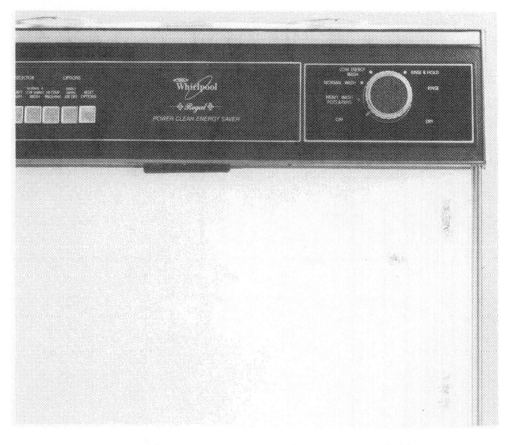

Figure 6.4 An appliance electrical control housing demonstrates the
use of a higher melt flow rigid PVC compound in an application requiring an 80°C UL Recognized Component listing. (Courtesy of Whirlpool
Corporation, Benton Harbor, Michigan.)

conductive carbon-filled compound requires extreme cleanliness, thermal stability during processing, and the toughness to withstand a multitude of repeat plays without damage to the disk (Fig. 6.7). Another
example of a consumer home entertainment application for injection-
molded PVC is a 19 in. (48 cm) television back. The application requires UL 94V-0 flame retardance, high impact resistance, high melt
flow, and appearance gloss (Fig. 6.8).

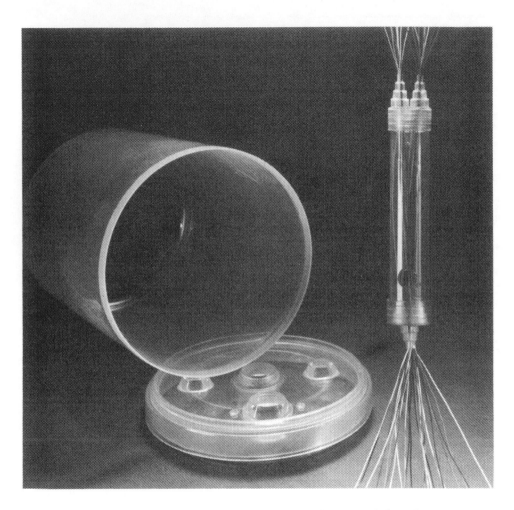

Figure 6.5 A transparent lead-acid battery jar and rigid PVC cable splice housing demonstrate the capabilities for use in communication applications.

Figure 6.6 A UL 94V-0 high-flow, high-impact rigid PVC can be used in telephone terminal connector applications.

Figure 6.7 An ultraclean version of a low-molecular-weight PVC homopolymer is utilized in the CED video disk. (Courtesy of RCA Corporation, Indianapolis, Indiana.)

6.2.5 Medicine and Health

PVC homopolymer is classified under an FDA Prior Sanction as GRAS
(generally recognized as safe). It is possible to formulate rigid PVC
injection-molding compounds that satisfy FDA guidelines for food-contact and medical applications.

6.2.6 Specialty Plumbing

Because of the excellent chemical resistance of PVC, which includes
resistance to the attack of toilet tank chemicals, rigid injection-molded
PVC has qualified for use in intricately designed toilet valves. Other
specialty plumbing applications include two-way and four-way flow
reversal valve applications, which require outdoor weatherability and
chemical resistance (Fig. 6.9).

6.2.7 Furniture

Furniture has been made from conventional PVC pipe formulations for
a number of years. Today, new attractive designs are available. Weatherable, UV-resistant, color-stable, high-impact rigid PVC injection-

Figure 6.8 An injection-molded rigid PVC television back examplifies good toughness, gloss, and inherent flame retardance, along with excellent processability. (Courtesy of N.A.P. Consumer Electronics Corporation, Knoxville, Kentucky.)

molding compounds are useful in furniture fittings and accessories.

6.2.8 Summary

These examples of current products demonstrate that rigid PVC injection-molding compounds now offer many opportunities to diversify PVC away from the dependence on the building and construction markets. Technological advances are continuing. Perhaps future copolymers, alloys, blends, or grafts, which combine the benefits of PVC with those

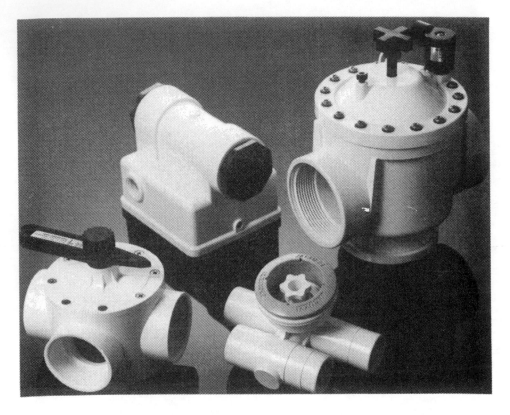

Figure 6.9 The assortment of specialty plumbing parts shown here require excellent chemical resistance. Several also require outdoor weatherability.

of other polymers, will one day be injection molded into even more demanding applications.

6.3 ECONOMICS

Economics have been a significant contributing factor to the slow development of rigid PVC in the United States. Rigid PVC gained a reputation among injection molders during the 1950s and 1960s for poor processability. During this same period, other thermoplastic materials were rapidly gaining wide acceptance in the injection-molding industry. General-purpose ABS, for example, was introduced to injection molding in 1950. During its early years as a physical blend of three polymers, ABS was also characterized as a very tough but difficult to process material with poor appearance and poor stability. However, in 1955, a graft polymerization breakthrough vastly improved its moldability, low-temperature impact, and appearance gloss. General-purpose ABS rapidly became recognized as the workhorse thermoplastic material for injection molding, bridging the gap between commodity resins like polystyrene and polyethylene and high-performance products like polycarbonate, nylon, and acetal.

Later, during the mid-1970s, rigid PVC made a similar breakthrough in moldability. However, by this time other materials with similar performance properties (i.e., GP-ABS) were already well established among designers and molders. Although rigid PVC was priced competitively with GP-ABS on a price-per-pound basis during the later 1970s, rigid PVC suffered from a 28% specific gravity disadvantage (1.04 versus 1.33).*

Designers and molders were simply not willing to evaluate rigid PVC, a material with physical properties comparable to those of ABS, because of a higher cost on a unit volume basis, and a wide reputation for poor processability (Fig. 6.10). Thus, although the significant technical breakthrough allowing rigid PVC to be injection molded into larger parts, with intricate appearance took place during the mid-1970s, growth in PVC injection molding was hampered because of a higher cost relative to that of ABS.

However, the technological breakthrough to a high melt flow rigid PVC during the mid 1970s was indeed timely. Because thermoplastics are products of petroleum chemistry, plastics pricing was dramatically affected by the world energy crisis of the 1970s. Although crude oil had become a critically important fuel, it served a dual role. Only approximately 4% of the crude oil consumed in the United States goes into petrochemicals and polymers, but energy costs have had a profound effect on thermoplastic supplies and prices in recent years (Fig. 6.11). Polymers that are 100% hydrocarbon-based have been and will continue to be vulnerable to escalating crude oil prices. PVC, while 57% inorganic chlorine by weight, is less dependent on hydrocarbon feedstock prices, and thus has been and should continue to be less affected by the inevitable price escalation of crude oil (Fig. 6.12).

An environmental factor has also recently had a major effect on the economics of thermoplastic injection-molding materials in the United States. The shift toward unleaded gasolines has placed a growing demand on the higher octane aromatic refinery products, benzene, toluene, and xylenes. Benzene is the basic building block for poly-

---

*When comparing the cost efficiency of any two injection-molding materials, it is essential to compare costs on a unit volume (dollars/cubic inch) basis rather than a weight basis (dollars per pound). The price in dollars per pound can be converted to cost in dollars per cubic inch by the following equation: $/lb X specific gravity X 0.03618 = $/in.^{3}.

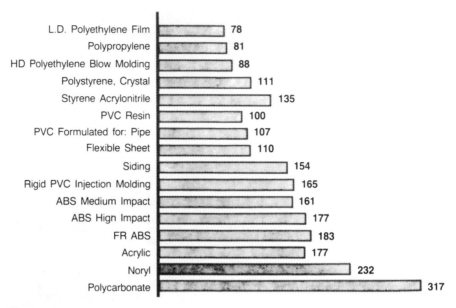

Figure 6.10 First quarter 1980 cost per cubic inch of rigid PVC injection molding resin relative to other injection-molding thermoplastic materials. (Courtesy BF Goodrich Chemical Group, Cleveland, Ohio.)

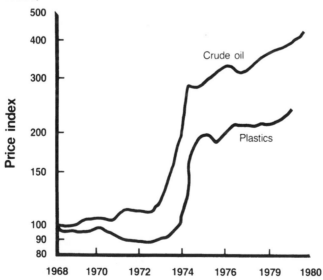

Figure 6.11 Thermoplastic material pricing has closely paralleled crude oil pricing during the energy crisis of the mid-1970s. (Courtesy Society of the Plastics Industry, Inc., New York, New York.)

	Type of Energy	
	Crude Oil/ Natural Gas	Electricity/Coal
Polyvinyl Chloride	69%	31%
Polyethylene		
High Density	88%	12%
Low Density	87%	13%
Polypropylene	93%	7%
Polystyrene	94%	6%
Acrylonitrile-Butadiene Styrene Polymer	92%	8%

Figure 6.12 A comparison of thermoplastics indicating the energy dependency on crude oil or natural gas versus coal. (Courtesy BF Goodrich Chemical Group, Cleveland, Ohio.)

styrene and ABS. Thus, the recent increase in demand for unleaded gasoline has had a dramatic effect on the price of benzene and consequently contributed to price escalation of polymers based on aromatic monomers (Fig. 6.13).

During early 1980, the U.S. market price of rigid PVC injection-molding materials, in spite of the specific gravity disadvantage, crossed under general-purpose ABS on a cost per unit volume basis. Continued aromatic hydrocarbon price escalation since 1980 has resulted in a significant 20 to 30% cost per unit volume advantage for rigid PVC relative to general-purpose ABS molding compounds. The more stable pricing of rigid PVC molding materials has been and will continue to be a driving force behind its growth (Fig. 6.14). Designers, specifiers, and molders are now willing to evaluate rigid PVC, and to their surprise, are finding that the new materials can be easily processed into demanding, value-added, appearance parts. Although thermoplastics as a whole have recently escalated in price due to the increased costs of hydrocarbon feedstocks, plastics still remain very energy efficient compared to the traditional die-cast and stamped metals they have replaced over the last several decades (Fig. 6.15).

Processing the new PVC injection-molding compounds also competes on a favorable economic basis with the established "workhorse" injection-molding material (ABS). At higher melt temperatures and mold temperatures, ABS develops outstanding surface gloss. However, these higher temperatures can result in a productivity loss due to extended cooling cycle time. On the other hand, rigid PVC can now be processed with gloss comparable to ABS at more moderate temperatures. This results in faster cycle times and higher output rates.

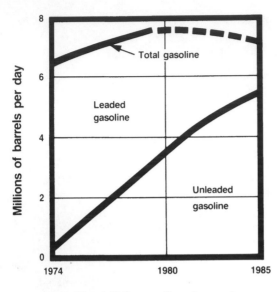

Total U.S. gasoline demand

Figure 6.13 Environmental concerns have caused the shift toward unleaded gasoline. Increased demand for aromatic hydrocarbon anti-
knock gasoline additives will likely result in higher prices of styrenic-
based thermoplastics. (Courtesy Society of the Plastics Industry,
New York, New York.)

   In summary, recent economic trends have placed rigid PVC injection-molding materials in attractive cost and pricing positions relative
to other long-established, 100% hydrocarbon-based thermoplastic
molding materials. It is anticipated that, in the long term, the hydrocarbon pricing trend will continue to favor rigid PVC (Fig. 6.16). In
addition, environmental pressures will likely result in increased demand
for aromatic hydrocarbons. This will enable rigid PVC to compete
even more favorably with styrenic-based thermoplastic injection-molding materials.

6.4 PROPERTIES AND CHARACTERISTICS OF CURRENT PRODUCTS

The new breed of low-molecular-weight, high melt flow compounds
evolved out of the earlier functional pipe-fitting compounds and the
need for improved processability for specialty applications. Logically,
one would expect that, in significantly reducing the polymer chain
length to improve processability, corresponding compromises would occur

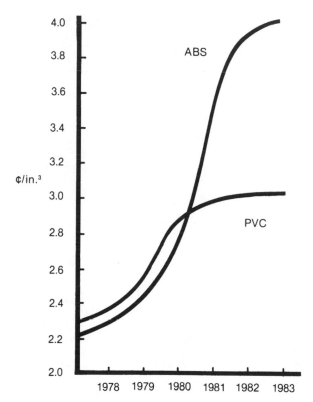

Figure 6.14 List price history of high-impact ABS versus rigid PVC compound on a cost per cubic inch basis. PVC became less expensive than ABS in 1980. Since then, the margin between the two materials has continually increased. (Courtesy BF Goodrich Chemical Group, Cleveland, Ohio.)

in physical properties, especially impact strength. All things being equal, this could be true. However, new state-of-the-art compounding technology is available that minimizes these effects. A comparison of typical physical property data is shown in Table 6.1 It is notable that, while achieving a dramatic 180% improvement in spiral flow length, the only significant trade-off is a modest 5 to 7°C heat distortion temperature.

While maintaining this excellent blend of physical properties listed in Table 6.1, the new generation of high melt flow rigid PVC compounds may also possess a number of the following general characteristics.

Table 6.1 Physical Properties of Pipe Fitting versus Custom Molding Rigid PVC

Property	Pipe fitting Type I, Grade I, normal impact	High flow, normal impact	Pipe fitting Type II, Grade I, high impact	High flow high impact
Specific gravity	1.41	1.36	1.35	1.33
Tensile strength, psi MPa	7,000 49	6,700 46	6,700 46	6,200 43
% Elongation	—	140	—	130
Flexural strength, psi MPa	13,000 90	12,900 89	13,400 92	12,000 83
Flexural modulus, psi MPa	— —	470,000 3,238	— —	410,000 2,825
Notched Izod impact strength (72°F) ft lb/in. (J/cm)	1.0 0.5	1.0 0.5	7-12 3.6-6.4	12 6.4
Heat deflection temperature (@ 264 psi), °C	73	68	75	68
Relative spiral flow, in. cm	24 61	47 119	24 61	43 109
Hardness, Rockwell R points	112	112	111	106

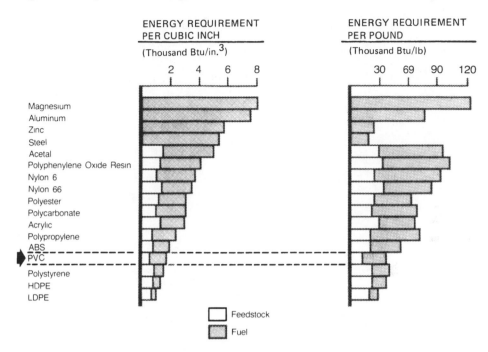

Figure 6.15 Relative energy requirements of various materials are compared. Of the plastic materials listed, PVC energy requirements are less influenced by feedstocks and plastic materials are generally less energy dependent than metals. (Courtesy B F Goodrich Chemical Group, Cleveland, Ohio.)

> Inherent flame retardance (UL 94V-0, UL 94-5V)
> Excellent chemical resistance
> Transparency, good clarity
> Good surface appearance (gloss)
> Outdoor weatherability
> Colorability in opaques
> Good processability
> Excellent electrical properties
> 90°C UL thermal-index
> NSF and USDA listings
> FDA sanction
> Low cost
> Good resistance to stress-cracking agents

Several of the above characteristics are inherent in almost any rigid PVC formulation, but those of any specific formulation should be obtained from the material supplier. For economic reasons, one may wish to emphasize only one particular characteristic. Many combinations of

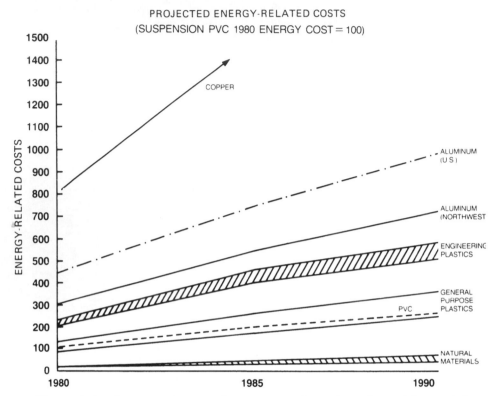

Figure 6.16 Because they are less energy dependent than either metals
or engineering thermoplastics, general-purpose plastics will likely
remain favored economically through the balance of the decade. (Courtesy BF Goodrich Chemical Group, Cleveland, Ohio.)

these physical properties and characteristics can be obtained in a single formulation. A few characteristics, however, cannot be combined
in the same single formulation (e.g., transparency and weatherability).

6.5 FORMULATIONS AND THEIR EFFECT ON MATERIAL PROPERTIES

The formulation of rigid PVC injection-molding compounds involves a
great deal of constantly evolving technology and could very well be
the subject of an entire book rather than, as presented here, a portion
of a chapter. Much of today's technology in this area is proprietary
to various suppliers, but considerable information is generally known
and available to provide the basis for a reasonably thorough discussion.

The purpose of this section is to show the wide range of properties that may be achieved by formulation of rigid PVC compounds for injection molding.

Pure polyvinyl chloride homopolymers and copolymers cannot be processed without the addition of other ingredients. After these ingredients have been mixed with PVC, either as a dry powder blend or fluxed to a molten state and cooled, the resultant mixture is a PVC compound. Rigid PVC powder molding compounds are used almost exclusively in the PVC pipe-fittings industry, which still represents the single largest market for injection-molded PVC. There has been a steady trend by large manufacturers of PVC fittings to reduce costs by mixing powder molding compounds in-house. Molders of PVC compounds in nonfitting applications typically are custom or captive molding operations that use a wide range of thermoplastics. They do not wish to get involved with the complexities of formulating PVC compounds or the housekeeping problems associated with handling powder compounds. Generally, it has been estimated that an annual usage of 10 to 15 million pounds of PVC compound is required to justify in-house compounding.

In order to discuss formulations and properties, we first need to review certain test methods used. This is followed by a review of the additives used to prepare rigid PVC compounds for injection molding. Stabilizers, lubricants, and process aids are necessary in all formulations. Other additives, such as impact modifiers, fillers, and pigments, are optional, depending upon the final properties desired. The purpose of each family of additives is described, together with characteristics, alternatives, effects on processing, and physical properties of the molding compound.

6.5.1 Testing

Thermoplastic formulations are usually characterized by tests that measure physical properties. Except where noted, physical properties discussed in this chapter have been obtained by test procedures according to the current ASTM Standards published by the American Society of Testing and Materials. A general ASTM Standard, D-1784, covers the testing of PVC compounds for selected properties detailed in other ASTM Standards.

Two properties are of prime importance when characterizing rigid PVC compounds for injection molding: melt flow and melt stability. Reference is frequently be made to these important characteristics, and time is devoted to how these may be tested. Later, in the section on processing, their relationship to molding characteristics becomes abundantly clear.

One of the most popular methods of characterizing PVC formulations, which has evolved in the industry, is the use of a torque rheometer, such as the Plasti-Corder from C. W. Brabender Instruments, Inc. The

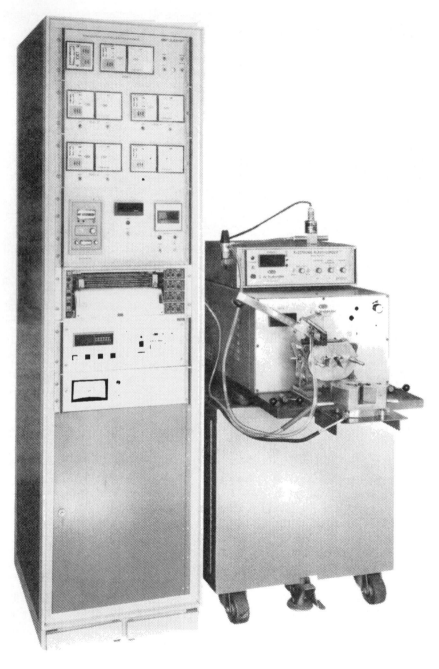

Figure 6.17 Typical torque rheometer apparatus. (Courtesy of C. W. Brabender Instruments, Inc., Hackensack, New Jersey.)

apparatus (Fig. 6.17) provides the user with information about the melt viscosity and melt stability of a PVC compound at any temperature within its processing range. These are critical properties for all formulations intended for injection molding. For the uninitiated, the apparatus can be run at a variety of conditions, selected by the user. As yet, no standard test exists for characterizing PVC molding compounds. Generally, the unit consists of a heated mixing chamber, which is controlled to a preselected temperature, containing two roller-type mixing elements, which are counterrotating. Figure 6.18 shows the disassembled unit. One of the mixing elements is driven directly by the drive motor and is rotated at a preselected speed. The second element is gear driven by the first element at a ratio of either 3:2 or 2:3. Torque required to maintain the preselected speed while mixing the material is continuously recorded as a measure of the melt viscosity of the material.

When the torque rheometer is used to evaluate the characteristics of a rigid PVC power compound, a curve much like that illustrated in Fig. 6.19 is obtained. The rise in torque to point 1 is associated with the ram compacting the powder compound in the chamber. The drop in torque to point 2 relates to the breakdown and melting of the resin grains and distribution of the lubricant. The rise in torque to point 3 represents fusion melting of the primary particles that compose the PVC grains. The drop in torque to point 4 is the reduction in melt viscosity of the compound as temperature equilibrium with the chamber is reached. The compound typically maintains this minimum torque until point 5, where cross-linking of the polymer, due to degradation, results in a sudden rise in torque. This point is often referred to as the point of dynamic thermal stability, and the time required to reach this point after fusion (point 3) is called the DTS time. Point 6 represents the termination of the test. A gradual upward slope of the curve from 4 to 5 indicates a gradual increase in cross-linking with heat history, whereas a gradual downward slope of the curve from 4 to 5 indicates a breaking of the polymer chains. Neither of these characteristics is desirable in molding compounds, since variability in processing results.

When a preprocessed rigid PVC compound, in the form of pellets or cubes, is tested in the torque rheometer, a curve more like that shown in Fig. 6.20 is obtained. The maximum torque is achieved almost immediately (dependent upon the conditions selected), and the stability or DTS time is measured from the beginning of the test. Various methods of interpreting torque rheometer curves are found throughout the PVC literature. It is important to choose one method to be consistent when comparing formulations.

The temperature of the melt should also be continuously recorded when using the torque rheometer. This is measured by a thermocouple mounted directly in the mixing chamber with the tip in the melt between

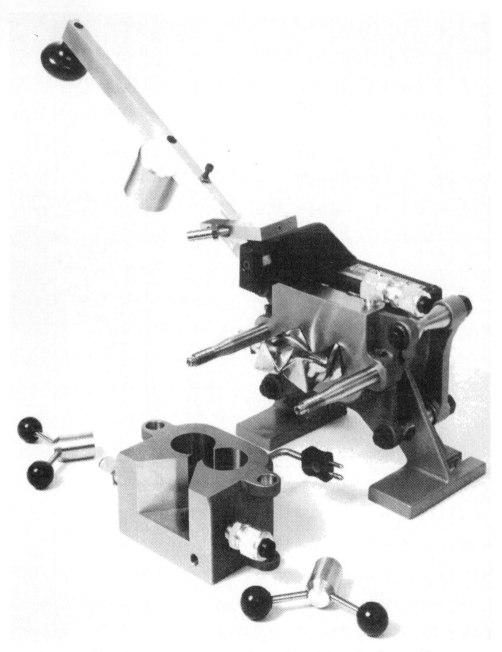

Figure 6.18 Disassembled torque rheometer mixing chamber. (Courtesy of C. W. Brabender Instruments, Inc., Hackensack, New Jersey.)

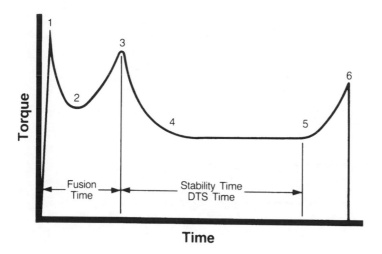

Figure 6.19 Typical torque rheometer plot of torque versus time for a rigid PVC powder compound for injection molding.

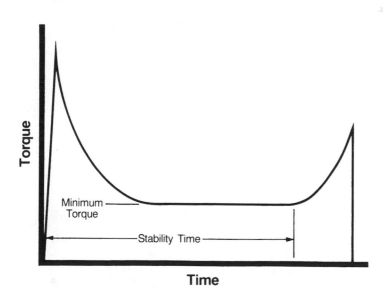

Figure 6.20 Typical torque rheometer plot of torque versus time for a rigid PVC pellet or cube compound for injection molding.

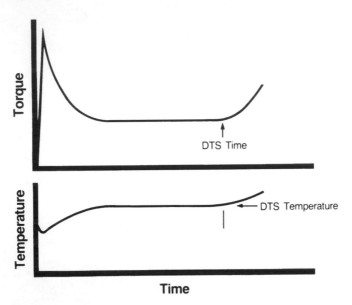

Figure 6.21 Typical plot of melt temperature in the torque rheometer
versus time with a companion curve of torque plotted versus time for
a rigid PVC cube compound for injection molding.

the two mixing elements. Figure 6.21 illustrates a typical temperature
curve with its companion torque curve. As the cool PVC compound is
added to the mixing chamber, a sudden reduction in temperature occurs. As mixing progresses, shear heating causes a rise in temperature until equilibrium is reached. Another rise occurs when cross-
linking starts or the DTS time is reached. Many compounds never
really level off, due either to a high material melt viscosity or relatively
cool test conditions.

    Figure 6.22 illustrates the changes in the torque curves when the
test temperature of the torque rheometer is varied. As the temperature rises from T to T1 and T2, the viscosity of the melt decreases, as
expected, but the stability time is also reduced. For a given PVC
compound, higher flow may be achieved with increased temperature
at the cost of a reduction in melt stability time. The thermal stability
of PVC formulations is time and temperature dependent. This relationship should always be kept in mind during processing by injection
molding.

    Although the torque rheometer has been used extensively in the de-
velopment and characterization of PVC compounds for many years,
other tests are used to measure melt flow and thermal stability. Measurements of melt flow using a capillary rheometer yield results that
correlate with torque rheometer tests and actual molding performance.

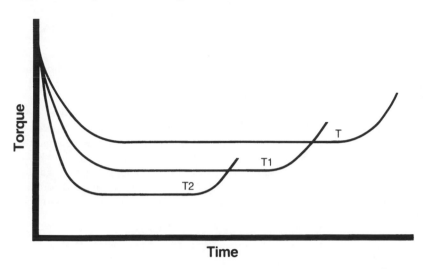

Figure 6.22 Typical torque curves obtained for a rigid PVC injection-molding cube compound when the temperature of the torque rheometer is raised from T to T1 and T2.

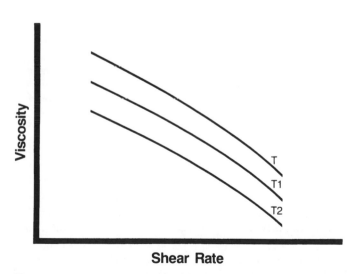

Figure 6.23 Typical plots of viscosity versus shear rate obtained from a capillary rheometer for a rigid PVC injection-molding compound as the temperature of the rheometer is raised from T to T1 and T2.

Testing the same PVC compound, as in Fig. 6.22, in the capillary rheometer at the same temperatures yields a family of curves for melt viscosity over a range of shear rates, as shown in Fig. 6.23.

In the capillary melt flow test, determination of the melt or thermal stability of a compound is more subjective since it depends upon visual changes in the color and uniformity of the melt extrudate. At some critical shear rate and temperature combination, the melt exhibits severe discoloration in the center of the extrudate, often accompanied by surface roughness or fracture. Using the same capillary rheometer, the observer may choose to maintain a preselected shear rate and introduce time delays between extrudate samples. In this case, a value for the relative thermal stability time is the residence time or dwell time, which results in a sudden change in color or physical appearance of the melt extrudate.

This section on testing has focused on the two critical process related properties, melt flow and thermal stability, for characterizing all rigid PVC compounds for injection molding. Other tests for characterizing performance-related properties (e.g., tensile strength, heat deflection temperature, and impact) are used as for other thermoplastics.

6.5.2 PVC Resins: The Starting Component

All rigid PVC compounds start with the polymerized polyvinyl chloride homopolymer or copolymer resin as the basic component. Polymerization processes that produce these resins include suspension, mass, emulsion, or solution polymerization. Injection-molding compounds use the resin products of either the suspension or mass polymerization process. Resins from both processes are in the form of a free-flowing white powder or grains normally about 80 to 150 μm in diameter. Examination of these PVC resins grains under high magnification shows each of the grains is actually an agglomerate of primary particles approximately 1 μm in diameter [3]. The resin grains obtained from the suspension and mass polymerization processes differ principally in the structure of the grain surface. However, these differences have little effect on injection-molding compounds and are not discussed here.

Primary particles 1 μm in diameter are considered the smallest (primary) flow units for rigid PVC when it is processed at low melt temperatures. Models have been described for the structure within the primary particles [4]. This microdomain structure is believed to consist of regions of amorphous PVC with infrequent small domains of more ordered crystalline structures forming a three-dimensional network throughout the primary particle. Processing PVC at higher melt temperatures, as in injection molding, breaks down this microdomain structure by melting most, if not all, of the crystallites. Re-formation of the crystalline structures occurs on cooling but is less perfect than the original structure due to the quenching effect of the cooling cycle

in most commercial molding processes. Copolymers of PVC have far
less crystallinity, since the molecules are far less ordered and, therefore, are less likely to form regular crystalline regions. The crystalline structure of homopolymer PVC provides unique properties, including toughness.

PVC resins, by themselves, are practically worthless as thermoplastics. They are very susceptible to degradation by dehydrochlorination
when exposed to elevated temperature. A PVC resin experiences a
significant color shift when pressed into a film at 374°F (190°C) for
times as short as 30 sec. Higher temperatures or longer exposure time
increases the color change and the amount of degradation. The degradation mechanism of polyvinyl chloride is simply the stripping off of
adjacent chlorine and hydrogen atoms from the chain to form hydrogen
chloride.

$$
\begin{bmatrix}
& H & H & H & H & \\
& | & | & | & | & \\
- & C & - C & - C & - C & - \\
& | & | & | & | & \\
& H & Cl & H & Cl &
\end{bmatrix}
\implies
\begin{bmatrix}
& H & H & H & H & \\
& | & | & | & | & \\
- & C & = C & - C & - C & - \\
& | & & | & | & \\
& H & & H & Cl &
\end{bmatrix}
+ \; HCl
$$

The released hydrogen chloride molecules catalyze further degradation of the PVC polymer. In the atmosphere, hydrogen chloride vapor
combines with moisture to form corrosive hydrochloric acid. This chemistry is the source of the processing problems encountered with this
polymer. We discuss later how this reaction can be minimized by the
use of thermal stabilizers and by control of the process.

The primary characteristic of a PVC resin to be used in a compound
for injection molding is the molecular weight of the PVC polymer. This
characteristic is often denoted as the resin's K value, a relative measurement of viscosity. Another measurement of molecular weight is
inherent viscosity (IV) as measured by dilute solution viscosity.

Higher melt flow can be obtained by reducing the molecular weight
of the PVC polymer or by copolymerizing it with another monomer.
Two copolymer systems that have found some acceptance are PVC-
vinyl acetate (VA) and PVC-propylene (PP). These copolymers do
not provide all the characteristics desired—the PVC-VA copolymers
have less thermal stability and lower heat deflection; PVC-PP copolymers have even lower heat deflection temperatures. The need for a
higher melt flow homopolymer PVC resin led several manufacturers to
undertake development of resins with K value 50 (IV 0.52) for injection-molding compounds.

Considerable technology has been developed to manufacture low-
molecular-weight PVC resins. Special compounding technology is also
required to formulate them into useful thermoplastic compounds for injection molding.

6.5.3 Heat Stabilizers

Heat stabilizers must be added to all rigid PVC formulations in order
to inhibit the thermal decomposition of the polymer during processing.
Injection-molding compounds are no exception; in fact, they are even
more critical, due to the high temperatures and shear rates involved.
It has been demonstrated that processing PVC compounds depletes the
ultimate stability of the compound based upon the heat history (time
and temperature) of the process [5]. The selection of the stabilizer
type and the amount depends upon the process intended as well as the
effect of the stabilizer itself upon the other properties of the composition.

Hundreds of heat stabilizers may be used with PVC. Most of these
have been organized into the following families by McMurrer [6].

Lead stabilizers and multifunctionals
Barium-cadmium, calcium-zinc, and other stabilizers
Tin stabilizers and multifunctionals
Antimony stabilizers

Today, most rigid PVC compounds intended for injection molding utilize either tin or lead stabilizers. The other systems have certain de-
ficiencies and are only used in very special cases.

Tin stabilizers comprise a wide varity of chemical compositions,
generally known as organotins. They are alkylated tin derivatives with
various organic groups. Three alkyl groups are common—methyl,
butyl, and octyl. The organic groups may be either carboxylates, such
as maleates and laurates, or (sulfur-containing) mercaptides. Generally,
the tin mercaptides find the widest use in compounds for injection
molding with the selection of either the methyl, butyl, or octyl version,
depending upon the desired properties. All three tin mercaptide types
find utility in transparent applications with octyltin stabilizers sanctioned to meet FDA requirements.

Organotin stabilizers are expensive ingredients, ranging in cost up
to $10.00 per pound. Therefore, the type and amount used in a form-
ulation must be balanced with the economic limitations of the ultimate
application. Generally, the cost of an organotin stabilizer is related to
the tin content and the complexity of the stabilizer structure. Thus,
a simple methyltin stabilizer with 15% tin content could be considerably
cheaper than a complex octyltin stabilizer with maybe a 20% tin content.
Usually, there is a fairly direct relationship between stabilizer efficiency and tin content: You get what you pay for. The effect of
stabilizer upon a compound's stability can be shown by the torque rheometer test described earlier. As the level of an organotin stabilizer is
increased in a formulation, the ultimate thermal stability, or DTS, increases fairly proportionally, as shown in Fig. 6.24.

When a PVC formulation is prepared for melt processing into cubes
or pellets for later injection molding, the type and amount of stabilizer

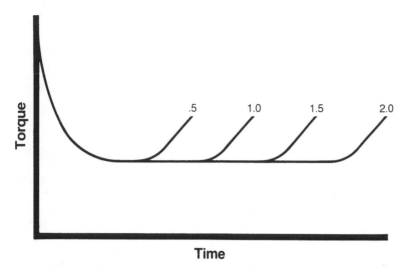

Figure 6.24 Typical torque rheometer curves showing the proportional increase in the dynamic thermal stability (DTS) time as the level of organotin stabilizer is increased from 0.5 to 2.0 parts in a rigid PVC injection-molding compound.

used must take into account the total heat history, including the com- pounding step, the compound will experience. The formulation must incorporate enough stabilizer to provide for the range of heat histor- ies to which it may be subjected by the wide variety of molding machines in common use. At the same time, the amount must be economically balanced, assuming good work practices by the molder.

PVC powder compounds for the manufacture of pipe fittings usually contain much lower levels of stabilizer for two reasons: (1) theoreti- cally, the elimination of the melt-compounding step reduces heat his- tory, and (2) the severe cost restrictions of this business. Recently, there has been considerable activity among stabilizer suppliers to pro- vide lower cost, multifunctional, "one-pack" systems for the PVC com- pounder-molder. These combination lubricant-stabilizer systems re- portedly provide improved cost performance. Although these systems do provide the custom or captive compounder greater simplicity of formulation, the trade-off may be less control over product properties and processability.

Because of their desirable electrical properties, lead stabilizers have been used in flexible PVC wire and cable formulations since these products were first made. Other uses include audio record compounds and opaque flexible vinyl film and sheet. In the United States, lead stabilizers have not been used for pipe and fittings, due to NSF regu- lations prohibiting lead in PVC for potable water applications. Rather

than processing lead-stabilized formulations for nonpotable water applications in the same plant as potable water applications, fittings producers evidently chose to avoid lead altogether. Other parts of the
world use lead stabilizers much more extensively because they impart
high thermal stability to PVC, and their cost is lower than that of tin
stabilizers. Other factors that should be considered are: (1) typical
lead loadings are two to three times higher than tin loadings, (2) the
resultant formulation has a higher specific gravity (or higher cost per
cubic inch), (3) lead-stabilized compounds generally do not flow quite
as well as tin-stabilized compounds at the same temperature and pressure, and (4) lead stabilizers cannot be used to achieve transparent
formulations, since they opacify the compound.

Figure 6.25 compares the torque rheometer curves for a rigid PVC
formulation with tin and lead stabilization. The composition containing
lead stabilizer has significantly more stability at slightly higher torque.
Generally, if the torque levels are nearly equal, a check of the melt
temperature will likely show that the lead-stabilized compound has
shear heated to a slightly higher melt temperature, which results in a
lower apparent viscosity.

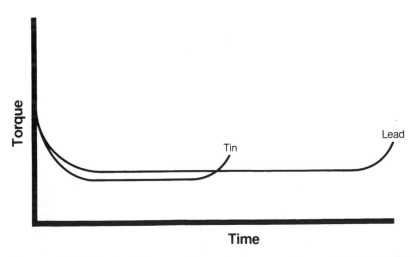

Figure 6.25 Typical torque rheometer curves for a rigid PVC formulation for injection molding with either a tin or a lead stabilization system.

Although lead stabilizers are not used for pipe fittings, several commercially successful PVC compounds for custom molding applications use lead stabilizers. Alternate running of lead- and tin-stabilized compounds in the same equipment should be accompanied by a complete purging of the machine to avoid discoloration or staining.

One final issue that should be addressed in the discussion of lead stabilized compounds is that environmental precautions should be taken by any compounder who wishes to produce lead-stabilized formulations. These materials can present an environmental problem in the workplace with respect to worker exposure to lead. After December 1983, OSHA requires the lead time-weighted average exposure limit to be 50 $\mu g/m^3$ of air. Suppliers of lead stabilizers now offer nondusting grades, special packaging, and material-handling equipment and techniques to meet this regulation.

6.5.4 Lubricants

Lubricants, like heat stabilizers, are necessary additives for the preparation of rigid PVC compounds. Thousands of pages concerning the function and selection of lubricants for PVC may be found in the literature. Lubricants are added to PVC to achieve better control over two important melt characteristics—melt viscosity and metal release. The complexity of using lubricants in PVC compounds arises since these additives influence the properties of the final rigid PVC product. Achieving a properly balanced lubrication system for processing and final properties can be a very long and involved effort.

As a matter of convenience, lubricants have been generally classified as either "internal" or "external," relative to their function in the PVC formulation. Internal lubricants are generally those that demonstrate compatibility with or solubility in the molten PVC polymer, whereas external lubricants typically are less soluble in the molten PVC polymer and function by providing release at the metal surface in processing machinery. Other additives, such as process aids, impact modifiers, and fillers used in PVC compounds, may change the effect of the lubricants.

The solubility or compatibility of lubricants in PVC is related to the chemical structure of the lubricant. Lubricants that are polar in nature are much more soluble in the polar PVC melt than nonpolar lubricants; the chlorine atoms attached to the PVC chain give it a higher polarity than a polymer such as polyethylene. Lubricants that have both polar and nonpolar functional regions in their structure have a dual internal and external lubricating capability. Large amounts of internal lubricants can demonstrate external lubricating characteristics as the polymer network becomes saturated with the internal lubricant. Thus, more is not necessarily better.

McMurrer [7] has categorized lubricants as follows.

Function	Group	Example
Internal	Fatty acid esters	Glycerol monostearate
	Fatty alcohols	Stearyl alcohol
	Fatty acids	Stearic acid
	Fatty acid amines	Ethylene bis-stearamide
	Metallic soaps	Calcium stearate
External	Paraffins, hydrocarbons	Paraffin waxes, oils
	Polyethylenes	Low-molecular-weight polyethylene waxes

This reference also contains the names of commercial products, technical information, and suppliers for several lubricants in each of the categories. Another source of information is the *Encyclopedia of PVC* [8].

When choosing lubricants for a rigid PVC injection-molding formulation, the compound should be carefully tested to ensure that the effects of the lubricant on properties and processing are understood. The torque rheometer and capillary rheometer tests mentioned earlier characterize the influence of the lubricant upon melt fusion, melt flow or viscosity, and melt stability of the formulation. Such properties as impact, tensile, and flexural strengths, modulus, and heat deflection temperature also need to be evaluated. The selection of lubricants can greatly influence special properties, such as light transmission, haze, and color of transparent compounds. Lubricants may even affect the aging and chemical resistance of rigid PVC compounds. Proper selection of external lubricants can play an important role in the heat stability of a compound during the molding process. A compound that does not have appropriate metal release characteristics can hang up in the injection unit and lead to localized melt degradation, which in turn can give color shift, streaking, or specks in the final molded part.

6.5.5 Processing Aids

An almost classic laboratory test or demonstration has been used many times to show the reason for using processing aids in PVC formulations. When a PVC resin containing both stabilizer and lubricant is brought to a melt state on a heated two-roll mill, the resultant thermoplastic melt is typically observed as having:

Dry and crumbly appearance
Little or no hot strength
Lack of adhesion to the hot metal surface

If a sufficient quantity of a processing aid is added, the melt becomes much more fluid in consistency, develops considerable hot
strength, and bands on the metal mill-roll surface. These characteristics are essential both in melt compounding and later for the uniform
remelting of the formulation in the injection machine. Basically, the
processing aid promotes homogeneous fusion at a lower temperature.

Typical processing aids for rigid PVC formulations include acrylic
polymers and copolymers, acrylic-PVC graft polymers, styrene-acrylonitrile copolymers, and α-methylstyrene polymers. Generally, higher
molecular weight versions of these polymers have found the widest acceptance for PVC compounds. The use of higher molecular weight process aids in lower molecular weight PVC resin-based formulations for injection molding raises some complex issues. The processing aid is needed
to achieve proper melt characteristics but tends to raise the viscosity of
the final melt, which in turn can lead to reduced melt stability. Today, several lower viscosity processing aids are available; although
they may contribute to better processing flow, they may be detrimental
to impact properties. The selection of the appropriate processing aid
is very dependent upon the requirements of the final application.
Acrylic process aids are much more suitable for color and clarity in
transparent products. Subtle effects, such as weld line performance
and gate blush, may depend upon process aid selection. Finally, the
type and amount of other additives may alter the characteristics of
the process aid in the compound. A detailed evaluation of processing
aids to ensure optimum product properties is essential.

6.5.6 Impact Modifiers

Many applications for injection-molded thermoplastics require resistance
to fracture when struck by another object or dropped onto a hard surface. This fracture resistance is commonly characterized as impact
strength, and tests, such as notched Izod impact and falling dart impact, are used to compare materials.

Although a rigid PVC compound with the proper stabilizer, lubricants, and process aid exhibits greater ductility than glassy polymers,
such as styrene, styrene-acrylonitrile (SAN), or polymethylmethacry-
late (PMMA), an elastomeric component is often used to achieve even
higher impact strength. Copolymerization is one approach, but the
most common method is to incorporate additives called impact modifiers
in the compound formulation.

It is important to understand the impact modification of PVC. The
key to achieving the required impact in the PVC compound is understanding how the modified PVC goes through what is called the brittle-

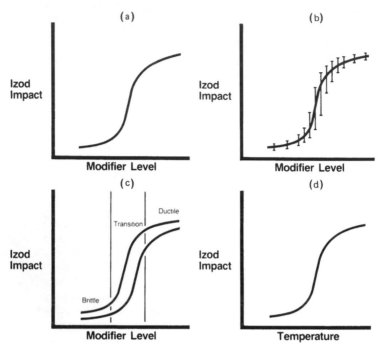

Figure 6.26 Typical plots of Izod impact versus modifier level and test temperature for rigid PVC injection-molding compounds, (a) classic S-shaped curve for average Izod impact versus modifier level when tested at a given temperature, (b) Ranges of individual Izod impact data points with considerable deviation in the steep, central portion of the curve, (c) The brittle-ductile transition region for Izod impact versus modifier level, (d) The typical S-shaped curve obtained when a rigid PVC injection-molding compound with a fixed level of modifier is tested at various temperatures.

ductile transition. A series of rigid PVC compounds with increasing levels of impact modifier yield notched Izod impact values that, when plotted against modifier level, yield the classic S-shaped curve shown in Fig. 6.26a. An understanding of this phenomenon is very important. Each point on the curve represents the average impact values obtained by testing several specimens. If one were to look at enough individual data points, the observer would see the steep central portion of the curve represents a region where considerable deviation in the test results occurs, as in Fig. 6.26b. The graphic representation of impact modification might be better represented by the regions in Fig. 6.26c. To avoid brittle fracture under these test conditions, a compound must have sufficient impact modifier to keep the test values in the ductile region, or upper portion of the curve.

Changes in Izod impact performance as a function of the level of impact modifier in the PVC compound have been shown at one test temperature (Fig. 6.26a,b). If a compound with a fixed amount of modifier is tested over a range of temperatures, the S-shaped curve appears again, as in Fig. 6.26d.

Other variables, such as the PVC molecular weight, modifier type or efficiency, selection of other additives, and sample preparation affect the impact performance of a rigid PVC compound. Therefore, a very general curve of impact strength versus temperature may be drawn (Fig. 6.27) to show that brittle-ductile transition curve moves up or down the temperature scale depending upon many factors.

Testing for impact performance of a rigid PVC compound must be done under conditions that closely simulate the end-use conditions of the product. Defining these test conditions is often difficult and time-consuming. Standard (ASTM) conditions are useful for ranking mater-

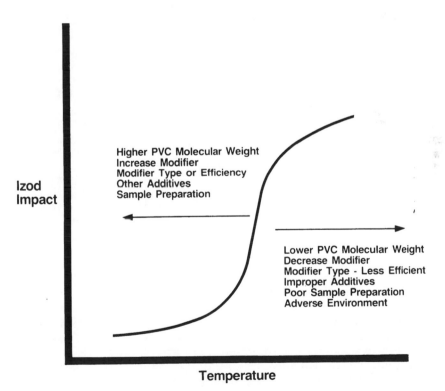

Figure 6.27 General curve for Izod impact versus test temperature for rigid PVC injection-molding compounds showing the factors that can cause the S-shaped curve to move up or down the temperature scale.

ials but only at the test conditions specified. Other conditions, perhaps closer to end-use conditions, may rank the same materials differently.

ABS graft copolymers have been used as PVC impact modifiers for many years. They are normally free-flowing powders of relatively small particle size for good dispersion during mixing and contain a high percentage of polybutadiene compared to typical ABS compounds for molding or extrusion. Rigid PVC compounds using ABS modifiers vary considerably in transparency, depending upon how close the refractive index matches that of PVC. Most ABS-modified PVC compounds are opaque. The amount of ABS modifier needed to achieve high impact with a lower molecular weight PVC resin is dependent upon the efficiency of the modifier, which in turn is dependent upon the structure of the complex multiphase ABS graft copolymer. ABS modifiers may be used anywhere from 5 to 20 parts per hundred resin (phr), depending upon the degree of impact resistance required. Increasing levels of ABS modifiers in lower molecular weight PVC tends to increase the compound's melt viscosity, which in turn affects the overall melt stability. These modifiers are only used in PVC compounds intended for interior applications, due to degradation of the polybutadiene by the UV component of solar radiation.

MBS (methacrylate-butadiene-styrene) and *MABS* (methacrylate-acrylonitrile-butadiene-styrene) *graft copolymers* have similar characteristics to ABS graft copolymers as impact modifiers for PVC, with the exception of optical properties. MBS and MABS modifiers are formulted to have refractive index values at room temperature almost identical to those of PVC. These modifiers yield, with other appropriate additives, highly transparent rigid PVC compounds. Their natural color exhibits some yellowness, which may be balanced to a more satisfactory tint by pigments and dyes. Since these high-impact transparent rigid PVC compounds are complex systems incorporating several additives and polymeric phases, they do not have the water-like clarity seen in crystal styrene or acrylic. They tend to be more translucent in thicker sections. Many applications can compromise some clarity to obtain impact, chemical resistance, or other desirable properties. Generally, the most ductile modifiers tend to have the least clarity, and careful evaluation is needed to optimize an impact-clarity balance.

Acrylic and modified acrylic impact modifiers are preferred for opaque rigid PVC formulations intended for exterior applications, due to their resistance to UV degradation. Selection of the modifier and of the amount should be based on outdoor aging studies.

CPE (chlorinated polyethylene) *modifiers* also have good weathering characteristics, as well as good retention of properties upon heat aging. They yield opaque compounds.

EVA, (ethylene vinyl acetate) *copolymers* are used to some extent as modifiers for rigid PVC where high melt flow is important.

The selection of an appropriate impact modifier for a rigid PVC injection-molding compound is complex. Since this additive is often
used at the highest level, compared with other additives, it usually
has the greatest effect upon properties. Generally, as the level of
modifier increases, the compound's specific gravity, tensile and flexural properties, and hardness all decline. With high amounts of impact
modifier, the combustion characteristics and chemical resistance of a
PVC compound also decline.

6.5.7 Fillers

Readily available low-cost inorganic materials are commonly added to
PVC compounds, primarily to reduce costs, and also to obtain certain
properties. These fillers comprise a wide range of compounds, including alumina trihydrate, calcium carbonate, talc, mica, clay, asbestos,
diatomaceous earth, certain sulfates, glass, and various flours. Of
these, the PVC industry has found calcium carbonate to be particularly
suitable for many applications, with clay, talc, and mica also finding
uses.

Compounds for pipe fittings utilize calcium carbonate primarily for
cost reduction, whereas higher flow PVC compounds use it to achieve
specific properties as well.

Calcium carbonate is available in a variety of forms, from ground
limestone to very pure precipitated grades. The average particle size
ranges form 0.07 to over 50 μm, depending upon the manufacturing
process. Many grades have been chemically treated to improve the
surface characteristics for both handling and processing. The cost
of a particular calcium carbonate grade is dependent upon the source
and treatment, with a general rule that finer grades and treated grades
cost more. Very fine particle size treated precipitated calcium carbonate is considered by some [9] an impact modifier for PVC since it increases ductility as measured by the Izod impact test. This can allow
reduction of polymeric impact modifiers in opaque formulations. Generally, the smaller particle size grades do not adversely affect flow
and seem to aid melt stability, probably acting as an HCl absorber.

The economics of using a filler for cost reduction should always be
looked at closely. The formulation costs of compounds based on high-
priced resins is reduced when fillers are used, but this is not necessarily true for PVC because of the lower cost of the resin.

6.5.8 Colorants

Rigid PVC molding compounds for interior applications can be colored
or tinted to almost any color by incorporating the appropriate pigments
or dyes. Sulfur-containing pigments should be avoided with lead-stabilized compounds, and lead-containing pigments should be avoided with
sulfur-containing tin stabilizers to prevent lead sulfide staining. Nat-

ural PVC compounds may be colored by the molder provided the color-
ant is compatible with PVC. Many color concentrates are based on
polyethylene and polypropylene, and these "carriers" tend to act as
lubricants to such an extent as to cause delamination of molded parts,
and, therefore, should not be used. Color concentrates based on PVC
carriers can have some dispersion problems if the carrier PVC resin
has a higher molecular weight than the PVC in the compound.

For exterior applications, the coloring of rigid PVC with good weathering characteristics requires a much more sophisticated approach.
Titanium dioxide (TiO_2) has been shown to significantly enhance the
weatherability of rigid PVC formulations, and typical exterior formulations usually contain greater than 10 phr TiO_2. The particular TiO_2
type selected, either anatase, rutile, or nonchalking rutile, depends
on many factors, such as desired color, chalking requirements, color
fastness, and even dispersibility. High levels of TiO_2 alone make the
manufacture of dark exterior colors very difficult. Dark colors of
rigid PVC face another problem when used outdoors. If the parts are
exposed directly to sunlight, they build up heat from solar radiation,
depending upon their color and the colorants used. If incorrect pigments are used, dark colors can absorb enough radiation to raise the
temperature of the part enough to cause significant changes in dimensions. Recently developed pigments allow many dark colors to be formulated with much less heat buildup due to solar radiation.

6.6 PRODUCT DESIGN AND MOLD CONSTRUCTION

The ability to produce large quantities of attractive complex, functional,
low-cost parts from various thermoplastics by the injection-molding
process has forever altered our daily lives. One only need to look
around any setting, whether it be home, office, factory, or while traveling, to see the abundant use of injection-molded plastics in our
society. Much of the enhanced life-style currently enjoyed would not
have been possible had not injection-molded plastics become common-
place.

Successful use of a plastic injection-molded part in any application
depends upon correct design and careful development. An ideal design
and development program consists of

1. Conception of the application idea
2. Definition of functional end-use requirements
3. Preliminary product design with desired aesthetics
4. Adaptation of the preliminary design to incorporate features
   needed to enable the part to be injection molded.
5. Selection of a plastic material to satisfy end-use requirements
6. Modification of the product design to allow for the characteristics
   of the chosen plastic material.

7. Prototyping in the chosen plastic (if possible) and testing the part
8. Preliminary mold design
9. Modification of the mold design to incorporate features needed to incorporate features needed for the selected plastic
10. Building the mold
11. Preproduction molding trial
12. Testing of parts to confirm performance
13. Commercial production

Obviously, a discussion of all the steps in a product design program goes well beyond the scope of this section. A recent publication, *Plastic Product Design*, by Beck [10], reviews most of the good design practices for injection-molding thermoplastics. Numerous other literature sources are also available, including the bimonthly publication, *Plastics Design Forum* (Industry Media, Inc., Denver, Colorado).

6.6.1 Product Design

Although most of the product design concepts in the literature are valid for rigid PVC injection-molding compounds, certain design features should be emphasized.

To take advantage of the excellent properties of rigid PVC, the designer must consider the influence of the molding process itself on achieving optimum part performance properties. The two most essential points to remember when designing a part to be injection molded from rigid PVC are as follows.

1. The melt is fairly viscous at all practical molding temperatures and, consequently, is prone to shear heating when injected through a narrow section too fast.
2. The melt degrades rapidly above 205°C (400°F).

Although the higher melt flow formulations of today process much easier than did earlier rigid PVC materials, these new compounds still require moderate heat history and injection velocities. A product design appropriate for rigid PVC must allow ample unrestricted flow to all areas of the part without using excessive temperature or extreme injection velocity. Part design must enable sufficient pressure to be maintained on the molten plastic during the mold-filling portion of the cycle to achieve uniform packing of the part. Excessive packing in any region creates areas of high localized frozen-in stresses that diminish properties. Insufficient melt pressure at extremities of the flow path result in poor appearance, weak weld lines, and possible reduction of impact strength.

Rigid PVC is primarily amorphous in flow behavior, so it does not have a temperature at which low viscosity is immediately achieved. It gradually softens and develops lower viscosity (higher melt flow) as the temperature is progressively raised. Since rigid PVC compounds achieve their ultimate properties at their optimum processing temperature of 390 to 400°F, the product designer can be assured of good performance when the design allows the part to be filled properly with the melt temperature in this range.

The following items should be considered by the designer of any product to be molded from a rigid PVC compound.

Wall Thickness

The overall dimensions of a part (length, width, and height) are usually dictated by the function of the application. However, most applications have considerable latitude in the wall thickness that may be incorporated.

Over the last few years, there has been a general trend to reduce wall thicknesses to counter the rapid increases in thermoplastic material prices and to reduce the cooling segment of the molding cycle. This trend has, in many instances, resulting in going from parts with uniform wall thickness that can be easily molded, to parts with nonuniform or thin-walled thickness in which the mold can barely be filled. This yields parts that lack the desired performance characteristics. The excellent economics of rigid PVC compounds relative to many other thermoplastics allow an ample wall thickness to be used to ensure proper melt flow and consistent optimum part properties.

As the flow path length increases, the wall thickness should be increased to prevent a serious pressure drop in the melt. The new higher flow rigid PVC compounds do not require the very thick walls once associated with the earlier PVC compounds, but attention still should be paid to this important design dimension. Typical wall thicknesses for center sprue-gated parts with weights in the range 2 to 5 lb are usually from 0.125 to 0.150 in. Parts with weights between 10 and 15 lb are being molded with wall thicknesses ranging from 0.180 to 0.225 in. Thinner walled parts requiring considerable flow length are still difficult for rigid PVC compounds, since higher than desired injection velocities are required.

Flow Restrictions

Every possible effort should be made by the designer to minimize flow restrictions in the melt path. There is considerable latitude in the use of slots, grill sections, and holes, for example, but care should be taken to ensure that ample melt flow through these areas is maintained. Transitions from thicker sections to thin should follow good design practices. A thick section fed from a very thin section should be avoided due to the pressure drop in the thin section. Rib running across the flow path reduce the available pressure within the melt and adversely affect mold filling. Ribs running in the flow direction actually help channel the melt by acting like a runner. To promote flow, every effort should be made to provide an ample radius wherever the polymer melt changes direction and section. This also improves part toughness by eliminating areas of notch sensitivity.

Gate Location

Too often, gate locations for injection-molded parts are decided either late in the design or purely based on appearance criteria. The designer should not view this critical feature as just a location for the molten plastic to enter the part, but as the point from which to supply uniform melt pressure throughout the part or the volume that the gate is designed to fill. A single sprue-gated part often has the gate located in the exact center of the part with little or no attention given to the product design features in the other areas. The sprue location should be such that all extremities of the part fill simultaneously. Part designs that use multiple gates should be carefully evaluated to ensure that both volumetric and pressure requirements are balanced. Well-designed parts do not have isolated areas that are filled and overpacked while other areas of the part remain unfilled.

Gate Dimensions

Gate dimensions are often best adjusted following mold construction, but the product designer may have certain appearance criteria that dictate the use of a particular gate configuration and location: gate dimensions may be severely limited. Often, small pin gates or tunnel gates are planned for a part with little thought about the volume of molten plastic that will travel through the gate in a very short time. The gate cross-sectional area and land length determine the melt pressure loss as well as the amount of shear heating to which the material is exposed. Rigid PVC injection-molding compounds require, as do other amorphous thermoplastics, ample gate area to ensure a good molding.

6.6.2 Mold Construction [11]

Either two- or three-plate molds may be used for rigid PVC injection-molding compounds. Prehardened, conventionally hardened, nitrided, or stainless tool steels are satisfactory. Steels other than stainless and nitrided steels should be protected by a corrosion-resistant finish on all surfaces exposed to the plastic and/or vapors that might be generated during processing. Triple-plate hard chrome and electroless nickel have been found suitable. Electroless nickel has been especially useful in molds with deep recesses, where the desired uniform thickness of chrome is difficult to achieve. Three hard chromed layers, 0.001 cm (0.0005 in.) each, provide maximum protection.

The required mold draft angle for the rigid PVC compounds is dependent on part size and dimensions. A draft angle of 0.5° is usually

sufficient unless there are textured finishes that would interfere with
part ejection.

Rigid PVC compounds generally exhibit fairly low mold shrinkage,
usually 0.004 to 0.006 cm/cm or in./in. shrinkage. Actual shrinkage
is dependent upon the pressure achieved in the cavity during processing. Higher cavity pressure produces more tightly packed molecules
and results in lower shrinkage upon removal of the part form the mold.
Parameters affecting cavity pressure are melt temperature, injection
and hold pressures, injection speed, gate size, and wall thickness.
Shrinkage is also affected by the flow length and direction in the cavity. The shrinkage is greatest in the flow direction and increases
slightly as the distance from the gate increases due to lower packing
pressure.

Localized burning and short shots can result when trapped air and
other gases become superheated under compression in the cavity ahead
of the molten plastic. The mold should be adequately vented to allow
for gas escape. Vents should be placed near weld lines, as well as in
the last areas of the cavity to be filled. Typical vents are slots 0.6
to 1.3 cm (0.25 to 0.50 in.) wide by 0.001 to 0.003 cm (0.0005 to
0.0010 in.) deep, located on the mating surfaces of one of the mold
halves. Venting may also be accomplished by grinding small flats on
core or knockout pins. In general, vents should be cut to a minimum
depth initially, then increased in depth as necessary.

Other features of mold construction—sprue bushing, runners, cold
slug wells, and gates—are discussed in the next section. Various literature sources are available to the reader for even more detailed information on mold construction. Two books that should help the reader
are Rubin's *Injection Molding Theory and Practice* [12] and the *Plastics
Mold Engineering Handbook,* by DuBois and Pribble [13].

6.6.3 CAD/CAM

A tremendous amount of technical effort has been expended during the
last few years in the area of CAD (computer-aided design) and CAM
(computer-aided manufacturing) relative to the design and building of
injection molds. Advances in the ability to predict melt flow in a mold
design have occurred rapidly. This allows product designers to incorporate the flow characteristics of a given material into their design
at a very early stage. Most of this work is based on melt viscosity
and heat-transfer characteristics of the plastic. Implementation of
CAD techniques for rigid PVC injection-molding compounds requires a
means of incorporating melt thermal stability characteristics as well.
It is anticipated that melt thermal stability will become an important
part of these CAD techniques as rigid PVC compounds evolve into wider
usage in custom applications. Much of the CAD/CAM techniques for
mold-cooling analysis and mold construction should be as appropriate
for rigid PVC compounds as any other commercial amorphous thermoplastics.

6.7 PROCESSING RECOMMENDATIONS [11]

Successful processing of rigid PVC by injection molding is dependent
upon a wide range of variables, which include machine type and size,
screw, nozzle, runners, gating, and mold design. The exact machine
conditions for optimum processing must be determined by the processor
for any chosen system. The following discussions review the general
conditions a molder should observe when processing the new high melt
flow rigid PVC compounds to develop ultimate properties in the molded
part. These same processing recommendations are useful in processing
the higher molecular weight, lower melt flow PVC polymers, but additional concern should be given to shear sensitivity and flow limitations
of these earlier compounds. More liberal flow channels, lower compression ratios, smear-tip screws, higher mold temperatures, greater
clamp capacity, and slower injection speed should be considered.

   Although the injection molding of rigid vinyl has progressed from
the possible, through the practical, to the desirable stage; certain
basic precautions still must be observed in processing thermally sensitive PVC. "First-time" processors should thoroughly familiarize themselves with the material suppliers' processing recommendations prior
to start-up.

6.7.1 Equipment

Only properly equipped reciprocating screw injection-molding machines
are recommended for injection molding rigid vinyls. Because PVC compounds are largely amorphous in their melt characteristics (no sharp
melting point), a plasticating screw is required to prepare homogeneous
melt for injection into the mold cavity. Plunger-type machines allow
melt nonhomogeneity, material stagnation and degradation. They
should not be used for rigid vinyls.

6.7.2 Machine Size

Barrel Capacity To obtain the widest processing latitude and optimum
physical properties of rigid PVC, an appropriate match of shot size,
i.e., volume of cavities, runners, and sprue, to barrel capacity is very
desirable. A shot weight of 60 to 75% of barrel capacity rated in PVC
is recommended. This minimizes melt residence time in the barrel, enabling processing at higher stock temperatures with optimum melt flow
and avoiding degradation.

   Since the optimum match of barrel capacity is not always practical
due to clamp requirements or machine availability, shot sizes as low as
30 to 32% of the barrel capacity may be used with the understanding
that the processing latitude of the material may be significantly reduced. As a result, the physical properties of the plastic material
may not be fully developed. When utilizing the lesser barrel capacities,

lower stock temperatures are often required to prevent thermal degradation due to the longer residence time in the barrel. Lower stock temperatures mean higher melt viscosity and more resistance to flow. Greater injection pressures are needed to fill the part, and molded-in stresses may result. It is likely that these molded-in stresses could adversely affect impact, dimensional stability, and other properties of the finished part. Because of less available processing latitude in back pressure and screw rpm, the use of higher barrel capacities is recommended to reduce residence time.

When calculating optimum barrel capacity, always consider the specific gravity of the rigid vinyl versus the specific gravity of the material for which the machine was rated. The injection capacity of molding machines is normally rated in ounces (a unit of weight) of general-purpose polystyrene.

Example Given that specific gravities of PVC and general-purpose polystyrene are 1.35 and 1.05, respectively, a 1.7 kg (60 oz. avoirdupois) barrel rated in general-purpose polystyrene delivers 2.2 kg (77 oz.) PVC.

A targeted minimum PVC shot weight, including sprue, runners, and parts, would then be 1.6 kg (58 oz.) on this machine (2.2 kg X 75% capacity = 1.6 kg, or 77 oz. X 75% capacity = 58 oz).

Clamp capacity A machine with a minimum clamp force of 300 to 400 kg/cm^2 (2 to 3 tons/in.^{2}) of projected part area, including runners, is recommended. The area of runners in a three-plate mold should be included.

6.7.3 Screw Design

Screws with a compression ratio (see Fig. 6.28) of 1.5:1 to 2.0:1 and a L/D ratio of 16/1 to 24/1 are recommended for molding rigid vinyls. Higher compression screws of 2.5:1 to 3.0:1 have been used, but they provide narrower processing latitude because they cause excessive shear-generated heat in the plastic.

To reduce the tendency of PVC to adhere and degrade on the screw surface and to provide protection against chemical attack, the screw should be deep nitrided to 67 Rockwell C. To ensure maximum corrosion protection, do not remove more than 0.005 cm (0.002 in.) in depth of the nitrided surface during grinding and polishing.

Flame-hardened screws are not as resistant to wear, fatigue, and corrosion. Although uncommon, the screw should be multiplated with hard chrome.

A screw tip of the smear-head design is preferred for medium-flow rigid PVC injection-molding compounds and has been used successfully with the higher flow materials. Sliding check rings having ample unrestricted flow channels to minimize shear heating and hang-up can be used successfully with higher flow rigid PVC injection-molding com-

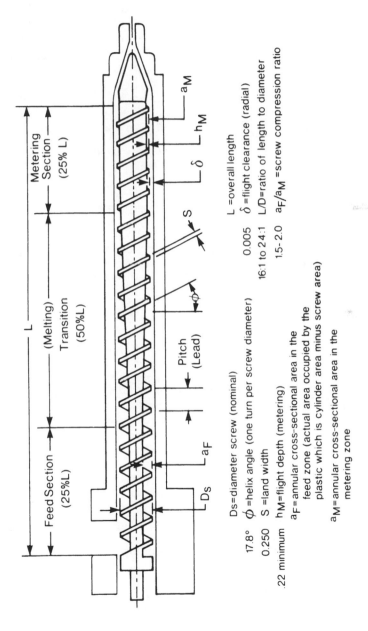

Figure 6.28 A typical injection molding screw design for rigid PVC materials (for illustration only). (From Ref. 11.)

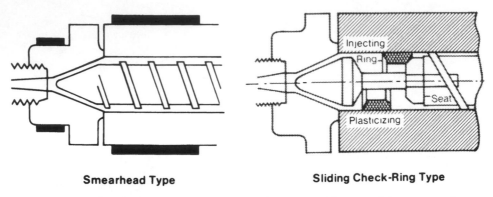

Smearhead Type   **Sliding Check-Ring Type**

Figure 6.29 Smear-head and check ring screw tips used in processing rigid PVC injection-molding materials. (From Ref. 11.)

pounds. Both 1.5:1 and 2.0:1 compression ratio screws with a sliding check ring have proven to be a very workable combination for easy-flow PVC. Ball-check tips are not generally recommended. Again, nitriding or multiple hard chrome plating should be used. Illustrations of smear-head and check-ring tips are shown in Fig. 6.29.

6.7.4 Nozzle

The nozzle length should be as short as possible. Regardless of length, the nozzle should be equipped with a separate heater control. A provision for thermocouple monitoring and control of the nozzle temperature is highly recommended, but a thermocouple that actually projects into the melt stream should not be used. Proportional, integrating, derivative (PID) solid-state, temperature controllers are strongly recommended. Depending on heater band power requirements, a silicon-controlled rectifier (SCR) or triac thyristor circuit may be used. The usual Variac or on-off relay controls are not as effective for maintaining the uniform processing control required for rigid PVC.

The reverse-taper nozzle design shown in Fig. 6.30 is preferred. Ideally, shear effects on the material are minimized by a "zero" land nozzle. The restriction created by the reverse taper promotes sprue break-off farther back in the nozzle, thus avoiding a potential source of cold slugs or stagnated, degraded material. If a nozzle with a finite land is used, the land length should not exceed 0.5 cm (0.2 in). An acceptable alternative is a commercially available full-taper nozzle.

The nozzle discharge-orifice diameter should be at least 0.6 cm (0.25 in.). If the orifice diameter is too small, the flow restriction is

Optimum Nozzle Design

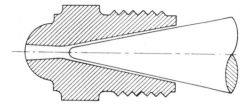

Figure 6.30 A reverse-taper nozzle with a short land length is recommended for use with rigid vinyl. (From Ref. 11.)

too great, and short shots, sink marks, shear burning, or other defects may occur. In all cases, the orifice diameter should be slightly
smaller than the entrance diameter to the sprue. If the orifice diameter is larger than the sprue diameter, the sprue does not release
properly during part ejection, and a cold slug of plastic is left in the
nozzle. Also, if the sprue diameter is smaller than the orifice, an undesirable shear edge is created.

6.7.5 Sprue Bushing

A sprue bushing with a standard 2.5° included angle, approximately
42 mm taper per meter (0.5 in. taper per foot), should be used. The
entrance diameter of the bushing should always be slightly larger than
the nozzle exit orifice. To promote a balanced pressure to the runners
and cavities, the exit diameter of the sprue bushing should be larger
than the diameter of the main runner. The use of a heated sprue bushing is not generally recommended when molding rigid PVC compounds.
Also, an excessively recessed sprue bushing should not be used because it usually requires a nozzle length unsuitable for molding PVC.

6.7.6 Runners

Conventional cold-runner systems are recommended for injection molding rigid vinyl. Although hot-runner systems are feasible, special
design and equipment considerations are necessary. Insulated runner
systems should never be used with PVC. Full-round runners are preferred because they provide the highest volume-surface ratio and the
least pressure drop and are the easiest to eject from the mold. Depending on the part size and weight, typical full-round runner diameters are 0.6 to 1.0 cm (0.25 to 0.4 in.). Because of the excessive
flow restriction, small-diameter runners, less than 0.6 cm (0.25 in.)
in diameter, should be avoided. Excessively large diameter runners
offer little advantage and contribute to longer cycle times and greater
material usage.

If a three-plate mold is used, full-round runners are still preferred, but trapezoidal runners can be used. Half-round and rectangular runners are not recommended for rigid PVC.

To maintain pressure and balanced flow during injection into a multiple cavity or multigated mold, the secondary runners should be slightly smaller in cross section than the main runner. Secondary runners should be perpendicular to the main runner, and the runner junction should be polished to remove burrs and sharp edges.

In addition to proper runner sizing, the layout of the mold is also an important consideration. A runner system should be designed to give balanced flow to all gates, ideally designed so that the melt reaches all the gates simultaneously.

6.7.7 Cold Slug Wells

During injection, the initial surge of material is generally cool, since it has remained dormant in the nozzle while the previous shot was being ejected from the mold. To prevent this cold material from entering the cavity and causing a visual defect, cold slug wells or runoffs should be incorporated into the runner system before material is allowed to enter the cavities. Properly sized runner systems designed for balanced flow, which incorporate cold slug wells, are shown in Fig. 6.31.

6.7.8 Gates

Rigid PVC compounds have been satisfactorily molded through a wide variety of gate designs, including fan, tab, edge, submarine, and sprue. In general, the gates should have a generous cross-sectional area to allow the material to flow freely with a minimum of pressure

Runner Systems with Balanced-Flow
Cavity Layouts and Cold Slug Wells

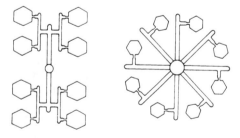

Figure 6.31 Appropriately positioned cold slug wells allow material runoff and enable a homogeneous hot material to enter the cavity. (From Ref. 11.)

Gate Designs

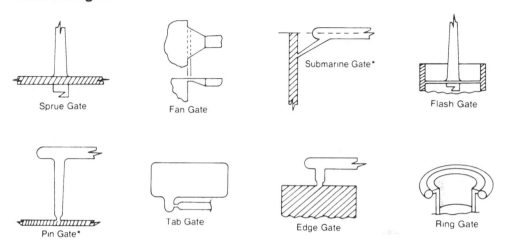

Figure 6.32 A variety of gate designs can be used effectively with the higher melt flow rigid PVC injection-molding compounds. (From Ref. 11.)

loss. The gates should be polished smooth, with all rough edges and sharp corners removed. Figure 6.32 illustrates several acceptable gate designs with rounded corners for minimum restriction.

The land length of a gate should be kept as short as possible to 0.08 to 0.1 cm (0.03 to 0.04 in.). This reduces shear heating and promotes the best combination of cycle time and injection speed.

In multigated cavities, the gate location and number of gates are very important in relation to the appearance and performance of the molded part. Since gate areas are almost always more highly stressed due to orientation, gates should be located in noncritical sections of the part.

Gating in thick sections of the part, allowing the material to flow to the thinner sections, keeps sink marks to a minimum. When gating into a thick section, the flow should be directed toward a cavity wall or deflector pin to break up the melt entering the cavity and to prevent a condition calling "worming." Worming is a random pattern of weld lines opposite the gate caused by the rapid cooling of the injected melt. If the design of the part requires a split in the flow front coming from the gate, a weld line results when the flow fronts meet. Care should be taken in designing parts to keep the number of gates and weld lines to a minimum. Multiple weld lines weaken the part and can detract from the surface appearance.

6.8 PROCESSING TECHNIQUES

6.8.1 Drying

The moisture sensitivity of rigid vinyls is much lower than that of most
molding materials. Normally, they do not need to be dried prior to
molding. However, should the products be exposed to high-humidity
conditions for extended periods, an accumulation of surface moisture
can result in processing problems and part blemishes. The inability
to attain a consistent reproducible cycle, intermittent short shorts,
localized burning, poor weld line strength, and/or silver streaking at
the gate (splay) may be indicative of a moisture problem.

If it becomes necessary to predry rigid vinyls, dry them at 60 to
65°C (140 to 145°F) for 1 to 2 hr. Prolonged or overnight drying is
not recommended. A hopper-mounted, dehumidified air dryer is preferred. If it is necessary to use a tray oven, the depth of the cubes
on the tray should not exceed 5 cm (2.0 in.).

6.8.2 Mold Temperature

Molds should be provided with good temperature control to obtain
optimum appearance and production rates. Inlet water temperatures
of 21 to 38°C (70 to 100°F) are normally used, depending on the size
of the part, wall thickness, and required flow length. The ejector
side of the mold is usually maintained 5°C (10°F) lower than the stationary side to facilitate part removal.

6.8.3 Stock Temperature

The stock temperature can be controlled by a proper combination of
the heater band settings, screw back pressure, and screw rpm. To
develop ultimate physical properties, it is imperative that recommendations for stock temperature be followed when molding rigid vinyls.

The recommended stock temperature for processing rigid vinyls
is 196 to 204°C (385 to 400°F). In no case should the stock temperature exceed 216°C (420°F).

To measure stock temperature, use an accurately calibrated needle-
probe pyrometer. When making a temperature measurement with a
needle pyrometer, the molten material should be injected directly from
the nozzle onto a piece of heavy cardboard or some other insulating
material that will not absorb heat from plastic. The injection pressure,
injection speed, and back pressure are normally at a lower setting for
taking these air-shots than when at normal cycle; therefore, a stock
temperature of approximately 5 to 10°C (10 to 20°F) lower than the
recommended range is a good objective when starting. The needle
should be jabbed into the molten plastic successively four or five times
in different locations before the actual reading is taken.

Occasional wiping of the needle probe with some mold release agent
helps prevent "freezing" of plastic on the probe during the initial por-

tion of the reading. If material freezes to the probe on the first insertion, it acts as an insulator on the probe's surface and erroneously
low values for stock temperature are obtained.

If gassing or bubbling of the hot plastic is observed during the air-
shot, it generally indicates a higher than recommended stock temperature is being achieved. Stock temperature should be rechecked. The
molten plastic rope should appear smooth and reasonably glossy if the
stock temperature is near optimum.

6.8.4 Heater Band Settings

To achieve a given stock temperature, heater band settings depend
greatly on machine size, screw design, and other settings, such as
back pressure and screw rpm. Large machines typically yield stock
temperatures higher than heater band settings.

For the initial trial of rigid vinyls for large machines (over 450
ton clamp), a uniform barrel temperature profile of 160°C (320°F) and
a nozzle temperature of 166°C (330°F) can be used; a uniform barrel
temperature profile of 171°C (340°F) and a nozzle temperature of 177°C
(350°F) for smaller machines, less than 450 ton clamp, should be satisfactory. These settings should be adjusted to achieve an air-shot,
stock temperatures 5 to 10°C (10 to 20°F) less than the final desired
temperature (more heat is generated once the machine cycles continuously). Since heat is generated by the screw within the material, it
is quite normal for the middle and front barrel temperature zones to
override the set point. As long as the machine is cycling regularly,
these set points do not need adjustments. Carefully monitor stock
temperature during initial start-up and after any condition changes.

Once exact heater band settings have been established and the de-
sire stock temperature for the material is achieved, subsequent runs
may be started at these conditions. Close monitoring of stock temperature is still recommended.

6.8.5 Screw Back Pressure

The proper value for screw back pressure varies from machine to
machine, but generally the back pressure should be in the 0.35 to 0.7
MPa (50 to 100 psig) range. Low compression ratio screws may require
higher back pressures. Sufficient screw back pressure is needed to
continuously provide melt homogeneity.

6.8.6 Screw rpm

For a screw of recommended geometry (see Fig. 6.2.8), a rotating
speed of 40 to 50 rpm should be satisfactory. Large machines generally
require lower rpm at optimum conditions. Due to increased diameter,
a larger screw has greater circumferential velocity than a smaller

screw at a given rpm. The greater velocity promotes more shear heating of the molding compound.

6.8.7 Injection Speed

A slow to moderate injection speed should be used at the start of the
molding run and increased to the point at which the part fills and no
signs of weld lines or sinks exist. If the injection speed is too fast,
excessive frictional heat buildup can result in velocity burning as the
material flows through restrictions or over sharp edges. This frictional heat can result in surface appearance problems, brittleness, or
even degradation of the material. Injection speeds for air-shots should
be relatively slow, since there is very little resistance to the material
flow.

6.8.8 Injection and Holding Pressures

The amount of first-stage injection pressure (booster pressure) required to fill the mold cavity depends on the stock temperature, injection speed, mold temperature, and mold design. Generally, pressures in the range 50 to 75% maximum available offer the best consistency and processing latitude. It is advisable to start with lower pressures and increase to the desired pressure to avoid flashing the mold.
The timer for the first-stage injection pressure should be set to switch
to holding pressure just as the part is completely filled. This should
coincide with the moment the screw completes its relatively fast forward travel leaving a 0.3 to 0.6 cm (0.125 to 0.25 in.) cushion.
Smear-tip screws often continue to slowly creep forward and may eventually bottom out.

  The second-stage injection pressure (holding pressure) should be
just enough to maintain a full part as the part cools and shrinks in the
cavity. This pressure is typically 1/2 to 2/3 of the first-stage injection pressure. Parts with thicker sections usually require greater
holding pressure. A cushion is needed throughout the holding stage.

  Overpacking the part by excessive time on either the first-stage
injection pressure or holding pressure increases molded-in stress and
is detrimental to properties. Generally, sink marks opposite the gate
indicate that more injection pressure or time is needed; sink marks
near the gate indicate that more hold pressure or time is needed. Once
it is apparent that gates are frozen off, hold pressure can be reduced
to save on energy consumption.

  A small cushion must be maintained ahead of the screw to compensate for part shrinkage as it cools under holding pressure, thus preventing sink marks. Ideally, the screw reaches the full forward position when material flow to the cavity has ceased.

6.8.9 Processing Summary

In summary, to develop the ultimate physical and appearance properties of rigid vinyls, the material should be at the maximum allowable
stock temperature that avoids burning. It should be injected at a
moderate speed, packed at the minimum pressure required to fill out
the mold details, and allowed to relax during the cooling stage.

6.8.10 Machine Start-up Techniques

Thoroughly clean the injection unit by either physically dismantling
and cleaning, or purge the barrel with polystyrene, general-purpose ABS, or acrylic.

Set temperature controllers and reduce injection pressure settings,
back-pressure setting, and screw rpm to the lower end of their operating ranges.

After temperature zones have stabilized, introduce the rigid PVC
to the machine.

Take air-shot stock temperatures and make adjustments to temperature settings and screw rpm to approach the desired stock temperature. Observe the appearance of the molten plastic very carefully at
this stage. A smooth, glossy surface is indicative of a good homogeneous melt. Another evidence of good melt temperature is the ability
to draw down the hot rope into a thin monofilament. A bumpy melt
and matte surface indicate nonhomogeneity and low melt temperature.
A smoking or froth melt suggests that the stock temperature is too
high.

Spray some mold release in the cavity and sprue bushing, and move
the nozzle into position against the sprue bushing.

Start molding parts in the semiautomatic mode of operation while
adjusting screw travel (feed), injection pressure, and injection speed
to obtain a full part.

If a sprue should happen to hang up in the sprue bushing, never
try to shoot through the hung-up sprue to remove it. This can cause
extensive shear heating of the material, which could lead to degradation of the PVC.

6.8.11 Purging and Shutdown Techniques

Rigid PVC is susceptible to thermal degradation upon prolonged exposure to high temperatures. If there is an interruption in the molding
cycle, the barrel should be pulled back form the mold and the rigid
vinyl should be purged from the barrel at 2 to 3 min intervals by making air-shots. If the delay is lengthy the PVC should be completely
purged from the barrel with polystyrene, general-purpose ABS, or
acrylic. At the end of a molding run, the injection-molding machine
should not be shut down with rigid PVC in the barrel; it must be

purged from the barrel with another material. The carriage (barrel) of an injection-molding machine should never be left in the forward position with the machine idle.

If rigid PVC is accidently overheated in the barrel, both the screw and barrel may have to be cleaned. If the condition is not severe, this may be accomplished by purging the barrel with acrylic, ABS or polystyrene at very low temperatures. The "cold" acrylic, ABS, or polystyrene scours the barrel and removes any degraded material. If this method does not work, remove the screw from the barrel and clean mechanically.

In the event that a power failure occurs during the molding operation, and the PVC cools and solidifies in the barrel, special start-up procedures should be used when power is restored. Initially, the heater bands should be turned on low heat, 93 to 121°C (200 to 250°F), and held there until the material in the barrel has had time to warm through. The heater band setting should then be increased to slightly below normal operation temperatures. As soon as the machine reaches the higher temperatures, the screw should be manually jogged; and as soon as the material moves, the temperatures should be raised to those for normal molding conditions. The remaining material is then purged from the barrel with one of the other thermoplastic materials recommended above. At this time, the rigid vinyl being used can be reintroduced into the barrel and production resumed.

Care should be taken when putting a mold into storage to minimize the possibility of corrosion. The mold should be thoroughly cleaned with a neutralizing agent or washed with a baking soda solution to neutralize any acids present, dried, and then sprayed with a commercial rust inhibitor, lubricant, or mold release agent. At the end of a workday or over a weekend, it is also advisable to spray or brush a thin layer of commercially available mold protective solution on the mold core and activity. This temporary protective layer is automatically removed by the first production shots.

REFERENCES

1.   B. A. Jacobs, The wacky world of small computers, Industry Week *April*: 1982.
2.   C. W. Behrens, Materials searching for superiority, Appliance Manufacturer *May*: 1980.
3.   E. B. Rabinovitch and J. W. Summers, Poly(vinyl chloride) processing morphology, *SPE 38th ANTEC*, New York, May 1980.
4.   J. W. Summers, The nature of poly(vinyl chloride) crystallinity—the microdomain structure, *SPE 39th ANTEC*, Boston, May 1981.
5.   R. J. Brown, The additive heat history of PVC stability, *SPE 39th ANTEC*, Boston, May 1981.
6.   M. C. McMurrer, Update: Heat stabilizers for PVC, Plastics Compounding, 5(7):Nov./Dec. 1982.

7. M. C. McMurrer, Update: Lubricants for PVC, Plastics Compounding 5(4):July/August, 1982.
8. E. L. White, Lubricants, in *Encyclopedia of PVC*, Vol. II, Marcel Dekker, Inc., New York, 1977.
9. C. F. Ryan and R. L. Jalbert, Modifying resins for Polyvinyl Chloride, in *Encyclopedia of PVC*, Vol. II, Marcel Dekker, New York, 1977.
10. R. D. Beck, *Plastic Product Design*, 2nd ed., Van Nostrand Reinhold, New York, 1980.
11. *Injection Molding Guide—Rigid Geon Vinyls*, BF Goodrich Chemical Group, Cleveland, Ohio, 1981.
12. I. I. Rubin, *Injection Molding Theory and Practice*, John Wiley and Sons, New York, 1972.
13. J. H. DuBois and W. I. Pribble, *Plastics Mold Engineering Handbook*, 3rd ed., Van Nostrand Reinhold, New York, 1978.

7

Injection Blow-Molding of Rigid PVC Containers

ROBERT DELONG/Captive Plastics, Inc., San Bernardino, California

I. LUIS GOMEZ/Monsanto Company, Springfield, Massachusetts

7.1 Introduction 363

7.2 The Injection Blow-Molding Process and Its Variables 364
    7.2.1 Blow Pressure 365
    7.2.2 Blow Rate 365

7.3 The Parison Injection-Molding Process 366

7.4 Cavity Filling in Multicavity Molds 366

7.5 Parison and Blow Core Temperature Control 367

7.6 Parison Blowing 368

7.7 Injection Blow-Molding of Rigid PVC Containers 369
    7.7.1 Background 369
    7.7.2 The Resin-Compound Picture 369
    7.7.3 Process Machinery for Injection Blow-Molding of
           Rigid PVC 370
    7.7.4 Process Elements 371
    7.7.5 Biaxial Orientation via Injection Blow-Molding 379

7.8 Alternate Processes 380

Bibliography 383

7.1 INTRODUCTION

Injection blow-molding is a two-step process for producing completely finished plastic containers. In the first step, the plastic melt is injection molded into a cavity where the parison is formed. The neck finish of the container as well as the shape of the parison are molded as the plastic under pressure is injected through a gate, around a core pin, and into the preform cavity. This injection-molding process is distinguished from conventional injection molding in that temperature conditioning of the parison, i.e., cooling the parison to a temperature needed for expanding into the finished shape, takes place at this stage. This process is also known as the conditioned injection blow process (CIBM).

The temperature-conditioned parison is then transferred via the core pin to the blowing station. At this second stage, air is introduced through the core pin to blow the parison into the shape of the blow mold. The expansion of the parison is distinguished from conventional blow-molding in that there is a blow core inside the parison at the start of the blowing operation and the conventional parison is unsupported. The complete container is then transferred to the ejection station.

7.2 THE INJECTION BLOW-MOLDING PROCESS AND ITS VARIABLES

In general, one can state that the processes of injection and blowing are unrelated. Although blowing does not influence the injection step, the process variables of injecting, however, do affect the blowing process. For example, injection pressure affects the packing and the tendency for the parison to swell upon removal from the mold or to distort upon reheating. The higher the injection pressure, the greater the tendency for the parison to swell, due to elastic recovery. Low injection pressures thus cause the least amount of swell. Another example is that of the injection rate: the slower the injection rate, the harder it is to control the temperature distribution of the parison; therefore, higher injection rates should be used. Very fast injection rates, however, produce higher melt temperatures, which can be objectionable for heat-sensitive material like rigid PVC.

The higher the parison temperature at the start of blowing, the faster it expands, and the better it reproduces the blow mold surface; therefore, to obtain the best surface finish, the highest possible parison temperature should be used. The higher the parison temperature, however, the less orientation is imparted to the finished bottle in the course of blowing and the lower the impact resistance of the containers. Therefore, to obtain the greatest amount of orientation, the lowest practical parison temperature should be used. The lowest temperature above the glass transition temperatures (for most rigid PVC compounds for injection blow-molding applications, $T_g \leqslant 70°C$) gives the greatest orientation at a given rate and percentage of stretch and rate of injection. It thus becomes apparent that if both high orientation and a good detailed surface are required, which is the most desirable case, a balance of processing variables, such as stock temperature, injection speed, parison temperature, blow rate, and blow pressure speed, are necessary. This balancing, however, should be achieved while running the parison indexing as fast as possible.

Generally, the key step in injection blow-molding is the parison injection, since the blown bottle is as good as its parison. The overall parison geometry and the accuracy and the alignment of the parison mold components, together with the temperature distribution in the parison, determine the wall thickness distribution in the blown articles. Therefore, all these variables control the amount of plastic sufficient for a given product. Consistency of the parison-making process determines production yield and, consequently, the amount of scrap produced. Furthermore, control of the parison means control of the product properties, including physical properties given by blow orientation and the amount of residuals, i.e., vinyl chloride monomer (VCM) and other extractant.

For a given parison geometry, the characteristics of the parison are determined by the injection-molding process variables and by the accuracy of the parison mold set. Assuming that the parison comes from a well-aligned and well-designed mold and exhibits minimal swelling, good homogeneity, and uniform temperature distribution, the blowing step is influenced by the following process variables.

7.2.1 Blow Pressure

The lower the blowing pressure, the less the expanded parison conforms to the surface of the blow mold for good surface definition and finish and the less is it kept in contact with the blow mold walls for efficient cooling. Thus, the highest possible blow pressure should be used. Although not absolutely necessary, the blow pressure may also be programmed to hit the thickest walls of the parison first.

Note: Blowing defects, such as "blowout" can always be avoided by means other than lowering blow pressure, i.e., by adjusting the wall thickness distribution.

7.2.2 Blow Rate

The slower the parison expands, the more it cools before reaching the blow-mold walls, and the less it conforms to the mold surface, therefore, the highest blow rate should be used. Note: For any given blow orifice, higher blow pressure coincides with higher blow rate.

In general, one can state that any plastic like rigid PVC that can be pressure molded can also be injection blow-molded; in other words, if a stress-free parison with uniform temperature distribution can be made, it can also be blown. The blowing step is a minor phase of the process and a rather easy one.

The same statement is also valid for any parison injection-molding process like the one used in the single-stage injection stretch-blow process (ISBM) or in the two-stage injection-molding reheat stretch-blow process, where preforms are made on one machine and then blown later on in another machine. This two-stage process is also known as the cold-parison process.

It is important to note that experience and expertise in extrusion blow-molding do not qualify for injection blow-molding; in fact, they provide a completely opposite, at times hindering, background.

7.3 THE PARISON INJECTION-MOLDING PROCESS

Today, reciprocating screw injectors are almost universally used for melt preparation.

Although a parison is actually nothing but an injection-molded piece, it differs from the normal kind in numerous important aspects, as follows.

1. It is injected around a temperature-controlled blow core and into a mold that is heated, except for the open-end forming section of the mold, the neck ring, and the area surrounding the gate, the so-called gate pad. The gate pad is normally cooled to avoid drooling and to obtain clean separation of the parison from the injection nozzle. The neck ring area is also cooled to set the threads to prevent deformation upon transfer into the matching blow-mold neck ring.
2. It is transferred from the original mold while hot, its temperature ranging from near the melt temperature to approximately 100°F below it, depending on the deformation, behavior, and finished product properties desired.
3. It must be produced with minimum cavity packing to avoid swell at and after removal from mold.

Generally, there are two schools of thought regarding the process of injecting a parison: low(er) or high(er) injection pressures. Most of the processors advocate injection blow-molding at low injection pressure in the 2000 psi range. The advantages are stress-free and low-swell parisons, lower clamping tonnage, reduced core deflection, and, as expected, less wear and tear on the machine. Few other processors advocate higher injection pressures in order to obtain faster cycles due to faster injection speeds. Higher pressures allow the use of lower melt temperatures, thus faster cooling and possibly higher "orientation" and stronger containers. The low injection pressure and longer cycle system lend themselves particularly well to low-capacity machine moldings, and consequently, fewer numbers of cavities.

7.4 CAVITY FILLING IN MULTICAVITY MOLDS

In multicavity molds, cavity filling uniformity is of extreme importance to assure good production quality and yields. In order to balance the flow, i.e., to fill cavities at the same time under equal pressures, two major approaches have been used by the processors: either increasing the nozzle diameter from the center of the plane to the outside and/or using individually heat-controlled nozzles to balance the flow by adjusting the temperature. Contrary to normal injection molding, the size of gate is not that critical, meaning that the usual pinpoint gate may be increased substantially to facilitate rapid injection at low or high pressures. This is good news for shear-sensitive materials like the rigid PVC melts, where large gate are very welcome. Large gates, however, are prone to drooling. The usual concern of a visible gate stub does not apply to rigid PVC parisons, since the gate area can be postformed in the blow mold.

To avoid drooling and to obtain clean separation of the parison from the injection nozzle, the gate should be cooled by means of a chilled gate pad.

Like injection molding, injection blow-molding depends heavily on the quality and sophistication of the tools used. The really significant features of the process must be built into the tools. They are as follows:

1. Alignment of the blow core, neck ring, parison mold, and blow-mold with each other.
2. Temperature control of the cold mold parts—gate pads, neck ring, and blow-mold—and the hot parts—blow core and parison mold.
3. Proper sealing of the blow-slot and of the several movable connections.
4. Hot-runner configurations and design for accurate, low- or high-pressure metering and drool-free separation.

7.5 PARISON AND BLOW CORE TEMPERATURE CONTROL

The blow core and the parison mold must be kept at slightly different temperatures. Heat transfer fluid is circulated to equalize their temperatures over the length of the parison mold, since otherwise the hot plastic overheats the mold and causes unequal temperatures over the length of the cavity. To provide heat transfer fluid circulation in the parison mold is not much different from the conventional cooling of an injection mold. Temperature control for the core rod, on the other hand, presents special problems because the fluid supply and return connections must accommodate both the extensive blow-core movement and the blow-air supply. Furthermore, temperature control of blow cores is particularly difficult for long and slender cores, such as those needed for long narrow-necked bottles. The tendency is for the blow core to overheat at the tip where the hot plastic impinges on it directly and to cool too much near the blow-slot (normally located in the neck area) because of the air expanding during blowing. For these reasons, circulating fluid-cooled core rods are not in wide use.

Instead, air is blown over the core rods after the bottle is stripped off at the ejection station. This approach, although not too precise, works.

Even with the most careful engineering of the temperature controls, there is always a temperature gradient between the neck and the gate. This is a fact that the process engineers learn to live with and to use to influence the wall thickness distribution during blowing. Consistency from one parison to the next is of extreme importance. If this consistency does not exist, irregular blowing, production interruptions, and rejects multiply rapidly.

The usual design practice is to place coolant lines at or near the ends of the preform (neck and nozzle) and to regulate these at a somewhat lower temperature than those in the body or expansion area of the preform. With block-style preform molds, such channels are easily gun drilled across the entire block. Unavoidably, parison to coolant distances are varied. Fortunately, parison diameters rarely exceed 1 in. for bottles up to 4 oz. capacity, and this variation can be process tolerated. As the process expands into larger sizes of containers and accompanying parisons, it may become necessary to revert to individual cavities and V-drill the coolant passages to obtain uniform parison temperature control.

7.6 PARISON BLOWING

In order to obtain controlled and reproducible expansion of the parison, blow air should be admitted rapidly, at full pressure, as soon as the parison is located in the blow mold. The rate of air flow through any valve should be restricted enough to preclude exceeding the maximum permissible expansion rate of the plastic. Therefore, the problem is to increase the blow rate as much as possible without the danger of overdoing it.

For two reasons air pressure should be high enough to accomplish rapid blowing and to promote good contact between the blow mold and the expanded product: (1) the surface finish of the product depends on it, and (2) unless there is enough pressure, the blow parison separates from the chilled mold wall, causing the cooling rate after blowing to suffer.

Quick blowing also results in causing the plastic to reach the blow mold wall while it is hot and deformable, contributing to good surface finish reproduction. The blown air should be dry and clean. It is occasionally desirable to add moisture to the air blast to obtain evaporative cooling on the inside of the blown product. This must be done with great care and always with moisture deliverately metered for the purpose. For rigid PVC, blow pressures should generally exceed 80 psi, the range of 125 psi being preferred.

The preceding discussion on the injection blow-molding process and its variables has been made to prepare the readers for a more detailed discussion on the injection blow-molding of rigid PVC containers that follows.

7.7 INJECTION BLOW-MOLDING OF RIGID PVC CONTAINERS

7.7.1 Background

The properties of PVC make it a material that lends itself well to a variety of rigid packaging applications. Within the container segment, utilization by toiletries, cosmetics, and food end-users constitute the major application areas. Currently, market needs have focused attention on injection blow-molding of rigid PVC containers for these markets. Some of the advantages of injection-blown PVC containers over extrusion-blown containers include gloss, clarity, freedom from die lines, and precisely molded finishes. However, despite the premium cost of injection blow PVC compound (about 10% higher than extrusion grade), the bottom line is currently economics. Alternate-side PVC extrusion blow machines (Bekum, Fischer, Kautex, and Hoover; see Chap. 5) equipped with triple-parison PVC heads, are just now emerging as a factor in the market. These six-cavity machines still are not cost effective against 8-, or 10-cavity injection blow, especially in sizes of 125 ml and under.

The future growth of injection-blown PVC containers will depend on several factors, including regulatory, market acceptance, and economics. Several observers believe the Bureau of Alcohol, Tobacco and Firearms (BATF) will reallow the test marketing of PVC for liquor now that they have allowed polyethylene terephthlalate (PET) as a packaging medium and the FDA situation is resolved for food. The airline liquor miniatures market is one target. Problems that will need to be resolved include reduced size impression versus glass, plus potential taste shifts in this high surface to volume ratio size. The current cooking oil conversion from glass to PVC will enhance the PVC image in the marketplace. Whether this new image will carry over to other food areas, such as flavorings, extracts, and sauces, is strictly conjecture at this juncture. Watching consumer acceptance closely is the liquid cough and cold preparations segment of the over-the-counter (OTC) pharmceutical market. At least one manufacturer, Vicks, is test-marketing Nyquil in a plastic bottle, in this case PET. Interprocess competition between extrusion and injection blow in the 4 to 10 oz sizes will evolve only with the advent of new resins with improved impact strength.

7.7.2 The Resin-Compound Picture

To place the injection blow-molding PVC compound market, estimated at 7 to 8 million pounds in 1982, in perspective, one must consider that the extrusion-blow compound market is estimated to be 15 times that size or 105 to 120 million pounds. When contrasted to the multi-billion pound PVC resin market as a whole, one can understand the specialty nature of this end use. As new PVC polymerization capacity comes online, the reactors are invariably termed "world-class," another way of saying large and dedicated. As the size of the reactor increases, so does the transition loss between various grades of resin. Generally speaking, extrusion blow-molding grades of PVC have a K value in the low to mid 60s. Conversely, injection blow-molding PVC compounds usually are based on resins with a K value of 52 to 55. In the near term, industry sources suggest the premium (currently about 10%) charged for injection blow compound versus extrusion blow will not decrease. But it is this premium price that is attracting renewed interest from among the PVC compound suppliers, despite the relatively small size of the current market. Continued inroads of 8-track and cassette tapes for recordings is idling the compounding capacity formerly used for vinyl records. There are currently four suppliers of injection blow-molding PVC compounds—Ethyl, Tenneco, Hooker, and Keysor. All offer a competitive offering to Ethyl's 9275J, considered the industry standard and the volume leader. Ethyl's compound is based on an injection-molding resin (K value 53 to 55) stabilized with octyltins; hence, it is FDA sanctioned. Although it combines high flow with good heat stability, both necessary for this application, it is deficient in impact strength. Lack of polymer chain entanglement in the melt state aids flow, but lack of chain entanglement in the solid phase detracts from impact strength. Notched Izod strength, as published by Ethyl, is 0.7 for 9275J as compared with 20 for their 9105, a popular extrusion-grade compound. These lower K-value resins do not respond well to increased percentages of impact modifier. Beyond certain levels, additional quantities of modifier offer little or no improvement in bottle drop impact strength. The industry standard of 100% pass at 4 ft cannot be met consistently in the 125 ml cylinder round size with currently available compounds. All the suppliers reportedly have a higher impact strength compound in active development. Market expansion into container sizes through 10 oz. would roughly double the size of the current injection-blow PVC compound market. Technically, PVC containers as large as 32 oz. have been made via injection blow-molding. Industry observers suggest injection blow-molding is economically competitive with dual-parison PVC extrusion blow through 12 oz and triple-parison PVC up through 8 oz. However, until the resin is optimized to improve impact resistance and tailored to the specific demands of this process, this competition cannot commence.

7.7.3 Process Machinery for Injection Blow-Molding of Rigid PVC

It is probably safe to say that 95% of all injection-blown PVC is processed on Gussoni patent-style equipment (three-station horizontal

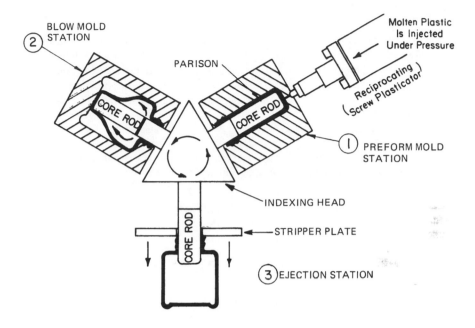

Figure 7.1 Three-station horizontal indexing injection blow process.

indexing—inject, blow, and strip format), as shown in Fig. 7.1 This style of machine is currently supplied by Rainville/Hoover and Jomar in the United States and Bekum in Europe.

By inserting one or more idle positions between the injection and blow stations, cooling of the parison results. This cooling to an optimum temperature orients the parison during blowing, yielding an oriented container. Four-sided machines have been built by Nissei in Japan and Rainville in the United States, and Fischer in Europe has produced a six-station format. None have achieved a wide commercial acceptance, although the Nissei ASB series is used for oriented PET bottles.

7.7.4 Process Elements

This discussion is applicable to Gussoni type three-station horizontal indexing injection-blow machines, inasmuch as these represent, by far, the majority in use. Essentially, the market can be summarized by two New Jersey machine suppliers, Rainville/Hoover in Middlesex and Jomar in Pleasantville.

Plasticator Types

Despite the similarity of machine operation, two divergent approaches
are employed toward melt introduction to the preform mold by Jomar
and Rainville. The Rainville machines use the Egan horizontal reciprocating screw plasticator as found on many injection-molding machines;
the Jomar uses a vertical stationary screw unit manufactured by Egar.
     Current practice in extrusion blow-molding of rigid PVC is to em-
ploy a pegged mixing screw, similar to those developed by Davis-Standard and Goodrich, with an extended tip, cored for air cooling, and to
employ barrel cooling, usually with an inert oil as the heat-transfer
medium. Somehow, this technology has not been transplanted to injection blow-molding.

Screw design: Because of the intermittent rotation and lower K-value
resins being processed, a "tighter" (than extrusion) compression section can be tolerated without excessive shear heating. At least three
rows of mixing pegs are advised, with the initial row well back in the
transition zone, approximately at the screw midpoint.

Compression ratios and lay-out: A compression ratio of 1.8 serves as
a good starting point. If Ethyl 9275J only is processed, a deeper
feed zone can be employed, necessarily increasing the ratio, perhaps
into the 2.0 range (see Fig. 7.2). Ethyl 9275J is a coarse granular
pellet instead of a die face cut pellet and some feeding problems are
encountered, especially on the Jomar. A L/D ratio of 24/1 is normally
employed, with the better mixing outweighing the shorter residence
time of a 20/1. Screw layout is usually a matter of personal taste,
prior exposure, or negative experience, often a blend of all. A feed-
transition-metering format of 0:16:8 or 4:12:8 should serve as a good
starting point.

Barrel and screw cooling: Coring of the screw for air cooling is recommended for the horizontal reciproscrew on the Rainville. Though
not impossible, it is impractical on the Jomar vertical screw. Air
blowers are used by both Jomar and Rainville to control barrel over-

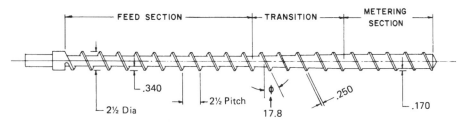

Figure 7.2 Metering-type screw: suggested dimensions of a 2.5 in.
screw for PVC.

ride. These are successful with low K-value resins. As more impact-resistant grades are developed, greater problems of dumping the frictional heat are envisioned. Hot oil will then become the accepted method, just as it has in extrusion blow molding.

Mold Design and Materials of Construction

Mold Design Considerations Probably the first design problem will be related to the position of the air entry in the core rods. One school subscribes to shoulder-blow or "top-opening" core rods, as the portion of the rod below the valve runs hotter (see Fig. 7.3). The alternative, bottom-blow core rods, also have their adherents, claiming cooler rods are easier to control (see Fig. 7.4). On one mold, changing from bottom-blow to top-blow core rods converted a problem mold into a smoothly running one. This decision can be delayed until after unit sampling without undue expense by including one of the alternate-style rods in the sampling trials.

The usual injection-blow design caveats should be observed—3.0:1 blow ratio, 10/1 L/D of core rod, and 2.5:1 bottle ovality ratio. In parison design for oval bottles, additional design freedom in parison ovalization is found with PVC. The pluglike flow of PVC allows a thick-thin ovalization ratio of perhaps 1.5:1 without the selective flow and/or knit line problems one would encounter with polyethylene at this ratio.

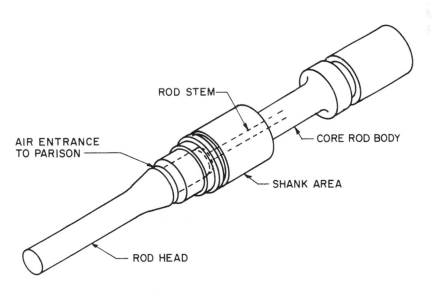

Figure 7.3 Top-blow core rod.

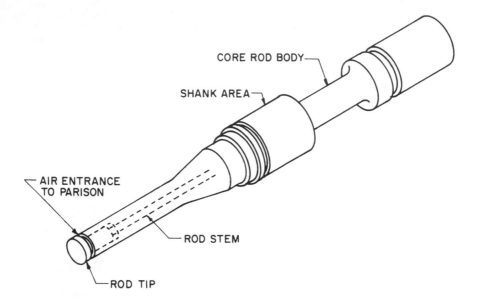

CORE ROD BODY—

SHANK AREA—

AIR ENTRANCE
TO PARISON

ROD STEM

ROD TIP

Figure 7.4 Bottom-blow core rod.

Mold shrinkage is low and predictable; common practice is to use
0.007 in./in. for length and 0.006 in./in. for diameters. Plastic bottle thread forms were designed around polyethylene in the early 1960s.
Rigid PVC does not require such massive profiles, and indeed some
molding sinks may be encountered. Experience dictates how much can
be shaved. Avoid all L-style profile if at all possible.

    Due to the relatively poor impact strength of available PVC com-
pound, as evidenced by very low notched Izod impact strength, blow-
mold alignment is critical. Any misalignment creates a notch and is
detrimental to drop impact performance. For this reason, block-style
mold construction is suggested, as opposed to the normal procedure
of individual cavities keyed to the die shoe. Figure 7.5 illustrates
various sizes and styles of core rods, some of them used with rigid
PVC.

Mold Materials The blow molds should be BeCu, suitably hardened.
Aluminum does not hold up well, chrome-plated steel corrodes under
and peels the chrome, and stainless imposes a cycle penalty due to inferior heat-transfer properties. Unless a very shallow base push-up
is accepted, a retractable bottom push-up must be fitted. Again, the
conventional industry practice of spring retracting, upper mold half
camming a solid (not split) bottom plug into position on mold closing,
is recommended. Figure 7.6 represents a typical injection blow-mold

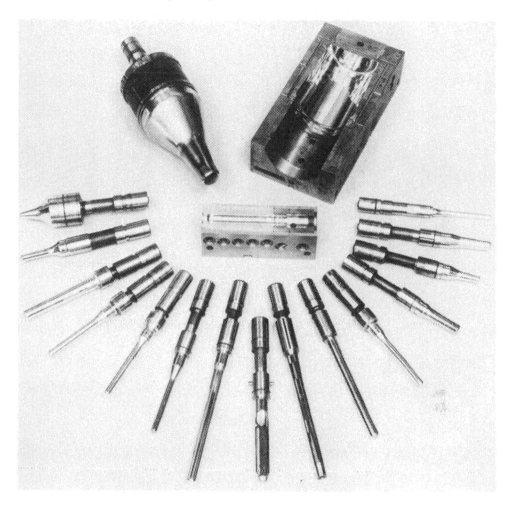

Figure 7.5 Various sizes and styles of core rods. Some can be used with rigid PVC. (Courtesy of Captive Plastics, Inc. Piscataway, N.J.)

parison mold half, and while Fig. 7.7 represents a typical injection blow-mold half. A view of a moving bottom plug blow mold is presented in Fig. 7.8.

Core rods of hardened stainless steel have not proved as satisfactory as conventionally hardened tool steel, polished with nickel or chrome plating. Undoubtedly some of the newer surface diffusion processes will eventually replace chrome plating. At present, the industry practice for parison molds is to use hardened 400 series stainless steel, as opposed to chromed tool steel.

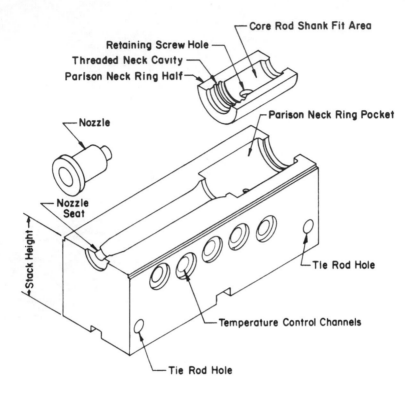

Figure 7.6 Typical injection blow parison mold half.

Manifold and Nozzle Design

The manifold and nozzle system accepts melt from a single entry point
(the plasticator) and distributes it uniformly to as many as 12 exit
points (parison molds). The operative word is uniform, in terms of
flow, temperature, and residence time. For ease of manufacture and
disassembly for cleaning, the manifold is undoubtedly made in two
halves and split either vertically or horizontally, with half the main
flow channel ball end-milled into each half. Both versions are in commercial use. Primary flow balance normally is by the use of variously
sized secondary runners from the main feed channel. Final equalization of cavity feed is by varying orifice sizes in the nozzle (see Fig.
7.9). Manifold heating is still the subject of experimentation. The
ease of electric core rod heaters may outweigh the uniformity of hot
oil, and each system has its adherents. Materials of construction
range from polished and hardened 400 series stainless steel to hardened tool steel with chromed flow passages.

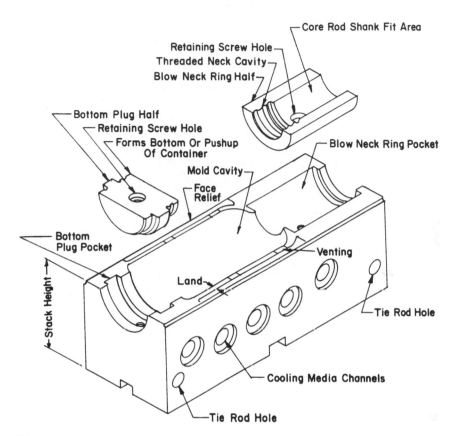

Figure 7.7 Typical injection blow mold half.

Figure 7.8 Moving bottom plug blow mold. (Courtesy of Captive Plastics, Inc., Piscataway, New Jersey.)

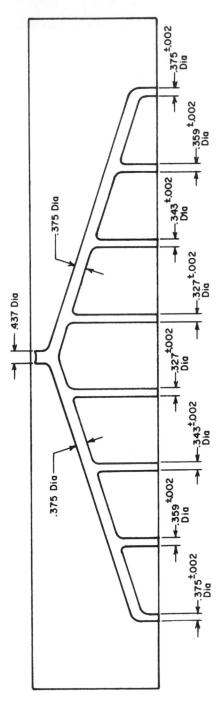

Figure 7.9 Bottom half of an eight-cavity manifold mold.

7.7.5 Biaxial Orientation via Injection Blow-Molding

The injection blow-molding operation imparts orientation in the circumferential direction as well as some orientation in the axial direction. In normal injection molding, most orientation is in the axial direction only, and additional stresses are frozen in due to shrinkage on the core and cooling under pressure. These frozen-in stresses are minimal in injection blow-molding, since the part is not cooled under injection pressure. This, plus the added biaxial orientation, produces blown articles that are not prone to stress crack and that exhibit substantially improved impact resistance. This applied only to those portions of the container that are indeed blown.

Most of the current work with the orientation of PVC, however, has been done with the extrusion-blow process, notably by Ethyl and Owens-Illinois on Bekum equipment. Orientation dramatically improves the impact strength of the container. This permits a reduction in the amount of impact modifier needed in the resin, thereby lowering the initial resin cost. Since most impact modifiers are several times as permeable as PVC, the permeation resistance of the bottle is enhanced, offering equivalent containment at reduced wall thickness (light-weighting) or superior containment for a wider range of products. Inherently, the neck and base portions of the preform receive minimal or no stretch during the biaxial expansion (orientation) operation. Hence, most work has been done with containers in the 16 to 48 oz. range, where the sidewall orientation benefits are most realized. There is one inherent advantage of cool-and-blow orientation, and that is reduced cycle times. This increase in speed is derived from *concurrent* pre-cooling to orientation temperature in one cavity along with the final blowing or orienting step in the finish cavity.

Thus, the driving forces for PVC orientation via injection blow-molding are (1) cycle reduction and (2) impact strength improvement. Two diverse machinery approaches are currently being pursued.

To effect the concurrent cooling to the optimum PVC orientation temperature (generally recognized as 190 to 200°F), at least one extra station must obviously be fitted, beyond the three normal (inject, blow, and strip) stations. Nissei and Rainville/Hoover use a fourth station; Fischer uses a six station format. None of these suppliers has achieved a measure of commercial success; what few PVC orientation machines have been sold are mainly in use for research and development at this time.

Impact strength improvement within the conventional three-station PVC injection blow segment is directed to the gate-scar area of the container base, as this is regarded as one of the weaker areas in drop impact strength. At present, development work is directed at orienting just the container base. The initial approaches involve shorter than normal parison lengths. The usual axial clearance between parison and blow mold is a 0.100 to 0.125 in.; in this instance clearance is increased

to a range of 0.500 to 1.00 in. To assure parison centering, long-
stroke, bottom-blow air valves are fitted to the core rods along with
appropriate air valve-actuating mechanisms.

7.8 ALTERNATE PROCESSES

Two alternate processes to the injection blow-molding (IBM) process
have been already mentioned throughout this chapter. These processes,
although more elaborate than injection blow-molding, offer some advantages worth noting in this work. Specifically, we are referring to the
injection molding reheat stretch blow (IMRHSBM) also known as the
cold parison and/or two-stage process and the injection stretch/blow-
molding (ISBM). All these processes, however, have in common a
parison-making step and some overall similarity in the reciprocating
screw unit used to plasticate and deliver the PVC melt to the preform
mold.

The major difference between the IBM and these two alternate processes is that, as shown throughout this chapter, there is no stretching of the parison in the IBM process, and stretching the parison to
improve molecular orientation is the key to these two alternate processes. The major difference between the ISBM and the IMRHSBM is
that, in the IMRHSBM process, the parison production is divorced
from the blowing operation; i.e., the parisons are produced in one
place, inspected, stored, and perhaps transported to a different location where they are finally reheated, stretched, and blown. In the
ISBM processes (see Fig. 7.10), all the operations are accomplished in
one machine. However, all these processes remove the preform from
the core rod, use traveling neck rings, and reheat the preforms.
Radiant heat is normally used to reheat the preforms. The penalty
these processes pay is extended cycle time; the main advantage is that
they produce bottles in one continuous process. Perhaps the Nissei
ASB process is one of the best known of the IBSM processes.* Cincinnati Milacron's RHB-5 and Corpoplast are perhaps the two best
known IMRHSBM processes available. In these two processes, injection-molded preforms are radiantly reheated, then stretched, blown,
and cooled (see Fig. 7.11). Although both the ISBM and IMRHSBM
have been designed and/or upgraded for the carbonated beverage
bottle industry and hence for high-volume throughout rates, they can
be easily modified for custom containers and short runs and can be
used for bottle runs that require changeovers. This is more so for the
IMRHSBM than for the ISBM process.

---

*As far back as the 1971 Dusseldorf Plastics Fair, Nissei ran PVC bot-
tles in a machine precursor of the present machine system.

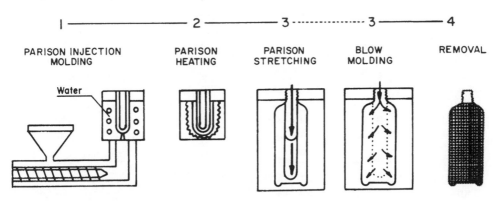

Figure 7.10 Injection stretch blow.

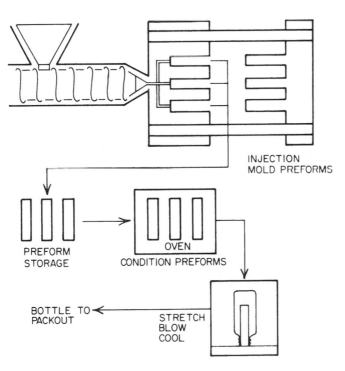

Figure 7.11 Reheat stretch blow.

Current state-of-the-art IMRHSBM processes use injection-molding machines with 8 to 16 cavities molded in a hot runner mode system. In a process variation that has been used with heat- and shear-sensitive materials like high-nitrile polymers, a cold runner, balanced mold system has been the choice. Runners are ground in situ and put back in the injection-molding machine hopper. There is great flexibility built around the IMRHSB process. For example, if bottle-neck finish is kept constant, no oven tooling changes are needed. Also, for the most part, there is no need to change the neck ring in the injection mold.

These two processes when compared to the straight IBM process produce containers that may show improved properties, mainly due to the increased orientation achieved through the stretching step. Some of these properties that may show improvement via orientation are as follows.

Clarity and gloss.

Impact resistance: Containers can be made almost unbreakable with reduced impact modifier. This has a positive economic implication.

Barrier properties: This is due in part to the impact modifier reduction and in part to the molecular rearrangement of the polymer.

Chemical resistance: Since the rubbers incorporated as impact modifiers are more sensitive to chemical attack than the PVC resin, the stretch-blow operation that contributes to impact modifier reduction enhances chemical resistance.

Taste and odor: Again, an increase in container wall barrier properties reduces the migration tendency of the taste and odor offenders, e.g., sulfur-bearing tin stabilizers, some antioxidants, and some rubbers. Furthermore, increased barrier properties reduce chemical attack or leaching of the container wall by the product.

The only two properties that have been found to be inferior to IBM bottles are heat shrinkage and abrasion resistance. The heat shrinkage comes into play when the temperature reaches about 130°F.

Stretch Ratio

Stretch ratio (bottle height to preform height) values ranging from 1.35 to 1.8 are commonly used to orient the rigid PVC containers normally used in these processes. By increasing the melt strength of the rigid PVC compound, i.e., increasing the concentration of the acrylic processing aids, higher stretch ratios can be used, which will improve the above-listed properties even further.

Blowup ratios (bottle diameter to preform diameter) from 2.0 to 2.8 have been commonly used with the above stretch ratios. Again, higher melt strength allows even higher blowup ratios.

Higher stretch and blowup ratios affect cycle time as follows: the larger the stretch and blowup ratios, the thicker will be the preform and, consequently, the longer the temperature conditioning time.

BIBLIOGRAPHY

F. Z. Bailey, and M. E. Bailey, Modern Plastics *April, 1968.*

A. Borer, assigned to T. C. Wheaton, U.S. Patent 2,789,312,

J. P. Burns, Plastics Design and Processing, May 1976.

R. DeLong, Modern Plastics *April, 1972.*

R. D. DeLong, Plastics Machinery Equipment *February, 1979.*

A. J. DeMatteo, assigned to USM Corp., U.S. Patent 3,100,913, July 20, 1963.

J. Farrell, Plastics World *April, 1976.*

A. Gussoni, U.S. Patents 2,853,736 and 3,011,216.

W. Kuelling, and L. Monaco, Plastics Technol. *June, 1975.*

Modern Plastics *October, 1978.*

Plastics Technol. *November, 1973.*

Plastics Technol. *December, 1973.*

Plastics Technol. *December, 1976.*

Plastics World *November, 1974.*

J. K. Presswood, *Practical Benefits of Orienting PVC Containers,* The Ethyl Corporation, Baton Rouge, Louisiana, 1981.

D. Rainville, *Modern Plastics Encyclopedia,* 1972-1973.

C. Strohecker, Plastics Design Processing *September,* 1980; *October,* 1980.

M. Swanzy, Plastics World *May, 1974.*

R. A. Treitler, Plastics Technol. *May, 1974.*

8

Calendered Rigid PVC Products

A. WILLIAM M. COAKER*/Plastics Consultant, Morristown, New Jersey

8.1 Introduction 386
   8.1.1 Definitions 386
   8.1.2 Historical Development 386
   8.1.3 Calender Roll Arrangements 387
   8.1.4 Controlling Web Thickness 387

8.2 Market Factors 393
   8.2.1 Formulas and Units of Market Volume 393
   8.2.2 Market Volume for Film and Sheet Products 393
   8.2.3 Rigid, Semirigid, and Flexible Vinyl Materials 397
   8.2.4 Specific Rigid Vinyl Film and Sheet Products 397
   8.2.5 Need for Market Planning 405

8.3 Manufacturing Operations 406
   8.3.1 Raw Material Handling 406
   8.3.2 Weighing and Mixing 410
   8.3.3 Fusion 411
   8.3.4 Feeding the Calender 412
   8.3.5 Solving Calendering Problems by Applying Theory 413
   8.3.6 Controlling Finish 432
   8.3.7 Drawdown and Cooling 433
   8.3.8 Trimming, Slitting, and Windup 434
   8.3.9 Use of Rework 436

References 436

Additional Reading 438

---

*Current affiliation: BF Goodrich Chemical Group, Avon Lake, Ohio

8.1 INTRODUCTION

8.1.1 Definitions

The word "calender" is derived from the Greek word for a cylinder, or roll, which is the essential working part of a calender.

The essence of plastics calendering is forming a sheet by passing thermoformable stock between corotating cylindrical rolls. The integrity and finish of the sheet or web may then be improved by passing it through successive nips between additional pairs of rolls.

"Film" refers to thinner products sold as roll goods. Heavier gage products cut into rectangles and sold in stacks are called "sheet."

8.1.2 Historical Development

Calendering was defined in 1755 by Samuel Johnson (*A Dictionary with a Grammar and History of the English Language*) as a process using rolls for smoothing and glazing textiles. In the metals industries, similar use of rollers to flatten and shape products is generally referred to as milling.

The earliest machines recognizable as forerunners of modern plastics calenders were described in patents taken out in 1836 [1,2], and were designed for making rubber-coated cloth.

Today, calendering is one of the most spectacular, highly instrumented, and expensive fabrication operations in terms of capital investment. Magram [3] stated that a capital investment of $8 million is required for a vinyl calender line capable of producing 20 million pounds (9074 metric tons) per year of film and sheet products, i.e., a capital cost of 40¢ per annual pound. This is conservative for a large, new rigid vinyl calendering line if the capital cost includes the site, buildings, and necessary peripheral activities.

Shortages of metals and rubber during World War II encouraged a great deal of work on development of replacements using rigid and flexible PVC and on calenders for these [4].

By the mid-1950s, the principal technical advances in calendering and associated equipment had been published in the literature [5]. By this time it was accepted that calenders used for the manufacture of rigid PVC require separate dc motors driving the rolls individually; peripherally drilled and specially crowned rolls; hydraulic loading cylinders to eliminate float and backlash on nip adjustments; flood lubrication of bearings; and power-operated nip adjustment and cross axis for adjusting web profile. Fine tuning of web profiles is conveniently done by hydraulic roll bending [6]. The feasibility of radiation gauging of webs with feedback control was also reported in those early years [7].

About 150 PVC calenders are installed in the United States, but 20% of them are virtually inactive due to overcapacity related to recession or cheap imports [8,9]. There are about 16 calenders, believed to be running mostly on rigid PVC, owned by five companies (or their successors): American Hoechst, Tenneco Chemicals, Klockner, General Tire, and Arlington Mills. Preliminary reports suggest that two other companies may start up new calendering lines in the United States on rigid PVC products during 1983.

8.1.3 Calender Roll Arrangements

A few roll arrangements have been found to work well on rigid PVC. Figure 8.1 shows four calender roll arrangements used for rigid PVC, with their associated feeds and web paths. Figure 8.1 a shows a four-roll L calender with cross axis on roll 3 and roll bending on roll 4. Figure 8.1 b illustrates a four-roll *inverted* L calender with cross axis on roll 3 and roll bending on roll 4. Note that the web follows the final roll for one quadrant in the L arrangement and for three quadrants on the normal inverted L calender.

The five-roll Luvitherm calender illustrated in Fig. 8.1 c makes an excellent product, but the Luvitherm process has not been commercially successful in the United States. This process requires emulsion PVC, and, historically in the United States, suspension PVC has been considerably cheaper than the emulsion form. Some time after its installation, one Luvitherm calender in the United States was converted to conventional hot melt calendering, and it does a fine job in this role.

The ascending, *inclined-Z* arrangement shown in Fig. 8.1d has the advantage on heavy-gage sheeting that it allows three quadrants of web contact on rolls 2 and 3. This helps in removing frictional heat generated in the nips while running a thick web from which heat is removed relatively slowly. On thin webs, the L or inverted L arrangement is generally preferred today.

The demands placed on rigid PVC calenders include being versatile enough to adapt to changing market situations, to produce film or sheet with uniform structure and surface, to hold thickness tolerances both in the machine and transverse directions (i.e., along and across the web), to keep recoverable strains consistent and low, and to run at highest speeds with greatest energy efficiency.

Lower capital installations, such as a sheet extruder followed by a three-roll calenderette, can be designed to do an excellent job on a narrow range of products. It has not yet been demonstrated in the United States that such an installation combines sufficient versatility to adapt to market changes while holding the very tight tolerances of this demanding and competitive industry.

8.1.4 Controlling Web Thickness

Changes in the web gage along the machine direction can be caused by many factors. However, good control of roll speeds is essential to gage uniformity.

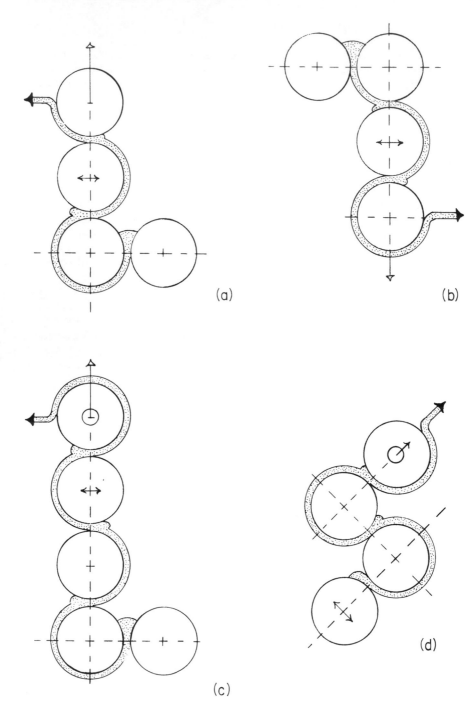

Figure 8.1 Feed and web paths for various calender roll arrangements.
(a) four-roll L; (b) four-roll inverted L; (c) five-roll Luvitherm; (d)
four-roll ascending inclined Z.

Defects caused by gage bands on the calender rolls, by metal marks (dents) on the rolls, or by using out-of-round rolls, stuttering pre-loaders, or defective bearings that chatter, must be corrected by re-pairing the rolls, preloaders, and bearings.

While calendering rigid PVC, the roll-separating forces applied by the polymer melt to the roll surfaces may be 6000 lb/linear in. (PLI) of roll face or higher, depending on the rheology of the stock, its tem-perature, the diameters and speeds of the rolls, and the gaps between

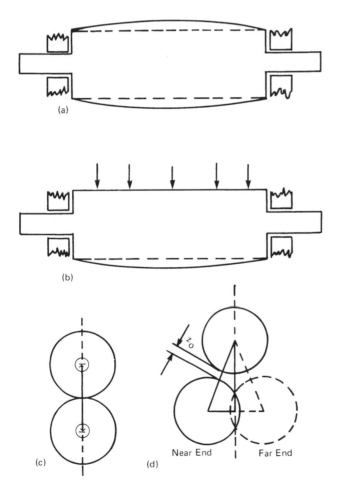

Figure 8.2 Use of roll crown and cross axis to compensate for deflec-tion during calendering. (a) Crown on an unloaded calender roll. (b) Loaded roll with crown compensating for deflection. (c) Calender rolls uncrossed and just touching. (d) Lower roll crossed with respect to upper roll.

rolls, among other factors. The resultant forces on the calender rolls cause them to deflect enough to produce a heavy center unless steps are taken to compensate for the roll deflection. Conditions are usually chosen to stay below 4000 PLI.

Figure 8.2 shows two ways to compensate for roll deflection during calendering. Figures 8.2 a and b illustrate a roll ground with "crown" on it, such that under the calendering load the face of the roll toward the nip is straight. This method lacks versatility, for if the crown is set to compensate for a separating force of 3000 lb/linear in. on a full-width web, the calender can only be run on products and at processing conditions that produce this separating force.

By trial and error, close but not exact compensation for roll deflection can be effected by cross axis. The details and quantitative aspects of this method are discussed in Sec. 8.3.5. The principle is shown in Fig. 8.2c and d.

Figure 8.3 illustrates the concept of compensating for roll deflection by applying "roll-bending" forces to the roll journal beyond the regular roll bearings. Positive roll bending can be applied to increase the effective crown on a roll, thereby eliminating a heavy web center. Negative roll bending is used to correct a light center in a calendered web. Again, compensation is close, but not exact, across the full width of the web. The details and quantitative aspects of roll bending are discussed in Sec. 8.3.5.

Figure 8.4 illustrates common defects in calender rolls caused by improper maintenance. Figure 8.4a and b show the results of bad roll grinding; Fig. 8.4 c shows the result of failing to grind down the roll after allowing a piece of hard metal to pass through the nip. Correcting such defects is expensive, but failure to do so is usually more expensive. Such defects cause the film and sheet products to be off-

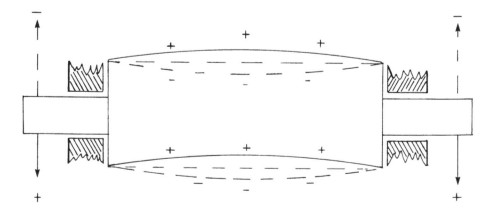

Figure 8.3 Effects of positive and negative roll bending on an unloaded roll.

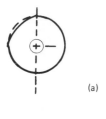

(a)

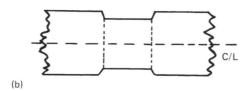

C/L

(b)

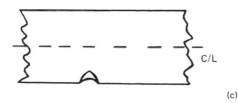

C/L

(c)

Figure 8.4 Common defects in poorly maintained calender rolls. (a) Roll out of round. (b) Gage band on roll. (c) Dent or "metal mark" on roll.

grade and unsalable, or only salable at much reduced prices. Figure 8.5 shows a partly stripped down calender.

For good gage control in a calendering line using significant draw-down from the calender, it is necessary to have the first pick-off roll close to the final calender roll, no more than 1 in. away, and prefer-ably only 0.5 in. distant.

The pick-off rolls are normally given a sufficiently rough matte finish to decrease the tendency for film being taken off to slip. This reduces necking down in width as the web is drawn down in gage, as well as preventing longitudinal slip.

For effective drawdown operations, the pick-off rolls must have good heat-transfer capabilities and be provided with good temperature-control systems. Most modern calenders for rigid PVC have from five

Figure 8.5 View of "heat" end of four-roll inverted L calender of type used for rigid PVC; calender partially stripped down for maintenance; cross-axis and roll bending mechanisms are visible. (Courtesy of Farrel CT. Div., Emhart Corp., Ansonia, Connecticut.)

to nine pick-off rolls, with groups of two or three provided with individual temperature-control systems and drives.

Most thin-gage rigid PVC film calendered in the United States is
used in widths much less than the width of the calender. Therefore,
the material is calendered at full width and slit inline or offline, depending on equipment availability and market requirements.

Inline tentering, which makes a film wider than the calender, has
not found favor in the United States. This is a difficult operation to
start up, and, hence, is economically sound only on long production
runs of film to be used at full width. To make biaxially or uniaxially
oriented film, offline tentering, or drawdown, is preferred over inline.

8.2 MARKET FACTORS

8.2.1 Formulas and Units of Market Volume

Table 8.1 gives some typical formulations for products made of rigid
vinyl calendered film or sheet. Statistics on the market volumes of
flexible and rigid PVC products are often published in terms of the
PVC content of the end products.

The Society of the Plastics Industry, Inc. (SPI) reports PVC resin
use in thousands of pounds [10]; *Modern Plastics* [11,12] and typical
European sources [13] generally report in metric tons, so attention
must be paid to units.

Usage statistics must also be interpreted with care because one
group of figures represents the sum of sales and captive use by domestic resins manufacturers, but another group gives estimates of
total market volumes, including imports. During 1979-1980 when PVC
prices were at their peak in the United States, imports of rigid vinyl
film and sheet may have been as high as one-half the total market
volume. Currency exchange rates and low duties have also favored
imports of rigid vinyl film into the United States.

8.2.2 Market Volume for Film and Sheet Products

SPI [10] does not break out rigid vinyl calendering as a separate PVC
use category. It constitutes a portion, perhaps one-fourth, of "all
other calendering" in their figures. The three PVC resin use categories SPI does report explicitly under calendering are "flooring," "tex-
tile," and "all other calendering uses." All other amounted to 436,
391, and 445 million pounds in 1979, 1980, and 1981, respectively.

Modern Plastics [11,12] presents a much more detailed breakdown
by market segments that does justice to flexible and semirigid calendered vinyl products without explicitly recognizing calendered rigid
vinyl packaging films. These are lumped together with extruded pack-

Table 8.1 Typical Rigid PVC Calendering Formulas

Ingredient	Formulation		Ingredient	Formulation	
Clear general-purpose rigid packaging film			Drug-packaging film		
PVC (K 60)	100	parts	PVC (K 55)	100	parts
MABS impact modifier	6	phr	MBS impact modifier	10	phr
Butyltin stabilizer	2	phr	Acrylic processing aids	5	phr
Acrylic processing aid	1.5	phr	Epoxidized soybean oil	5	phr
Costabilizer	0.5	phr	Ca-Zn soap plus pentaerythritol stabilizer	2.8	phr
Lubricants	1.8	phr			
Total	111.8		Total	122.8	
Food-packaging film			White credit card core stock		
PVC (K 64)	100	parts	P (VCl-VAc) copolymer (~8% VAc)	100	parts
MBS impact modifier	10	phr	MBS impact modifier	12	phr
Acrylic processing aid	2	phr	TiO_2 pigment	9	phr
Octyltin stabilizer	2	phr	Lead stabilizer	4	phr
Lubricants	1.8	phr	Acrylic processing aids	2.3	phr
Total	115.8		Siliceous scrubbing agent	1	phr
			Calcium stearate	0.4	phr
			Ester wax lubricant	0.3	phr
			Total	129.0	

Table 8.1 (continued)

Ingredient	Formulation		Ingredient	Formulation	
Homopolymer floppy disk cartridge film			Copolymer floppy disk cartridge film		
PVC (K 60)	100	parts	P (VCl/VAc) copolymer ($\sim$ 8% VAc)	100	parts
ABS impact modifier	10	phr	MABS impact modifier	12	phr
Acrylic processing aids	7	phr	Lead stabilizer	4	phr
Butyltin stabilizer	2	phr	Acrylic processing aids	2.5	phr
Lubricants	1.5	phr	Lubricants	1.5	phr
Flame retardant	1.0	phr	Flame retardant	1.5	phr
Carbon black	0.5	phr	Carbon black	0.5	phr
Total	122.0		Total	122.0	

Table 8.2 Markets for PVC Film and Sheet Products

Market segment	Metric tons (millions of pounds)		
	1979	1980	1981
Packaging film	114 (251)	101 (222)	125 (275)
Packaging sheet	48 (106)	48 (106)	51 (112)
Paneling	11 (24)	10 (22)	11 (24)
Credit cards	7 (15)	8 (18)	8 (18)
Tapes and labels	5 (11)	5 (11)	6 (13)
Stationery, novelties	6 (13)	6 (13)	6 (13)
Bookbinding	2 (4)	2 (4)	2 (4)
Totals	193 (425)	180 (396)	209 (460)

Source: From Refs. 11 and 12.

aging films: market statistics for *calendered* rigid PVC film and sheet are incomplete.

Adding the totals for the market segments into which most calendered rigid vinyls go, we arrive at the figures for calendered plus extruded film and sheet, which appear in Table 8.2. About one-third of the packaging film and two-thirds of the packaging sheet shown in Table 8.2 are calendered rigid PVC, and essentially all the paneling, credit card, and tape and label films are also calendered; stationery, bookbinding, and novelties are split between calendered and extruded products.

Most of the very thin (0.5 to 0.75 mil) semirigid packaging film is blow-extruded; a small amount is solution-cast. In the gage range 3 to 30 mils, more than half the straight rigid vinyl film and sheet, historically, has been calendered. Since BF Goodrich Company's vinyl film extrusion plant at Gloucester, Massachusetts, was closed down in 1981, one industry source suggests that about 80% by weight of the rigid vinyl film and sheet made in the United States is calendered, and only about 20% is extruded (private communication).

If the market for rigid vinyl film under 3 mils thick expands, the processes competing for the business will include calendering using inline and offline drawdown and blow extrusion.

Prices and profitability have held up better on the more specialized film and sheet products than on commodities such as industrial blister packaging film. Peaks and valleys have characterized pricing in the U.S. PVC business for many years. Recent figures [14] show pipe-grade suspension PVC has sold in the United States at prices varying between 17.5 and 32.5 ¢/lb during the period from April 1980 to September 1982. Such price changes in resin affect the pricing of those calendered rigid PVC film and sheet products on which competition is fierce.

8.2.3 Rigid, Semirigid, and Flexible Vinyl Materials

When classifying materials and items as rigid, semirigid, or flexible, it must be understood that these words are used to describe both the inherent properties of a material and also the behavior of particular pieces or parts, which are sometimes fabricated from composites. The term "rigid" PVC is used in this chapter to refer to the inherent property of the material, and *not* to the stiffness of thin parts made from it, which bend readily in flexure.

A rule of thumb commonly used in the industry for distinguishing between rigid, semirigid, and flexible plastics materials is based on their tensile moduli. Materials are called rigid if their tensile moduli are greater than 10^5 psi (703 kg/cm^2); semirigid when the tensile moduli fall between 10^4 and 10^5 psi (70.3 to 703 kg/cm^2); and flexible when their tensile modulus is less than 10^4 psi (70.3 kg/cm^2) under ASTM standard conditions of 23°C and 50% RH.

Semantic confusion on this subject was graphically illustrated by a customs case, Sekisui Products, Inc., Plaintiff, versus the United States, Defendant, heard on January 4, 1968, and decided on September 17, 1969, by the United States Customs Court, First Division [15]. The Defendant failed to call adequate technical witnesses to refute the Plaintiff's contentions, which were accepted. In a courtroom "demonstration" a 5½ ft long PVC panel was bent with difficulty by a witness for the Plaintiff until its ends touched. On release the panel sprang back to its original shape. The court accepted the Plaintiff's erroneous assertion that this was evidence of flexibility, even though accepted definitions teach that the ability to spring back is resilience, and flexibility is the ability to bend (i.e., flex) under a low force. When the decision was rendered the duty on 771.42 items, flexible or other, was 12.5% *ad valorem*, and the duty on 774.60 items not specifically provided for items of plastics, was 17% *ad valorem*. As rigid items the panels would have been dutiable at the 17% rate. The court's decision allowed the rigid panels to be imported as flexible items with a 12.5% duty (which has since been lowered).

8.2.4 Specific Rigid Vinyl Film and Sheet Products

Important groups of rigid vinyl calendered film and sheet products include the following:

Industrial Packaging Film

Most of this sold in the United States is of the glass-clear, double-polished type, as it comes off the calender, not press-polished. A small percentage is opaque of various colors. It is used for thermoformed blister packs, display boxes, box lids, stationery supplies, and specialty packaging. Some users need the thermoformed parts to easily denest.

Sales of rigid PVC industrial packaging film approached 80 million lb/year in 1982. Table 8.3 shows typical ranges of properties offered by this film. For specific customers additional properties may be specified, for instance, thermoforming temperature range, printability, resistance to chemicals or solvents, and the transmission rates of water vapor, oxygen, carbon dioxide, or fragrances.

The heavier gages usually are sold as sheet goods, and intermediate gages are sold as roll goods on large cores. The lighter gages usually are marketed as roll goods on 3 in. i.d. cores. Table 8.4 shows typical gage and width parameters for sheet and roll goods.

Table 8.3 Typical Property Requirements for Clear Rigid PVC Packaging Film

Property & Unit	Range
Gage, mil	3-30 ±5% average
Maximum width, in.	40-72
Specific gravity	1.30-1.35 average
Tensile strength, lb/in.^{2}	5400-6700 average
% Elongation	100-220 average
Tensile modulus, lb/in.^{2}	355,000-450,000 average
Heat distortion temperature at 264 lb/in.^{2}, °C	62-72 average
% Haze (on clear film)	3-6 maximum
% Gloss	83-87 minimum
Dart drop impact strength at 23°C, ft lb/mil	0.75-1.8 average
Strain recovery, 10 min at 66°C	Machine direction: 0-7.5% shrinkage Transverse direction: 1% growth to 5% shrinkage

Table 8.4 Typical Gage and Width Parameters for Rigid PVC Sheet and Roll Goods

Parameter	Sheets	Rolls
Gage (thickness)	10-30 mil	3-24 mil
Maximum width	40-41 in.	71 or 72 in.
Tolerance on average thickness and yield	±5%	±5%
Tolerance on individual thickness readings	±10%	±10%

Roll goods may be offered on 3, 6, 8, or 12 in. inside diameter cores. For uses critically affected by warpage of the film, returnable metal cores with provision for warpage-free starts may be used. These minimize warpage in the first few plies, which is often caused by a fold at the start of the film or by attaching the film to the core with thick tape.

Appearance properties, specified by setting visual standards, include maximum allowable gels, flow lines, crease whitening, and/or "applesauce" (surface roughness). Punch-out or cut-through performance is also specified in terms of particular test or in-plant performance in the customer's factory.

Note that ASTM specifications D1927 for rigid vinyl sheet based on PVC homopolymer and D2123 for copolymer sheet refer back to D1784, which was written for profile extrusion compounds. All the tests in D1784 are run on compression-molded or extruded test specimens and ignore the effects of the calendering operation on properties. Among the most crucial effects of processing are gage control, lay-flat, strain recovery and thermoformability. It is misleading to try to measure these on test specimens other than those made on a production calender, or other full-scale manufacturing equipment.

Food-Packaging Film

This film may be double-polished clear or opaque, and is used for luncheon meat packs and individual portion packs for jellies, jams, syrups, condiments, and snacks. (Twin packs containing cheese or peanut butter in one compartment and crackers in the other are good examples of this use.)

About 15 million lb/year of food-packaging film sold is calendered rigid PVC, mostly between 7 and 15 mils thick. The large amount of very thin (less than 1 mil) semirigid vinyl food-packaging film is mostly blow-extruded, not calendered.

The market for calendered rigid PVC food packaging film was growing steadily prior to 1974, but received a severe setback from publicity given in that year to the alleged hazards of exposure to vinyl chloride monomer (VCM) vapors. Following public hearings in the United States (and other countries), the Food and Drug Administration (FDA) proposed regulations to restrict the uses of vinyl chloride polymers in contact with food [16]. The PVC industry then drastically reduced the VCM residuals in their PVC suspension homopolymers, the base resins for food-packaging films. Nevertheless, some large users switched to acrylonitrile-styrene (ANS) copolymers, thermoplastic polyesters, or copolyester films.

There are indications that, during 1982, some major U.S. food packagers have again accepted rigid PVC as a safe material for contact with many types of foods. Perhaps PVC will resume its growth as a food-packaging material. This will depend on maintaining the favorable price to performance ratio PVC has recently enjoyed in competition with styrenic, polyester, ANS and polyolefin food-packaging films. A factor in this advantage is rigid PVC's oxygen permeability, which is considerably lower than that of polystyrene (PS) or polyolefin (PE and PP) films. A second factor has been the low cost of chlorine compared to petrochemical feedstocks.

Although the general physical properties of rigid vinyl food and industrial packaging films are similar, most of their ingredients differ in detail because of FDA regulations concerning food contact. In addition to general physical and appearance properties, special requirements for food-packaging performance may include the following.

Low oxygen transmission rates, such as 6 cm^3 mil or less per 100 in.^{2} day atmosphere at 50% RH and 23°C
Low water vapor transmission rates, such as 0.9 g mil or less per 100 in.^{2} day at 90% RH and 37.8°C
High heat distortion temperatures, such as 71°C minimum (to allow hot fill without distortion)
Compatibility with adhesives used in laminating barrier films onto PVC
Good hermetic sealing, using resealable adhesives, normal heat sealing, and/or sonic sealing
Availability in colors, using FDA-regulated pigment systems
Low intermigration by components of the food and the film

Due to concern about possible migration of residual acrylonitrile (RAN) into foodstuffs, U.S. – made rigid PVC food-packaging film formulas normally use methacrylate-butadiene-styrene (MBS) impact modifiers instead of cheaper methacrylate-acrylonitrile-butadiene-styrene (MABS) or ABS types. Heat distortion modifiers based on styrene-acrylonitrile (SAN) are avoided for the same reasons. The U.S. rigid vinyl industry wants to avoid being threatened with restriction for potential residual acrylonitrile (RAN) migration into foodstuffs as they were for potential residual VCM migration from packaging films.

FDA-regulated octyltin stabilizers are preferred over calcium-zinc-epoxidized soybean oil (ESO) combinations, because the tin-stabilized formulas have better heat distortion and barrier properties, and also thermoform better in deep draws.

Drug-Packaging Films

This market consists mostly of double-polished clear, amber or pigmented, opaque film used for thermoformed, unit-of-use packages for drugs. They are commonly made with see-through rigid vinyl on the bottom and printed aluminum foil on the top of the package.

For over-the-counter sales of nonprescription drugs, the trend is toward tamper-resistant unit-of-use packaging, and away from glass or plastic bottles. This is a rapidly growing market segment estimated to be between 15 and 20 million pounds per year of rigid vinyl film in 1982.

However, this market is more difficult to break into than some others and resists film formulation changes because the components of a drug package (e.g., rigid vinyl film ingredients) must be registered in an FDA drug master file. To qualify as a drug package, the manufacturer must submit shelf-life data on the drug in the proposed package, and details of packaging components. New data are required when any change is made in the formula or the film's thickness.

About two-thirds of the rigid vinyl films used for drug packaging in the United States is stabilized with calcium-zinc-ESO; the other third uses octyltin stabilizer. Although some authorities consider calcium-zinc-ESO toxicologically cleaner than octyltin systems, its use slightly compromises heat distortion and barrier properties. The distortion temperature using Ca-Zn-ESO is from 4 to 6°C lower, and barrier properties are slightly poorer. Also, Ca-Zn-ESO stabilizer requires use of very low molecular weight PVC resin to keep processing temperatures down. The side effect of this is to slightly lower general physical properties.

Floppy Disk Cartridge Film

Floppy disk cartridges, made from opaque film with a matte or embossed suede finish, are usually black in color. As a means of storing computer programs, inputs and outputs, floppy disks are fast gaining ascendency over punched cards, tapes, and rigid metallic disks. Cartridge film is regarded as a glamorous market because of its association with "high technology" and the fact that it has been growing at about 30% per year. It will probably more than double before it stabilizes. Including both the jackets for full-sized disks and the smaller diskettes, an estimated 12 to 15 million pounds per year of rigid vinyl film is projected for 1982.

Film gage varies from 7.5 to 8.5 mil for diskette jackets and from 9 to 11 mil for full-sized floppy disk cartridges.

Initially, this market demanded the outstanding dimensional stability and lay-flat only attainable with pressed sheet. The matte or
suede finish was obtained from special press plates. As calendering
houses improved their strain recovery and lay-flat, the market grad-
uated to less expensive calendered sheet, and is now in the process of
converting to still cheaper roll goods. For this use, large-diameter
cores are mandatory. They vary in size from 6 to 12 in. i.d., which
give progressively less curl of the innermost plies near the core than
is found with 3 in. i.d. cores.

From customer to customer there are differences in details of the
requirements, but most have tight specifications for size, appearance,
lay-flat (warpage and cupping), curl, reverse curl, strain recovery,
deflection temperature, heat sealability, flammability, stiffness, toughness, dimensional stability after fabrication into jackets, humidity stability, accelerated aging, behavior in intensive use, and so forth.

Jacket manufacturers who heat-seal with hot bar or dielectric methods prefer lower melting copolymer formulas; those using sonic welding
opt for homopolymer-based film.

Credit Card Core Stock

Core stock for conventional financial transaction cards (credit and
debit cards) is sold in sheet form. Most of it is matte, white or off-
white, but an occasional order is placed for a mass tone, such as gold
for the well-known American Express Gold Card.

The market for various print specialties, such as bank identification
cards, and clear overlays is often lumped with credit card core stock,
giving a total volume of about 17 million pounds per year of calendered
sheet. However, the volume of core stock alone is only about 10 million pounds year, and has not grown during the period of high interest
rates in credit markets.

Most conventional credit cards are made with a sandwich construction in which clear overlays are laminated over the printing on the
core stock to protect it. Total thickness of the finished card typically
is 30 mil. This could be made up of 24 mil core stock with two 3 mil
plies of overlay, or 26.5 mil core stock plus two 1.75 mil plies of overlay.

The core stock may be 20 to 26.5 mil thick with both sides print-
able or 10 to 13 mil with only one side suitable for printing.

The front and back printing is done in register on the thicker core
stocks. The thinner core stocks are printed on one side only, with
the front and back printing for the credit cards laid up in register.
In either case, overlay is laminated onto both sides of the core stock,
with a slightly longer laminating cycle used with two-ply core stock.
This ensures that a good bonding temperature is reached where the
front and back layers of core stock touch each other.

It is relatively difficult to calender 20 to 26.5 mil core stock at a
commercially acceptable rate with essentially no air occlusions (these

give "windows," "pits," and "drags") and with equal printability on both sides. Control of rework ratio, mixing, fusion, size of the rolling banks on the calender, temperatures of the rolls, condition of the matte on the 3 and 4 rolls, takeoff procedures using correctly placed driven stripper rolls, and post treatment are entailed. Lubricants that interfere with printability must be omitted from core stock formulas or burned off the printing surfaces with corona discharge treatment before printing.

Calendering 10 to 13 mil core stock for printing on one side only is much easier, because the printing surface can be made by proper matteing of roll 4 of the calender, or by a combination of matteing on the calender and embossing. Also, air occlusion in the final calender nip is much less of a problem with 10 to 13 mil webs than with 20 to 26.5 mil webs.

Properties needed in credit cards, and for which core stocks must be formulated, include good embossability, character height retention after embossing, good cut-through giving smooth edges when the individual cards are dinked out of the master sheet, and adequate toughness and flex-life to give credit cards a 2 year service life before they crack.

Although copolymer with about 8% vinyl acetate content gives better ink adhesion than homopolymer in core stocks, homopolymer is sometimes specified for its superior solvent resistance or higher heat distortion temperature. Lead stabilizers give better ink drying with some inks than do tin stabilizers. Tin mercaptide stabilizers must be rigorously excluded from overlays designed for lead-stabilized core stock; otherwise, there is cross staining of the stabilizers during lamination.

For good printability, credit card core stocks should be wetted by inks and solutions with surface tension up to and slightly above 40 dynes/cm. Suitable test solutions can be prepared following the directions in test method ASTM D2578. The quality of the print should be confirmed using ink rolldown tests. However, surface tension is a good online quality control (QC) test for printing stocks.

Use of ultraviolet (UV)-cured inks in place of the traditional solvent-based inks is being studied primarily to reduce solvent emissions from printing plants.

Credit Card Overlay Film

This is a thin-gage type of film with a matte finish, made from an inherently clear formulation. About 2 million pounds per year is used in the United States as overlay for credit cards.

A clear, impact-modified copolymer-based formulation is used for this. It is calendered with a matte finish, because matte gives better laminating performance than does double-polished film. The smooth press plates used for laminating polish out the haze created by the matte on the film's surface.

To prevent smearing the printing on the core stock while still having good adhesion of the overlay to the core, it is necessary for the
overlay to soften at a lower temperature than the core stock does.

Overlay films from 3 to 5 mils thick are calendered directly, and
thinner gages down to 1.75 mils are made by drawing down inline or
by biaxially orienting offline. Crease whitening must be low and
strain recovery consistent in overlay film.

Film for Light-Gage Uses

Due to secrecy by manufacturers, it is difficult to determine the volume of light-gage rigid PVC tape film currently made in the United
States by calendering. The film is rolled to a thickness of 3 to 4 mils
and then drawn down to between 0.75 and 1.5 mils. Probably several
million pounds per year are going into pressure-sensitive tape applications and packaging, such as stretch wrap film.

A low-gel PVC resin must be used for this application, because
gels appear magnified after drawing down to light gages. Good,
modern fusion, calendering, takeoff and windup systems are necessary
for making this type of product. A β-ray thickness gage with feedback controls is advantageous for gage control on very thin films.
The original uses of this PVC were pioneered with film made by the
Luvitherm process or conventional biaxial orientation.

Cooling Tower Baffles

A heavy-gage heat- and water-resistant product is supplied for this
use. About 10 million pounds per year of calendered rigid PVC sheet
is estimated to go into this market.

Appearance is noncritical, so a fair amount (up to 50%) of rework
is used, and selling prices are correspondingly low.

The main technical requirements are good heat stability, a relatively high heat distortion temperature, sufficient impact strength to withstand abuse during installation or abnormal weather, such as hail-
storms, and sustained resistance to hot water. To achieve high heat
distortion temperature, use of copolymer and plasticizer is contrain-
dicated. Also, to minimize any tendency to support fungal growth,
biodegradable plasticizers such as ESO and lubricants like stearic acid
should be avoided.

Orange Mask Film

This film must effectively mask portions of a printing plate from light
used in offset lithographic processes. The market consumes about 5
million pounds per year of calendered rigid PVC in the United States.

The key properties of orange mask film are excellent dimensional
stability, outstanding lay-flat, and uniformly high opacity (absence of
pinholes or thin spots through which light can leak). To meet these

last three requirements a gage tolerance of ±5% is strictly enforced.

Insulation Facing for Pipe

Both matte and double-polished film and sheet, most of which is white or grey, are used for this. Calendered rigid PVC film is thermoformed to fit around the insulation on pipe and fittings. This is distinct from the use of semirigid PVC pipe-wrapping tapes. About 5 million pounds per year of calendered rigid PVC goes into this market in the United States. Volume is growing as confidence is generated in the performance of rigid PVC insulation facing and as its ease of installation is recognized.

One supplier codes his products by supplying a light-stabilized outdoor grade as double-polished and a less light-stable indoor grade as matte. It is also claimed that factory housekeeping is easier if the smooth-surfaced outdoor grade is installed, since dust washes off the smooth more easily than off a matte surface.

One of the functions of rigid vinyl facings is to protect pipe insulation during the frequent washdowns required in certain food and drug factories for hygienic purposes.

Key properties are good thermoformability, heat sealing characteristics, and, for the outdoor grade, adequate light stability to resist 3 to 5 years' weathering in the most aggressive climates of the United States, i.e., those exemplified by Florida and Arizona.

Established Specialty Uses

Identified specialty uses amount to at least 12 million pounds per year. Moderate volume specialty uses range from clear overlays for wood-grain laminate sheet, ceiling tile facings, light diffusers for lighting fixtures, aircraft interior panels, and venetian blinds for aircraft, buses, and trains, to film for inexpensive, flexible recordings (used in advertising promotions and recordings for the visually handicapped), engraving stock, lampshades, waterbed heaters, face masks, relief maps, thermoformed containers for tape casettes, tank linings, ductwork to handle corrosive vapors, and various layups for subsequent laminating.

There is no sharp cutoff between PVC-ABS and ABS-PVC film and sheet. Higher ABS content formulas are often used for panels of business machines, appliances, luggage, autos, machined name tags, and various trays. In addition, ABS-PVC is used for FDA-regulated uses, such as margerine packs. Together these constitute multimillion pound market segments not included in the volume figures quoted above for calendered rigid PVC.

8.2.5 Need for Market Planning

Successful commercial calendering of rigid PVC is characterized by sound, long-range planning. This involves analyzing market segments and determining which are, indeed, growing, and for which calendered rigid PVC vinyl film uniquely satisfies the cost-performance requirements.

Market planning should be done in a comprehensive way, taking into account raw materials need, the design of manufacturing equipment required for particular products, and potential problems in satisfying a market. New products should be designed to fulfill the technical requirements of a market segment. At the same time, they should be formulated and priced to earn a respectable return for the manufacturer.

Investment decisions on installation of new calendering lines or upgrading older lines should be based on a sound knowledge of the growth markets for calendered rigid PVC (and other products intended to be made) and the current and probable future technical requirements. Equipment should be selected with a view to what products it will be called upon to manufacture throughout its payout period and remaining useful life.

Offers for sale of several idle or underutilized calenders have discouraged startup of new types of lines. Those under construction are said to be based on conventional technology, generally using existing calenders.

Film market areas characterized by good product and market planning include floppy disk jackets, drug packaging, insulation facing, credit card core stock and overlay, and rigid vinyl food packaging.

8.3 MANUFACTURING OPERATIONS

8.3.1 Raw Material Handling

PVC resins polymerized in suspension or bulk are used in the United States for most calendering operations. These resins should be purchased and handled in bulk at calendering plants because bagged resin costs more, and also, its use entails a much higher risk of introducing dust, paper, and string contamination into products. Bagged resin should be confined to emergencies. (Reserve quantities of all resins used should be kept on hand in bags.)

One key design element of a rigid vinyl calendering plant is the number and size of the bulk resin hoppers. Adequate bulk storage is essential to low-cost, versatile operation of a rigid vinyl calendering plant.

Extensive tests for resin properties should be run before changing resin suppliers because some PVC suspension resins give tacky stocks that release with difficulty from calender rolls. Using the wrong resin in clear rigid formulas may result in plateout or crease-whitening problems.

Impact Modifiers and Processing Aids

Although small amounts of chlorinated polyethylene (ClPE) and ethylene vinyl acetate (EVA) are used in rigid PVC calendering in the
United States, ABS, MABS, and MBS impact modifiers and acrylic processing aids in powder form predominate. The latter, when dispersed
in air, produce highly explosive dusts that can be set off by static
sparks or other ignition sources. Bulk handling systems for these
materials require special provision for dissipating static electricity and
suppressing dust explosions. Such equipment is sophisticated and
costly, and requires frequent maintenance. Therefore, most rigid
vinyl calendering facilities order these materials in bags that are
broken into a weighing hopper from which the correct charge is added
to the blender with a minimum of dusting. Whenever there is a spill
it should be cleaned up promptly and completely, with operators trained
to keep dusting to a minimum.

Data are available from the manufacturers on the amounts to use to
achieve desired effects in terms of enhanced toughness and hot
strength of webs. Baek et al. [17] showed how to evaluate the rheological effects of additives to rigid PVC.

Fillers and Pigments

The two-thirds or so of rigid vinyl film and sheet marketed in the
United States as "clear" is actually slightly tinted with dyes or pigments. Customers want the film to have a bluish, amber, or reddish,
cast, for example, but often the hue is barely discernible, and these
products are not referred to as being "pigmented."

Fillers are usually incorporated in rigid vinyl calendering formulas
to improve processing or end-product performance, rather than to
cheapen products. Compared to flexible vinyls, relatively low levels
of fillers are used in rigids. Some fillers tend to smooth out rolling
banks on a calender, thus reducing air occlusion and flow lines.
Others act as scrubbing agents to keep the rolls free of plateout, or
act as denesting agents to help items in a nested stack of thermoformed
shapes to separate easily.

Fillers and pigments in the form of powdered solids are normally
incorporated at the primary mixer, where the necessary good dispersion can be obtained. Many systems for adding color at the fusion
step are available, but none gives the color accuracy and consistency
demanded by the majority of customers.

Historically, about one-third of the calendered rigid vinyl film and
sheet sold in the United States is filled and pigmented. About 90% of
this is ordered in large quantities of simple colors, such as black floppy
disk jacketing, white credit card core stock, or light gray or cream
pipe wrap.

Small orders for many different colors, some of which involve making
new color matches, result in a host of problems. If not tackled properly, these may cause a mountain of off-colored rework to build up.
This has to be sold at scrap value or dumped with severe profit penal-

ties if it is not worked off. Manufacture of these products should not
be started without employing a good color-matcher backed up with
proper instrumentation; using high-shear primary mixers, such as
Henschel, Littleford, or Papenmeier (Welex); getting as near as possible to plug flow through the fusion system; and developing expertise in scheduling colors through the plant.

To reduce dusting, pigments are often purchased as a dispersed
phase in low-molecular-weight PVC resin.

Fillers are often surface treated to reduce erosion of mixing and
fusion equipment. Finely ground calcium carbonate or silica is typical.

Stabilizers

The stabilizers used in rigid PVC calendering are organotin compounds,
proprietary mixtures of calcium and zinc soaps with costabilizers such
as pentaerythritol, ESO, and FDA-regulated phosphite chelators, lead
compounds, and antimony compounds. In the past, combinations of
barium, cadmium, and zinc compounds were popular in opaque and
pigmented rigid vinyl film and sheet. Today the overall cost effectiveness of the tin-, lead-, or antimony-based systems is considered better, even though the calcium-zinc systems have lower toxicity. Both
lead and cadmium compounds are considered relatively toxic. Hence,
regulations covering their manufacture and safe use are onerous and
force special handling to avoid problems in exposure of personnel,
discharge of effluents, and disposal of used bags or containers.

For clear industrial packaging film, the most popular stabilizers
are alkyltin alkyl mercaptoacetates, alkyltin mercaptides, tin maleate
esters, or mixtures of these. The volume in most calendering plants
is sufficient to justify buying liquid stabilizers in totes rather than
drums, and most of the tin stabilizers are liquids amenable to this.

Food-packaging film is stabilized in most cases with di-(n-octyl)tin-
S,S'-bis(isooctylmercaptoacetate), which is a liquid regulated by the
FDA for such use.

Drug-packaging films are stabilized with calcium-zinc soaps plus
pentaerythritol, which may be added to the blender as solids. Additional costabilizers are phosphite chelators and ESO in combinations
regulated by the FDA. Alternatively, FDA-sanctioned tins may be
used for this.

Lubricants

A lot of pertinent technical information can be obtained from reputable,
technically oriented suppliers.

The proper choice of lubricant systems is a key to minimizing rate
limitations in calendering and to minimizing subsequent product problems. They help to minimize the generation of excessive frictional heat
in calender nips producing thin film. Optimization of lubricants in
long runs of medium-gage films greatly delays the onset of plateout,

thus saving downtime for calender and takeoff roll cleaning. In heavy gages, appropriate lubrication greatly reduces the occurrence of air occlusion in the rolling banks ahead of the nips. This allows the calender to be run faster without objectionable bubbles and bubble trails in the finished film.

In spite of the positive role lubricants play during the calendering process, they have many potentially harmful effects on the properties of the finished film. For instance, they may inhibit heat sealing, reduce printability, induce objectionable-looking bloom on the film surface, interfere with adhesion, reduce light stability, provide a nutrient medium for ugly growths of fungus, and so forth.

The majority of current lubricant systems used in rigid PVC calendering contain carefully balanced ester-wax combinations or alkyl acrylate polymers. These form a strong protective layer between hot metal surfaces and the hot PVC matrix during processing, without causing excesive slippage or allowing objectionable sticking. Recent information [18,19] shows a small degree of slip is desirable in calendering. However, lubricants must not be sufficiently incompatible with PVC at film use temperature to hurt the performance of the film in most uses.

No single laboratory instrument or item of small-scale processing equipment (Brabender Plasticorder, Mechanical Spectrometer, Weissenberg Rheogoniometer, capillary viscometer, cone-and-plate viscometer, instrumented extruder, or instrumented laboratory mill) has, as yet, become a single final arbiter for lubricant systems.

Although much useful information can be gained in laboratory processing, a full-length commercial trial followed by end-use evaluations on the film is necessary to qualify a new lubricant system. Sustained commercial calendering generates a combination of higher pressures in the nips, more frictional heat, higher shear rates, and more exposure to oxygen than any of the test instruments does. Some of the properties of commercially calendered film (e.g., strain recovery, printability, and heat sealability) may be affected by the lubricant and most lab preparations do not show these effects as the film customer sees them.

Miscellaneous Ingredients

Fire retardants, antistatic agents, denesting agents, UV screeners, or optical brighteners or compounds that fluoresce under black light may be incorporated into rigid vinyl calendering formulas to achieve special properties' enhancement.

Scrubbing agents and plasticizers are added in some cases to improve processing. However, external plasticizers reduce the heat distortion temperature of the product. When used at low levels they often also make the film more brittle. This so-called antiplasticization effect can be avoided by careful choice of plasticizer, overcome by adding additional impact modifier, or nullified by adding just enough plasticizer to get beyond the antiplasticization range. However, the development of better polymeric and oligomeric processing aids has greatly reduced the use of external plasticizers and internally plasticized PVC copolymer resins in rigid vinyl formulas.

Radiation sterilization tends to cause yellowing of rigid PVC. Low levels of some plasticizers are used to minimize this effect.

Vinyl shrink-wrap films are uniaxially oriented after calendering to give the film the ability to shrink and tighten up in a heat tunnel. "Splittiness" in these uniaxially oriented films is overcome by incorporation of low levels of plasticizer. This is also an example of a good use for suitable plasticizers in calendered rigid PVC. Vinyl stretch-wrap films also generally contain some plasticizer.

8.3.2 Weighing and Mixing

A necessary condition for good performance of a rigid vinyl calendering line is that the ingredients charged to the mixing equipment be accurately metered. The most reliable systems (and the easiest to calibrate) depend on weighing the ingredients batchwise into a powder-blending unit. A good rule of thumb is that each weighing should be accurate to within ±1% of its nominal value. If a labor-saving, automated ingredient weighing system is selected, accuracy and versatility should not be sacrificed. To handle short or experimental runs in which each batch may be different, the system should readily accept changes. But, operators must keep a clear record of the changes actually made.

Charges of rework, if they have the same composition as the virgin formula to which they are being added, need not be weighed quite so accurately, but it is asking for trouble to allow the proportion of rework during a run to vary more than ±5% of a selected preassigned level. Normally, rework in the form of flake is charged to the cooling blender in the desired proportion each time a virgin batch is dumped.

Blenders should be sized to handle the maximum throughput anticipated on the calender line, with a little to spare. Equipment vendors usually recommend that the cooling blender have double to three times the holding capacity of the intensive blender. However, to allow for uniform dispersion of rework and time for always cooling the powder-rework combination to below 140°F (60°C), the cooling blender should have four to five times the capacity of the intensive blender. This becomes crucial when running the primary mixer up to the high dump temperature required for residual VCM removal. And if the blend from the primary mixer is not cooled enough, there is a danger of caking or discoloration prior to fusion.

Large horizontal blenders with low-horsepower motors and slow-moving ribbons or ploughs do not do as good a blending job for rigid PVC as do high-intensity mixers with fast-moving impellers mounted on vertical shafts that discharge into large cooling blenders. If abra-

sive ingredients are used in any of the formulations, the intensive
mixers should be provided with chrome-plated impellers and baffles;
otherwise, graying may occur from erosion of the metal of the working
elements.

Allowing about 1 min each for charging and discharging, 4 to 8
min for active mixing, depending on the type of mixer (single or twin
impeller), the turnaround time of the primary mixer is 6 to 10 min.
Note that copolymer formulas agglomerate at lower temperatures than
those based on homopolymer. For instance, vinyl acetate copolymers
need to be cooled to 120°F (49°C) or below, to avoid caking. This
total time will be slightly longer if residual vinyl chloride monomer is
being removed from the resin used in food- and drug-packaging formulas.

In that case a vacuum is applied to a suitable intensive blender
through a dust filter. The mixing cycle is extended until just before
the blend becomes hot enough to agglomerate into pea-sized lumps (usually about 275°F or 135°C). At that temperature very little residual
VCM is retained by the blend [20].

The cooled and blended powder-rework combination is normally discharged into one of two live-bottomed holding hoppers. The holding
hopper not in active use is locked and cleaned in preparation for the
next product change. From the holding hopper in use, the blend is
fed to the primary fusion device by a weigh hopper, which discharges
exact increments to a Banbury. Volumetric or gravimetric feeders service continuous fusion devices.

8.3.3 Fusion

Banbury fluxing units made by Farrel (or equivalent) find use in calendering lines dedicated to rigid vinyl. However, a fixed-speed Banbury is not an ideal fusion device in a line making both rigid and flexible PVC films. The cost of a versatile Banbury-type fluxing unit
with a continuously variable speed motor is relatively high because
of the size of the motor required for this unit's very high peak torque
on rigid vinyls.

In newer rigid lines the calendering industry has adopted the Farrel Continuous Mixer (FCM), Stewart Bolling Company's equivalent,
the Mixtrumat, or the Buss Ko-Kneader. A Werner Pfleiderer system
and planetary gear extruders (PGE) are not used extensively in the
United States although they are popular in Europe. The relatively
large working surface area in the PGE and the relatively thinner sections of polymer melt handled by it are claimed to give somewhat smaller
maximum-to-minimum temperature differentials in the polymer melt than
the other machines. However, PGE are subject to catastrophic damage
by tramp metal.

Maintenance costs on the fusion device during the useful life of
the calendering plant should be estimated. These are an important

factor in selecting the device to use, and also in setting product pricing for sustained profitability. Whichever device is adopted, it is wise to consult the manufacturer in depth about optimizing operating conditions for the range of formulas to be run. A visit to the machine manufacturer's processing laboratory is advisable when such a facility is available.

8.3.4 Feeding the Calender

Originally calenders were fed by manually putting large individual "pigs" of stock cut from a mill into the first nip. Today small pigs are guillotined out of an FCM extrudate. Alternatively a continuous strip from a mill or a rope or a thick narrow strip from a strainer die may be used. In Europe a PGE may be run in cascade with a short extruder with a slit die; the resulting sheet is then fed into the first nip [21]. In the United States the majority of calenders are more conventionally fed; e.g., a wig-wag distributes the pigs from an FCM or Buss Ko-Kneader, or the strip from a mill or the rope from a strainer in the first nip.

An outage of stock in the first calender nip is a potentially serious event because it signals an imminent loss of stock in the second and third nips. There the rolls are apt to run together and score each other if the web vanishes. For fear of this many calender operators keep an excessive bank of stock in the first nip at all times.

That practice in itself is harmful, because some of the stock resides in the bank too long. While there it cools down, and this produces cold marks and air occlusions when it finally goes through the overloaded first nip.

One typical arrangement for checking the amount of stock in the first nip is to scan the nip with a closed-circuit television camera that projects the image onto a screen at the control console. In addition, a metal detector is always placed just ahead of the calender to prevent accidental damage by adventitious hard metal inclusions in the feed.

Quick-opening devices fitted to the calender not only save damage in the event of a sudden power outage, but give operators the confidence to run with optimal bank size in the first nip. Also, a low-stock alarm alerts the crew as soon as a condition dangerous to the calender develops.

Other guidelines are the following. The width of the web formed in the first nip should not be a lot narrower than the sheet taken off the final roll. The stock guides should be run as wide as possible without creating excessive edge trim. When mill rolls are part of the feed system, it is wise to use bamboo stock knives for manually cutting and folding stock, thus eliminating the risk of a metal mill knife scoring the calender.

Note that the mixing action of a mill evens out the temperature gradients in a Banbury drop, but it does not produce stock of uniform temperature, because the frictional heat generated in the mill nip introduces new temperature gradients.

One of the goals of a good calender feed system is to supply stock as close as possible to its optimum processing temperature. It is undesirable to allow the fused stock to cool before it reaches the nip. If the physical layout in the plant necessitates long feed belts carrying pigs, strips, or ropes of fused stock, heaters should be provided to prevent it from cooling unduly along its way to the calender.

8.3.5 Solving Calendering Problems by Applying Theory

Background

Most calendering problems impede operations or result in production of off-grade, and, in this sense, are "practical." Formulas have been developed from experience or theoretical analysis to help in solving some of them. Examples are worked to show how selected formulas function, rather than to highlight typical conditions.

Many calendering difficulties are directly caused by variations in raw materials or problems with equipment in the film user's plant. An integrated company setting up a film or sheet line to make product for their own use normally makes a narrow range of products to definite specifications and plans to make the same products for many years. However, most of the rigid vinyl calendering in the United States is done by merchant sellers. They make a very wide range of products and need versatility to respond quickly to market forces. Skills in negotiating specifications with customers and developing new products are essential for them to stay in business.

In selecting calendering equipment, the need for versatility has favored conventional calenders over potentially lower capital installations.

Processing Problems in Calendering Rigid PVC

Problems include low quality and/or quantity yield caused by difficulties in processing; low throughput; excessive power consumption; inability to maintain processing conditions within desired ranges; discoloration or sticking; solid plateout on metal processing surfaces; liquid condensation on takeoff rolls or cooling cans; and objectionable dusting, fuming, or odor.

Typical Product Problems

Typical product problems whose etiology may involve equipment, formulation, or operating conditions, fall into several groups, including the following.

1. Appearance properties: gels; flow marks; bubbles and/or bubble trails; clarity or opacity; finish, including matte, embossment, or excessive "air-side" to "roll-side" differences
2. Physical or chemical contamination: black specks, color streaks, or bloom; toxic, smelly, or cross-staining ingredients
3. Problems specific to roll goods: winding tension, soft edges, rough edges, or winding shifts (blowouts); warpage, cupping, or stretched edges; curl or reverse curl; telescoping rolls; roll cores crushed or collapsed
4. Problems specific to sheet goods: edges out of square; beveled edge cuts; different sized sheets in a stack; "angel hair" from malfunctioning cutter
5. Problems in the user's plant: poor feed, adhesive coating, thermoforming, heat sealing, or decurling
6. Gage (thickness) control: poor lay-flat; excessive "race-tracking"; uneven thickness
7. Strain recovery: shrinkage or expansion in machine and/or transverse directions too great or uneven
8. Physicochemical properties: incorrect tensile strength, tensile modulus, specific gravity, impact strength, deflection temperature; gas or vapor transmission; poor printability, surface wettability, solvent and/or moisture resistance, environmental resistance (e.g., heat and light stability or cold crack); discoloration during heating or sterilization

Gage Problems in Finished Roll Goods

Changes in Thickness of Film Along the Web Most thermoforming and fabrication operations applied to rigid PVC roll goods are more tolerant to gage changes along the machine direction of the web than they are to gage changes in the transverse direction, i.e., across the web.

However, thin areas in the film tend to blow out in a thermoformer, and thick areas do not conform adequately to molds. Also, since stiffness varies as the cube of the thickness of the film, significant gage changes along the web produce discernible increases or decreases in stiffness.

Most gage specifications allow ±5% or ±10% in gage on individual readings; occasionally, ±20% is allowed on thinner films. For satisfactory performance, the changes in average thickness across the web must be much smaller than this (see sections below).

Causes of gage variation along the web include uneven takeoff from the calender, poor control of roll speeds along the line, bad bearings or malfunctioning preload devices on the calender rolls, out-of-round rolls, and uneven feed to the calender. These causes rarely occur to an objectionable degree in properly maintained modern calender lines, but they are found in older lines that would benefit from modernizing. It is cheaper to repair these defects than to live with them and produce off-grade goods.

*Difference in Film Thickness from One Edge to the Other on Roll
Goods* Film comes off a calender with orientation strains that tend to
"recover," making the film shorter, thicker, and, sometimes, wider.
If one edge of a roll of film is thicker than the other edge at the time
of windup, the thinner edge will be loose ("soft"). This film, although
it is the same length at the time of windup, will shrink (shorten and
thicken) on the loose side during strain recovery. If "lateral neck-
down" (narrowing) occurred at the time of takeoff, this film also gets
wider during the strain recovery process. Although some longitudinal
shrinkage is inevitable, lateral widening during strain recovery should
be minimized by controlling lateral neck-down during takeoff.

One way in which film may come to be wound up with one side
thicker than the other is for a web to be calendered with a heavy center and slit inline in the middle; it is then wound up as two rolls, each
with the film thicker on the side that was the center of the original
web.

The relationship between length of film rolled tightly on a core and
the film and core dimensions, provided there are no splices, is

$$L = \frac{\pi}{12} \left[t \left(\frac{D - d}{2t} \right)^2 + (d + t) \left(\frac{D - d}{2t} \right) \right] \qquad (8.1)$$

where

t = film thickness (gage), in.
D = outside diameter (o.d.) of the roll, in.
d = o.d. of the core; also, inside diameter (i.d.) of the roll, in.
L = length of film on the roll, ft

These dimensions are illustrated in Fig. 8.6.

Behavior known as race-tracking, or the "banana effect" occurs in
web-handling equipment when one side of the web is longer than the
other, as may be illustrated by the following. Consider a very large
calender (96 in. roll face width) making an 84 in. wide web with a
heavy center; this is being slit down the middle and wound up as two
42 in. wide rolls. The film is nominally 10 mils thick; core o.d. is
6.25 in., and nominal roll o.d. is 24 in. The thick side of each roll
is exactly 0.0101 in. thick, and the thin side is 0.100 in. in gage when
first wound up.

Substituting in (8.1), we find that the thick side of the roll reaches
an o.d. of 24 in. when L = 3481.8 ft. This side of the roll winds up
tight, leaving the other side of the roll loose and free to shrink. Using
(8.1) to estimate the shrinkage by successive approximations, it turns
out that the thin side of the film is 3469.0 ft long and 0.0100369 in.
thick after just enough strain recovery has occurred to pull it tight.
Roll o.d. on the thin side is then 23.89 in. The magnitude of the race-
tracking effect can be calculated using simple trigonometry.

Following the definitions shown in Fig. 8.7,

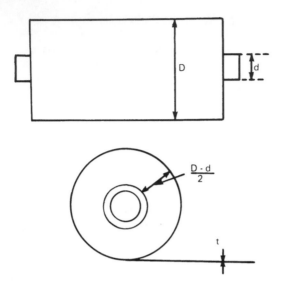

Figurę 8.6 Calendered film rolled up on core showing dimensions.

where

w	= width of web, ft
l	= nominal length of web, ft
GF	= longer edge of web when rolled out on flat surface
DEC	= shorter edge of web
BE	= amount of race-tracking, ft ($\equiv$ s, in.)
r	= radius of curvature of the race-tracking, ft
DAC	= angle subtended by l ft of web, θ, radians

Angle BAC = $\theta/2$ radians

Then,

$$\frac{\text{arc DEC}}{\text{arc GF}} = \frac{r\theta}{(r + w)\theta} \propto \frac{3469.0}{3481.8}$$

Whence, if w = 3.5 ft, r = 948.5 ft, and if l = 30 ft,

$$\theta = \frac{30}{r + w/2} = \frac{30}{950.25}$$

So, $\theta/2$ = 0.0157853. Now,

BE = $r(1 - \cos \theta/2)$ = 0.11817 ft

s = 1.418 in.

If race-tracking is less than 1 in. for a 30 ft rollout of the film, which in this case corresponds to a radius of curvature of about 1350 ft, the film handles well in critical web-handling equipment.

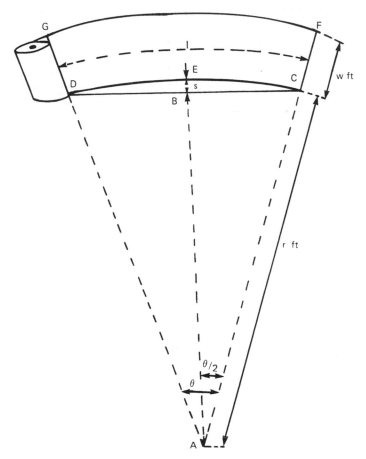

Figure 8.7 Race-tracking of roll goods laid out on a flat surface.

A "slide-rule" calculator with 10 significant figures, such as a
Texas Instruments SR-52, has adequate accuracy for such calculations.
   Race-tracking is less critical than indicated above in some modern
web-handling machines provided with heaters and accurate tension
controls. The limiting radius of curvature on race-tracking can be
determined for each customer's situation.

Edge-to-Center Thickness Differences on Roll Goods "Stretched edges"
are due to the edges being longer than the center; "belly" is due to
the center being longer than the edges of a web. Both cause severe
difficulties in web handling, and originate with edge-to-center thickness differences.

Consider nominal 4 mil film calendered in a 6 ft width and wound
up on 6.25 in. o.d. cores into rolls with a nominal 36 in. o.d., with
no splices allowed within a roll. When no compensation is used, assume
calender roll deflection makes the center of the web 10% (0.0004 in.)
thicker than the edges. Various amounts of compensation for roll
deflection are then applied, and the results tabulated, beginning with
no compensation, in Table 8.5.

An empirical formula for assessing the severity of web-handling
problems caused by film belly uses the minimum radius of curvature
permissable for race-tracking, where one side of the film is longer
than the other, is

$$\Delta t_{max} = t \frac{w}{r} \frac{D}{D-d} \tag{8.2}$$

where

Δt_{max} = thickness difference from edge to center, in.
w   = width of roll, ft
r   = minimum radius of curvature of race-tracking, ft;
D   = o.d. of roll, in.
d   = o.d. of core; i.d. of roll, in.

Numbers from the above example in (8.2) give

$$\Delta t_{max} = 0.004 \frac{6}{1350} \frac{36}{36-6.25} = 0.000022 \text{ in.}$$

which is very difficult to measure and control. By improving the web-
handling equipment, r may be reduced to 400 ft; D can be dropped to
24 in. to increase Δt_{max}, in which case Δt_{max} = 0.000081 in.

If the webs described in Table 8.5 were slit down the middle, the
race-tracking radii of curvature for the resulting rolls are

r_1 = 969 ft for Δt = 0.00003 in.

r_2 = 1459 ft for Δt = 0.00002 in.

This also shows the very accurate thickness control needed for r
$\geqslant$ 1350 ft.

In older plants lacking sophisticated control systems, it is impossible to hold film gage consistently within such close tolerances. A
good system for compensating for roll deflection is essential to the production of acceptable thin-gage calendered rigid PVC film. An automated measuring system, such as a Beta Gauge provided with computerized averaging capability, is also very helpful.

Compensating for Roll Deflection

Estimating Roll Deflection During Calendering A feel for the order of
magnitude of calender roll deflection during calendering, and how to

Table 8.5 Calculated Results for Nominal 4 mil Film Wound up with Various Center-to-Edge Thickness Differences[a]

t_1, in.	t_2, in.	t_3, in.	L_1, L_2, ft	L_3, ft
0.00440	0.00400	0.004160	18,701	17,982
0.00410	0.00400	0.004041	20,069	19,865
0.00405	0.00400	0.004021	20,317	20,213
0.00403	0.00400	0.004012	20,417	20,354
0.00402	0.00400	0.004008	20,468	20,426
0.00401	0.00400	0.004004	20,519	20,498
0.00400	0.00400	0.004000	20,570.5	20,570.5

[a]Where t_1 = film thickness at the center of the web; t_2 = initial film thickness at the edges of the web; t_3 = film thickness at the edges of the web after shrinkage; L_1, L_2 = length of film initially wound up on roll; and L_3 = length of film at edge of web after shrinkage.

estimate it quantitatively for various situations, can be gained by substituting realistic numbers in one of the published formulas for roll deflection.

Consider roll 4 of an inverted L calender as shown in Fig. 8.8, and apply the formulas presented by Stone and Liebert [22]. Use the following definitions, and substitute the actual numbers indicated.

where

q (lb/in.)   = total nip pressure on the roll, plus the weight of the roll divided by l; use 2176
l (in.)   = width of the web; use 78
E (lb/in.^{2}) = effective modulus of elasticity of the roll material; use 22 X 10^6
I (in.^{4})   = moment of inertia of the roll body; use 30,054
D (in.)   = diameter of the roll body; use 28
d (in.)   = diameter of the roll cavity; use 7
h (in.)   = roll neck overhang; use 15
y (in.)   = roll deflection at distance x (in.) from the roll center
y_{max} (in.) = maximum deflection at the roll center

$$y_{max} = \frac{5ql^4}{384EI}\left[1 + \frac{24}{5}\frac{h}{l} + 2\left(\frac{D}{l}\right)^2\right] \tag{8.3}$$

and

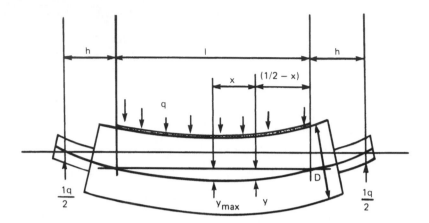

Figure 8.8 Dimensions used to calculate calender roll deflection under load (From Ref. 22.)

$$y = y_{max} - \frac{5ql^4}{48EI} \left\{ \left[\frac{3}{5} + \frac{12}{5} \left(\frac{h}{l} \right) + \left(\frac{D}{l} \right)^2 \right] \left(\frac{x}{l} \right)^2 - \frac{2}{5} \left(\frac{x}{l} \right)^4 \right\} \quad (8.4)$$

Making the numerical substitutions in (8.3), y_{max} = 0.00346 in., and values for y at intermediate points between the roll center and the edge of the web can be calculated from (8.4).

If the roll face width is increased to 126 in. and a 120 in. web run on it, with other conditions held constant, substitution in (8.3) shows: $y_{max}(120)$ = 0.01519 in., which is a very large increase from 0.00346 in. for a 78 in. web run on a roll with 84 in. face width, and illustrates a problem confronting calender makers.

Note that, in practice, 3 in. would be trimmed from each side of the 78 in. web, leaving it 72 in. wide at windup. A web of 120 in. width at the calender would not be wound up wider than about 112 in. Widths at windup would be less if neck-down occurs in the calender train.

Equations (8.3) and (8.4) can also be used for calculating the point at which the edges of the calender rolls are run together, if a thin and narrow web is run without using roll crossing or roll bending to compensate for deflection.

Use of Roll Crown to Correct for Deflection   It is normal operative practice to profile rolls on a calender with a crown adapted to average operating conditions. The crown is then supplemented with cross-axis and/or roll bending, to the extent that conditions depart from the average.

A former practice, that of trying to run many different formulations at a constant rolls' separating force by running stiffer formulas hotter and softer formulas colder, is now out of favor. Softer formulas are underfused and have poor physical properties; stiffer formulas are run too hot. This causes them to discolor and to have very low residual heat and light stability.

Roll crown is no longer commonly used as the sole means of compensating for deflection. However, calenders are crowned to diminish the extent to which cross-axis and roll bending have to be used. This reduces the "oxbow" effect, which results from nonquantitative compensation for deflection across the whole working face of the rolls.

A convenient equation for guiding the crowning mechanism of a roll grinder is [22]

$$z = y_{max} \left[\frac{\cos (2\alpha/l)x - \cos \alpha}{1 - \cos \alpha} \right] \tag{8.5}$$

where z is an estimate of y, the roll deflection calculated from (8.4), which is a more complex equation.

The use of (8.5) is illustrated in Table 8.6, in which z was calculated with $\alpha = 58.6°$.

The agreement between y and z is excellent, and perhaps could be made perfect using a slightly different value of α.

Roll crown is defined as the additional diameter required to compensate for deflection. A crown of 0.006918 in. is added to compensate for a deflection of 0.003459 (i.e., crown is twice the deflection).

Use of Cross Axis to Correct for Roll Deflection Compensation for deflection by roll crossing was analyzed mathematically by Carrier [23] and Gooch [24].

Using the definitions of dimensions shown in Fig. 8.9, and, if z is the additional gap due to crossing,

Table 8.6 Use of Equation (8.5) to Approximate Roll Deflection Calculated from Equations (8.3) and (8.4)[a]

x	0	6	12	18
y	0.003459	0.003370	0.003104	0.002669
z	0.003459	0.003370	0.003104	0.002669
x	24	30	36	39
y	0.002075	0.001336	0.000472	0.000000
z	0.002075	0.001337	0.000473	0.000000

[a]Where x = distance from roll center line, in.; y = roll deflection calculated from equations (8.3) and (8.4); z = roll deflection calculated from equation (8.5).

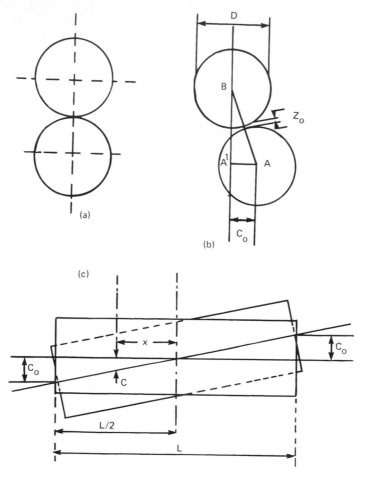

Figure 8.9 Dimensions used in analysis of cross axis as a means of compensating for calender roll deflection. (Adapted from Ref. 24.)

$$z = \left[c_0^2 \left(\frac{2x}{L} \right)^2 + D^2 \right]^{1/2} - D \qquad (8.6)$$

In an inverted L calender, the deflection of roll 3 is usually very little, because the forces from the second and third nips tend to cancel out. Assuming that all the deflection for which compensation is needed occurs in roll 4, we can use deflections previously calculated to compare against compensation possible with roll crossing.

The opening between the rolls is the initial opening, which we assume to be zero, plus the opening due to deflection y, plus the opening due to roll crossing z. At any distance x from the center of the rolls, the web thickness then is (y + z), if no allowance is made for puff-up.

In practice, if the web is to be trimmed to a finished width of 72 in., cross axis would be adjusted to give the same thickness 36 in. from the center as at the center. Values are shown in Table 8.7 which includes the theoretical oxbow effect.

In order to counteract this effect, it is common practice to have the final roll on a calender contoured with some degree of "reverse oxbow." This helps when running the most critical products. Note that real oxbow exceeds that calculated with the use of simplified assumptions. This is due to section changes in the roll shoulders and bodies, thermal gradients in the rolls and the fact that the loading is tapered, not constant.

Use of Roll Bending to Adjust Film Profile Adjusting web profile by applying an external bending moment to both ends of a roll shaft on a calender is called roll bending. Positive roll bending counteracts deflection caused by separating forces in a calender nip. Negative roll bending compensates for ground-on crown greater than needed for particular operating conditions.

Roll bending has been discussed by Jukich [25], Seanor [26], and Gooch [24]. Following the dimensional definitions in Fig. 8.10, the results of simple bending, ignoring the weight of the roll, are

Table 8.7 Use of Cross Axis to Compensate for Roll Deflection Showing Theoretical Oxbow Effect[a]

x	y	z	y + z	Theoretical oxbow effect
0	0.003459	0.000000	0.003459	—
6	0.003370	0.000083	0.003453	-0.000006
12	0.003104	0.000332	0.003436	-0.000023
18	0.002669	0.000747	0.003416	-0.000043
24	0.002075	0.001328	0.003403	-0.000056
30	0.001336	0.002074	0.003410	-0.000049
36	0.000472	0.002987	0.003459	—
39	0.000000	0.003506	0.003506	+0.000047

[a]Where x = distance from roll center; y = gap due to deflection; z = gap due to cross axis; and y + z = gap.

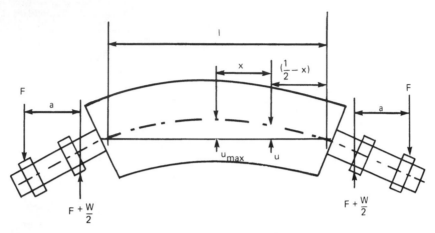

Figure 8.10 Dimensions used in analysis of roll bending as a means of compensating for calender roll deflection. (Adapted from Ref. 24.)

$$U_{max} = \frac{-Fal^2}{8EI} \qquad (8.7)$$

and

$$U = \frac{Fa}{2EI}\left(x^2 - \frac{l^2}{4}\right) \qquad (8.8)$$

where

U (in.)	= deflection due to roll bending at a point x inches from the transverse center line
U_{max} (in.)	= deflection due to roll bending at the transverse center line of the roll
W (lb)	= weight of the roll; use 13,710 or 175.8 lb/in
F (lb)	= force applied by the roll-bending cylinders
a (in.)	= moment arm between the main roll bearings and the point of applying the roll bending force; use 36
l (in.)	= active length of the roll face; use 78
x (in.)	= distance toward the roll edge from the transverse center line of the roll
E (lb/in.^{2})	= modulus of elasticity of the roll material
I (in.^{4})	= moment of inertia of the roll

For the case where it is wished to counteract the deflection of 0.003459 in. (previously calculated in Sec. 8.3.5) by roll bending, from (8.7):

$$-0.003459 = \frac{-F \times 36 \times (78)^2}{8 \times 22 \times 10^6 \times 30,054}$$

therefore,

$$F = 83,536 \text{ lb}$$

If the area of piston exposed to hydraulic pressure for positive roll bending is the rod end of a piston with o.d. 8 in. and rod diameter of 3.5 in.,

$$\text{Area of piston exposed to hydraulic pressure} = \pi(4^2 - 1.75^2)$$
$$= 40.64 \text{ in.}^2$$

Hydraulic pressure p corresponding to positive roll bending force F is

$$p = \frac{F}{40.64} = \frac{83,536}{40.64} = 2056 \text{ psi}$$

Another illustrative calculation, using the same hypothetical calender as an example, involves determination of the maximum crown for which the system can compensate. When force F acts in an upward (negative) direction, assume hydraulic pressure is applied to the cylinder ends of 8 in. o.d. pistons and that the maximum hydraulic pressure is 2200 psi.

$$-F \text{ (upward)} = -2200 \times \pi \times 4^2 = -110,584 \text{ lb}$$

therefore,

$$U_{max} = (-) (-) \frac{110,584 \times 36 \times 78^2}{8 \times 22 \times 10^6 \times 30,054} = 0.00458 \text{ in.}$$

which is equivalent to 0.00916 in. of crown, the maximum this negative bending mechanism can counteract.

Note also that adequate hold-down cylinders must be used on roll 4 of an inverted L calender when the latter has negative roll bending applied to it. This prevents rolls 3 and 4 from being run together by the force of the negative roll-bending mechanism in the absence of a web.

Separating Forces and Torque Between Rolls

Methods that provide accurate results in estimating roll separating forces and torques at practical running conditions on commercial calenders are very complex. They involve very sophisticated rheological measurements, followed by extensive iterative calculations using a computer.

A couple of simplified procedures that give approximate results for separating forces are reviewed. Selected modern studies are referenced as a starting point for those who wish to go into the subject in depth.

Ardichvili [28] developed a tractable equation for the separating force between corotating rolls, milling or calendering a newtonian fluid under isothermal conditions. Other simplifying assumptions implicit in this treatment include ignoring kinetic energy corrections, lateral movements, and elastic properties of the stock, and assuming that the rolls have equal diameter and run synchronously, and that there is no slip between the rolls and the fluid.

The conformation of this type of nip is illustrated in Fig. 8.11.a. The separating force is given by the expression:

$$F = 2\mu Vrw \left(\frac{1}{h_o} - \frac{1}{H} \right)$$ (8.9)

where

F = total separating force between rolls, lb
μ = viscosity of the fluid being calendered, lb sec/in.^{2}
V = peripheral speed of the rolls, in./sec
r = radius of the rolls, in.
w = width of the web passing through the nip, in.
h_o = separation between the rolls at the nip, in.
H = height of stock at the nip entrance, in.
q = separating force between rolls, lb/in.

As an example, making substitutions in (8.9), we get

$$F = 2 \times 0.02 \times 15 \times 14 \times 76 \left(\frac{1}{0.003} - \frac{1}{1} \right)$$

$$= 212,162 \text{ lb}$$

or

$$q = 2792 \text{ lb/in.}$$

Note that this applies to calendering about a 4 mil film (due to 4/3 springback of the stock coming out of the nip) at 25 yards per minute (ypm) on a calender with 28 in. diameter rolls using an active face width of 76 in. and stock with viscosity of 0.02 lb sec/in.^{2}.

Another tractable expression for estimating the separating force between calender rolls is due to Ancker [29]:

$$F = 4.47\mu \frac{D^2 Nw}{t}$$ (8.10)

where

μ = apparent melt viscosity, lb sec/in.^{2}
D = roll diameter, in.

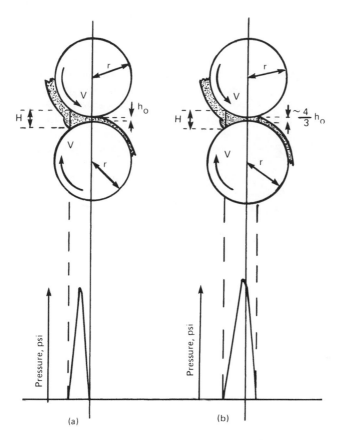

Figure 8.11 Dimensions used in calculations on movement of stock through calender nips. (a) Nip and nip pressure distribution, assuming stock exits at nip. (From Ref. 28.) (b) Nip and nip pressure distribution where stock exits beyond nip. (From Ref. 32.)

N = speed of the rolls, revolutions per second
w = width of web, in.
t = thickness of web, in.

Using the same example for (8.10) as for (8.9) and converting units, gives

$$F = \frac{4.47 \times 0.02 \times 28^2 \times 0.17052 \times 76}{0.004}$$

$$= 227,086 \text{ lb}$$

$$q = 2988 \text{ lb/in.}$$

which is in reasonable agreement with the result from (8.9).

Eley [30] and Gaskell [31] developed more elaborate equations. Bergen and Scott [32] experimentally confirmed Gaskell's conclusion that the zone of elevated pressure on the rolls extends beyond the nip, as shown in Fig. 8.11 b.

Dexter and Marshall [33] measured apparent viscosities with a capillary extrusion rheometer. They devised adjustment factors based on temperature, plasticizer content, and shear time. Using corrected apparent viscosities, they obtained good agreement between predicted and measured roll separating forces on a mill for plasticized PVC and PE. Interesting parameters they used to relate the behavior of stocks in a calender nip with those in a rheometer are

$$S = \frac{-2V}{h_o} \tag{8.11}$$

where

S = maximum shear rate in the nip, $\sec^{-1}$
V = roll peripheral speed, in./sec
h_o = nip opening, in.

using the numbers from the previous example,

$$S = \frac{-2 \times 15}{0.003} = 10,000 \sec^{-1}$$

$$\tau_s = \left[\frac{2.31}{S} \left(\frac{r}{h_o} \right)^{1/2} \right] \tag{8.12}$$

where τ_s is the shear time in seconds.
So, for the cited example,

$$\tau_s = \frac{2.31}{10,000} \left(\frac{14}{0.003} \right)^{1/2} = 0.0158 \text{ sec}$$

The shear time adjustment factor is determined from the dimensionless number $2.31 (r/h_o)^{1/2}$, which in this example is 158, giving the adjustment factor 0.56.

Kiparissides and Vlachopoulos [18] calculated nip pressure distributions for various situations. They reached the conclusion that effective speeds V and roll diameters D average out for newtonian fluids, if the rolls are run at different speeds, or different size rolls are used:

$$V_{eff} = \frac{V_1 + V_2}{2} \qquad (8.13)$$

and

$$D_{eff} = \frac{D_1 + D_2}{2} \qquad (8.14)$$

However, for power-law fluids, pressure profiles were higher for synchronous cases than for nonsynchronous ones.

Dobbels and Mewis [19] and Alston and Astill [34] studied calendering theory; they provide extensive references and insights into mathematical approaches, but do not give experimental verification.

Munstedt [35] showed that rigid PVC slips at the capillary wall during rheological measurements in an extrusion rheometer, except at very high stock temperatures.

Chaffoureaux et al. [36] developed a quantitative method for measuring stick-slip phenomena at interfaces between hot metal and lubricated molten PVC stock. They showed that the velocity of the PVC at and near an interface with a metal is a function of the shear stress, the temperature, and the amount and nature of the formulation additives, as well as depending on the presence or absence of grooves in the metallic wall. Vlachopoulos and Hrymak [37], using rheological data from Chauffoureaux, got fair agreement between theory and experiment on pressures in calender nips. If the slip correction was not used on a rigid vinyl calendering formula, theory predicted pressures 75% higher than those actually measured. Accuracy of torque predictions depended on the effective stock temperature used.

Agassant and Avenas [38] were able to predict separating forces and torques on calender rolls with moderate accuracy, using rheological data on the material being calendered. Their method shows promise, but more experimental verification along these lines is needed before it becomes a useful engineering tool. Their equations are in the forms

$$F = 2KRW \left[\left(\frac{2n+1}{n} \right) \frac{U}{h_o} \right]^n \lambda \left(n, \frac{H}{h_o} \right) \qquad (8.15)$$

and

$$C = 2KRW(2Rh_0)^{1/2} \left[\left(\frac{2n+1}{n} \right) \frac{U}{h_o} \right]^n \mu \left(n, \frac{H}{h_o} \right) \qquad (8.16)$$

where

F = separating force on the rolls
C = torque on the rolls
K = coefficient of power-law viscosity

W   = width of the web
n   = flow index for power-law viscosity, such that η = apparent
        viscosity = $K[\dot{\gamma}]^{n-1}$
$\dot{\gamma}$ = shear rate
U   = peripheral speed of the rolls
H   = one-half the bank height
h_o = one-half the nip opening
λ, μ = functions depending only on n, once H is large enough, that
        can be graphed or tabulated

Relating Processing Rheology to Film Quality

Due to the difficulties associated with accurately measuring the temperature of stocks in calender nips, relatively few experimental data
are available on viscous heating effects during calendering. Lyngaae-
Jorgensen [39] calculated that viscous heating effects in a capillary
rheometer used on rigid PVC may produce a 20°C rise in stock temperature along the capillary at shear rates of 1,000 sec^{-1}.

Power input in calendering rigid PVC on a medium-sized four-roll
calender usually is 2.5 to 3.5 hp/ypm (G.W. Eighmy, private communication). If all this power caused adiabatic heating of the polymeric
web, the temperature rise going through the calender could be very
substantial, namely, from about 100 to over 400°F (55 to 222°C). This
is contrary to experience. But, in steady-state running, the calender
normally takes heat out of the stock.

Perlberg and Burnett [40] estimated for a particular flexible PVC
that the stock temperature rose from 334°F going into the entrance
nip to 360°F in the final nip, due to viscous heating. Assuming input
of viscous heat at the rate of 2545 Btu/hp hr, they indicated this
stock would have risen to a temperature of 441°F, but for removal of
heat by conduction to the calender rolls and by radiation to the sur-
roundings. These factors kept the stock temperature from rising
above 360°F.

Kiparissides and Vlachopoulos [41] calculated that most of the temperature rise due to viscous dissipation occurs close to the roll surfaces in a web passing through a calender nip and that excessive local
heating effects, enough to cause defects in quality of the film or sheet,
can occur during commercial calendering.

A fruitful study was reported by Bourgeois and Agassant [42] on
calendering of rigid PVC. As the calender was opened up to make
heavier gages, the film became defective due to air bubbles. This defect could be reduced by increasing the maximum pressure in the
bank, either by increasing the molecular weight of the PVC used or by
cautiously speeding up the calender. As the nip was closed down,
surface matte showed up in the film, and the thinner the film, the
lower the speed at which unacceptable matte was experienced. For a
given speed, increasing calender temperature or lubrication of the

compound decreased the thickness where matte appeared. Smaller
calender roll diameters also allowed calendering thinner films faster
without matte.

Automated Control Systems

The trend in modern calendering facilities is to automate more of the
control functions than was possible a few years ago. The operator
today is more of a coordinator than formerly, and is freed from attention to some of the detailed control functions necessary in older calender lines. Weighing of ingredients, mixing, fusion, and delivery of
fused stock to the first calender nip at the rate required by the calender is semiautomated in modern plants, but still requires overall
operator surveillance.

Speed controls on the calender rolls, the pick-off rolls, and the
rest of the calender train are very critical. Means for speeding up
and slowing down groups of rolls are necessary to make speed adjustments efficiently. For instance, it should be possible to speed up and
slow down the pick-off rolls in concert with the downstream train
and hold them within a range of ±0.1% of the set speed. Note that
digital tachometers read more accurately than the older analog units
and are now used more.

Temperature controls on the heat transfer fluids used on the calender and pick-off rolls should be capable of holding the fluids within
±1°C (±1.8°F) of their set points, with only minor fluctuations after
making a change in set point. Rolls that are run cold (cooling cans,
embossing rolls, etc.) should have systems capable of holding their
heat transfer media within ±5°C of their set points.

To achieve product versatility in a calender line, it is necessary
to have individual roll drives on the calender rolls and to have well-
controlled roll crossing and roll bending available for maintaining gage
profile across the web.

Perhaps the most striking recent improvement in calender control
systems is closed-loop, computerized web gauge control, based on the
attenuation of β-ray emissions passing through the web. In the United
States several such systems have been made by LFE and Measurex.
Ohmart, Industrial Neucleonics, and Accuray make somewhat less
sophisticated β-ray thickness gauges. The sensors do a full-width
scan of the web in one or two places, depending on the design of the
system. Digital thickness data and other online information, such as
nip openings, calender take-off roll speeds, roll-crossing, and roll-
bending data, are fed into a computer. The thickness data are averaged and displayed on a screen and can also be printed out.

The operator can use the averaged data to make decisions and run
the line on manual control, or the operator can switch over to automatic thickness control. The computer then controls nip opening,
roll crossing, and roll bending to hold the preassigned web thickness.

In this case, the operator concentrates on maximizing throughput by optimizing roll speeds, maintaining quality parameters other than gage, and coordinating activities related to production, such as running on-line tests for quality assurance.

Part of the relatively high cost of the LFE and Measurex systems is for the capability to print out roll by roll information on products as they are made, which is compiled by the computer. These data can be furnished to customers as part of the quality assurance program.

Two of the early claims made for computerized control [43] tended to oversell its alleged benefits. One was that it allows film manufacturers to control gage so well that they can run very close to cheating their customers, and so maximize profits. It suggested that a product sold by the pound should be run on the maximum gage specification, and a product sold by the yard should be run on the low end of the gage range. However, suppliers facing competition find that their customers are dissatisfied when a technological advance is misused in this fashion and tighten their specifications or even switch suppliers. Another claim is that "smoothed-over" data can be generated by this system to calm customers.

The true and worthwhile advantages of computerized gage control include quick and quantitative diagnosis of gap control problems (e.g., unplanned roll speed fluctuations, a gage band in a roll, malfunctioning bearings, out-of-round rolls, instability of preloaders, and cross-axis or roll bending), ease in controlling the thickness of deeply embossed products, and raising critical customers' confidence in the reliability of calendered rigid PVC as an engineering material. Manufacturing efficiency and quality yields can be much improved.

8.3.6 Controlling Finish

Calendered rigid PVC is sold in double-polished, matte, or embossed finishes. Specialty finishes can be obtained by pressing or laminating and pressing. Chrome plating the final two rolls on a calender allows the highest gloss film to be made, but highly polished steel rolls today approach this finish.

Where matteing is critical, surface quality of the film is measured by profilometer traces and scanning electron microscope (SEM) photographs. Use of corotating driven stripper rolls [44] helps in obtaining uniform side-to-side texture along with minimum residual strain on heavily matted film and sheet.

Inline embossing of rigid PVC film and sheet is practiced much less, proportionately, than it is for flexible PVC. An essential requisite for good embossing is to have the film hot enough when it reaches the emboser. For thin film whose temperature drops rapidly on leaving the calender, it is necessary for good embossing to run the pick-off rolls (or stripper rolls) hot and to have the embosser close to these. The

steel roll of the embosser is kept cool, but the rubber backing roll
tends to heat up, so a high-temperature rubber must be used for the
rubber roll in the embosser.

An interesting application is the use of a light "kiss emboss" to
smooth down a heavy matte finish that is too rough for optimal printing.
This method can allow longer runs of printing stock. By this means,
one can start a run with too rough a matte and smooth down the early
production with a kiss emboss until the matte "wears in." At this
point one can back off on the kiss emboss.

Printing of rigid PVC film or sheet is always done offline. But
printing stocks may be given inline or offline corona discharge treatment to improve their wettability by burning off excess lubricant and
superficially oxidizing the printing surfaces.

A good method for measuring the wetting tension of film surfaces is
ASTM D2578. This test is useful for optimizing the severity of a corona discharge treatment on rigid PVC, even though it was initially
devised for use on polyolefin films.

8.3.7 Drawdown and Cooling

Drawdown

It has been stated that film and sheet extruders cover the thickness
range 0.001 to 0.125 in., whereas calenders can only handle the range
0.005 to 0.025 in. [45]. More recently, calendering was assigned the
thickness range 0.002 to 0.050 in., and blow extrusion was quoted at
0.001 to 0.003 in., flex-lip extrusion was quoted at 0.005 to 0.0125 in.
on vinyls [8]. However, Meienberg indicates that a calender may produce 0.004 in. flexible or semirigid PVC film at 300 ft/min, and the
finished product is wound up at 750 ft/min in a thickness of 0.0015 in.
[46]. Efforts to achieve such speeds on thin rigid PVC buck up
against the limitations cited by Bourgeois and Aggasant [42]. Calendering at 90 ft/min and winding up at 240 ft/min after drawdown are
typical for thin rigid PVC films.

Space is lacking to discuss drawdown procedures in depth. However, in properly engineered installations, it has become a routine
procedure to calender rigid PVC film to gages down to 4 mils and then
to draw it down to 1 mil, using a 4:1 drawdown ratio. Intermediate
draws of 3:1 and 2:1 can then be used for 1.3 mil and 2 mil film, respectively, without changing calendering conditions.

This is a far better procedure than trying to calender to 1 mil thickness directly, because drawdown gives a better product (smoother
surface at commercially interesting speeds), runs with less heat generation in the nip, much lower roll separating forces, entails less risk
of damaging the final two calender rolls by running them together, or
from small pieces of metal that get by the detector. It also uses less
power and thus saves money.

Current opinion favors using five or more pick-off rolls with diameters no greater than 7 in. The first pick-off roll should be located
within 0.5 in. of the final calender roll. Most of the drawing should be
done between the calender and the first pick-off roll.

On typical rigid PVC formulas, to avoid increasing strain recovery
in the course of drawing down, the final calender roll should be run
at or above 190°C, while the first pick-off roll is normally run at about
165°C, unless the film is to be postembossed. In that case the pick-
off rolls are run hotter, at a temperature dictated by the specifics of
the particular emboss.

Tentering rolls on each outside edge of the film on the roll 1 pick-
off may be used to restrain lateral neck-down (narrowing of the web)
during drawdown.

Relaxing, Tempering, and Cooling

The first essential for satisfactory performance of the cooling section
of the calender train is that all the driven tolls be properly synchron-
ized; otherwise, excessive recoverable strains or wrinkles are induced
in the film. Second, there should be no long (1 yard or more) horizontal or vertical runs where the warm film has to support its own
weight. Temperature control on tempering and cooling rolls should be
such that wrinkling induced by thermal shock can be avoided.

After initial relaxing and tempering, most modern calendering lines
include a long enough horizontal run to accomodate a β gauge (about
2 ft). Following this are several large-diameter cooling cans so
threaded that alternate sides of the film or sheet contact them over
180° or more of their cooling surface. The number of cooling cans has
to be selected for the heaviest gage products to be run, which cool
more slowly than thin-gage products. Film or sheet should be adequately cooled all the way through before it is wound up or stacked. Otherwise, numerous defects are courted.

In the production of rigid PVC film and sheet in the United States,
inline roller stretching equipment and inline tentering equipment have
not achieved acceptance. Such operations, if needed, are better done-
offline where they do not impact the operating of a calender. Tough-
ening due to balanced orientation and the ability to shrink in a heat
tunnel for shrinkwrapping require controlled stretching.

8.3.8 Trimming, Slitting, and Windup

The old adage, that the strength of a chain is that of its weakest link,
applies nowhere in calendering more forcefully than at the downstream
end of the line. Sometimes an older line has been modernized, with
drawdown capability added and fusion system capacity increased at
great cost. Cooling may become marginal, unless it, too, is upgraded.
All too often, however, inadequate trimming, slitting, and windup

equipment is left in place and the performance of the line, then, remains poorer than it should be. The inadequacy of cooling may not be manifest until the slitting and windup system is also upgraded.

If inline slitting capability cannot handle a particular job properly, it is usually better to run the line at full speed and width and to slit offline, than to struggle with an overranged inline slitter that causes the calender to run a narrow web slowly and with frequent off-quality excursions.

Score-type trimming equipment may be used on rigid PVC where the film or sheet is still warm and therefore soft, but these devices must be kept very sharp to avoid giving ragged edges to the trimmed film. Shear-type trimmers are more expensive, but do a more consistent job and help achieve high operating efficiency along with good-quality yields.

Before selecting slitting and trimming equipment it is important to understand the requirements of the markets to be served. If the majority of orders are less than 10,000 lb and require slitting, generally these should be slit offline. Also, if the widths are less than 9 in., they should be slit offline.

Some rigid vinyl customers want goods in widths of 27 or 54 in., which can be accommodated on a 66 in. wide line, but which are bad for an 84 or 88 in wide line, since neither of the latter can run 27 in. three-up. Widths of 39 and 40 in. run well on a 51 or 54 in. line, but require a 96 in. line to be run two-up. A width of 72 in. can just be handled on an 84 in. line, and more comfortably on an 88 in. line. Lines of these sizes are satisfactory for running 12 in. goods six-up.

Automatic cutting of rigid PVC webs only works up to about 20 mils on most slitter or winders. If a substantial portion of the business is for gages above that number, it may be necessary to have inline sheeting equipment. These heavy-gage stocks are more often sold as sheets than as rolls. On a sheeter, accuracy and cleanness of cut are important. Also, a vacuum device to remove angel hair should clean off each sheet prior to stacking. For the cleaner to function properly, charges of static electricity must be dissipated or prevented from occurring in the first place.

For many years, accumulators were installed on film lines to allow winding to be interrupted while a manual cut was made at right angles to the machine direction. Winders that cut the web and start a new roll automatically are more popular today. Accumulators are not being installed on many modern lines.

If the markets sought require it, provision should be made for winding up on 3, 6, 8, and 12 in. cores with no foldback at the start of the film on the core (to minimize warpage of the inner plies).

Tension should be controllable and sensibly constant throughout the winding process, to avoid the extremes of loose, sloppy rolls on the one hand, or telescoped rolls with crushed cores on the other. Surface winders are used for most flexible PVC webs and work also on

thin rigid PVC film. For thicker and stiffer rigid PVC roll goods,
center winders are generally recommended. A versatile line sometimes
is set up with both a surface winder and a center winder, each to be
used when appropriate.

8.3.9 Use of Rework

Appropriate use of film rework is a key element in optimizing the profitability of calendering rigid PVC. Edge trim is normally ground up
as it is produced and as much as possible put back into the run from
which it comes. Other film rework should be segregated by formulation,
color, and grade. If it is off-grade due to a materials problem, it
should be worked off only in noncritical items.

Best results are obtained with film rework when the grind is such
that the longest dimension is 3/16 in. or less. Rework in roll or pellet
form must be used very cautiously.

The ratio of rework to virgin stock should be controlled very
closely, based on experience and common sense. The system should
be such that it is virtually impossible by mistake (e.g., on a night
shift) to charge non-FDA-regulated rework into an FDA-regulated
formulation.

ACKNOWLEDGMENTS

Thanks are due to G. W. Eighmy, Jr., for his many helpful comments
on an early draft, and to my wife, Barbara, for her help with clarify-
ing the language, typing the manuscrift, and doing the artwork.

REFERENCES

1. E. M. Chaffee, U.S. patent, Making rubber fabrics, August
    31, 1836.
2. J. Pickersgill, British patent 7178, Preparing and applying In-
    diarubber to fabrics, 1836.
3. A. Magram, Panel on economics of the PVC industry, J. Vinyl
    Technol. 2(1):26-27, 1980.
4. British Intelligence Objectives Subcommittee, Final Report No.
    445, H. M. Stationery Office, London, 1946, pp. 32-4, 74-97.
5. F. B. Makin, British Plastics 28:500, 1955.
6. Goodyear Tire and Rubber Co., British patent 723,041, 1955.
7. J. Brown, Institution of Rubber Industries Annual Reports on
    the Progress of Rubber Technology, 19:135, 1955.
8. D. V. Rosato, Plastics World 37(7):64, 1979.
9. G. W. Eighmy, Jr., Modern Plastics Encyclopedia, 1982-1983,
    p. 220.
10. Ernst and Whinney (compilers), Monthly Statistical Reports, The
    Society of the Plastics Industry, Inc., 1979, 1980, 1981, 1982.

11. Modern Plastics, Materials 1981, *58*(1):67, 1981.
12. Modern Plastics, Materials 1982, *59*(1):77, 1982.
13. K. H. Michl, Kunststoffe *70*(10):591, 1980.
14. Modern Plastics *59*(10):16, 1982.
15. Decision of Customs Court in Sekisui Products, Inc., vs. United States, 63 Customs Court 123, C.D. 3885, September 17, 1975.
16. *Federal Register*, 40:171, 40529, September 3, 1975.
17. T-M. Baek, K-H. Chung, and R. Salovey, J. Vinyl Technol. *3*(4):208, 1981.
18. C. Kiparissides and J. Vlachopoulos, Polymer Eng. Sci. *16*(10): 712, 1976.
19. F. Dobbels and J. Mewis, A.I.Ch.E.J. *23*(3):224, 1977.
20. M. F. Saggese, F. V. Owens, and B. Fila, U.S.patent 4,219,640, Polymers of vinyl chloride having low VCM content, assigned to Tenneco Chemicals, Inc., August 26, 1980.
21. H. Kopsch, Kunststoffe *71*(10):743, 1981.
22. M. D. Stone and A. T. Liebert, TAPPI *44*(5):308, 1961.
23. G. F. Carrier, J. Appl. Mech. *72*:446, 1950.
24. K. J. Gooch, Modern Plastics *34*(7):165, 1957.
25. M. Jukich, Modern Plastics *33*(8):138, 1956.
26. R. C. Seanor, SPE Tech. Papers, *12th ANTEC*, Vol. II, 1956, p. 298.
27. R. C. Seanor, Mech. Eng. *79*:293, 1957.
28. G. Ardichvili, Kautschuk *14*:23, 1938.
29. F. H. Ancker, Plastics Technol. *14*(12):50, 1968.
30. D. D. Eley, J. Polymer Res. (renamed J. Polymer Sci.) *1*(6): 529, 535, 1946).
31. R. E. Gaskell, J. Appl. Mech. *72*:334, 1950.
32. J. T. Bergen and G. W. Scott, J. Appl. Mech. *73*:101, 1951.
33. F. D. Dexter and D. I. Marshall, SPE J. *12*(4):17, 1956.
34. W. W. Alston and K. N. Astill, J. Appl. Poly. Sci. *17*:3157, 1973.
35. H. Munstedt, J. Macromol. Sci.-Phys. *B14*(2):195, 1977.
36. J. C. Chauffoureaux, C. Dehennau, and J. Van Rijckevorsel, J. Rheol. *23*(1):1, 1979.
37. J. Vlachopoulos and A. N. Hrymak, Polymer Eng. Sci. *20*(11): 725, 1980.
38. J. F. Agassant and P. Avenas, J. Macromol. Sci. -Phys. *B14* (3):345, 1977.
39. J. Lyngaae-Jorgensen, J. Macromol. Sci. -Phys., *B14*(2):213, 1977.
40. S. E. Perlberg and P. P. A. Burnett, *Encyclopedia of PVC*, Vol. 3, (L. Nass, ed.), Marcel Dekker, Inc., New York, 1977, p. 1395.
41. C. Kiparissides and J. Vlachopoulos, *Polymer Eng. Sci.* *18*(3): 210, 1978.

42. J. L. Bourgeois and J. F. Agassant, J. Macromol. Sci.-Phys.
     B 14(3):367, 1977.
43. Plastics Technol. *25*(5):24, 1979.
44. F. D. Nicoll, U.S. patent 4,311,658, Manufacture of continuous
     plastic sheets, assigned to Tenneco Chemicals, Inc., January 19,
     1982.
45. J. Pegram and F. I. Warwick, SPE Tech. Papers, ANTEC,
     Paper XXX-1, March 1966.
46. J. T. Meienberg, *Modern Physics Encyclopedia*, 1981-1982,
     McGraw-Hill, Inc., New York.

ADDITIONAL READING

Farrel Machinery Group, Emhart Corp., Ansonia, Connecticut,
    Bulletin No. 233, Plastic Calender Lines, 1980.
Farrel Machinery Group, USM Corp., Ansonia, Connecticut,
    Calender upgrading and repairs, 1977.
Field Information Agency, Technical: FIAT Final Report No. 866,
    August 1946, pp. 1-17. Good description of the Luvitherm process
    for making thin rigid PVC film.
Hermann Berstorff Maschinenbau GMBH, Hanover, West Germany, Cal-
    ender plants for processing plastics, 1975.
Hermann Berstorff Maschinenbau GMBH, Hanover, West Germany,
    Technical development of plastics calender plants, by Willi Wocken-
    er, 1974.
EKK Kleinewefers Kunstoffmaschinen, GMBH, Bochum, West Germany,
    Calandrette, 1979.
EKK Kleinewefers Kunstoffmaschinen, GMBH, West Germany, Extru-
    ders (includes description of planetary roller extruder PWE), 1979.
R. J. Krzewki and E. A. Collins, J. Macromol. Sci.-Phys., *B 20*(4):
    443, 1981.
L. C. Uitenham and P. H. Geil, J. Macromol. Sci-Phys., *B 20*(4):593,
    1981.

9

Thermoforming of Rigid PVC Sheet

ROBERT A. McCARTHY/Springborn Laboratories, Inc., Enfield, Connecticut

9.1 Introduction 439

9.2 The Thermoforming Process 440

9.3 Thermoforming Techniques 443

9.4 Rigid PVC Thermoforming 449

9.5 Markets for Thermoformed Rigid PVC 451

References 453

9.1 INTRODUCTION

The shaping of heated thermoplastic sheet using vacuum and pressure dates back to the 1930s in the United States and a drape-forming process, in which the heated sheet is smoothed by hand over a form, was developed even earlier. Although thermoforming has been improved in speed and sophistication, it is still not considered a major plastic-fabricating process. Its main advantages are inexpensive molds and equipment, which make the process feasible for relatively short runs. In addition, there is little limitation on part size, multiple and differing molds may be used, thin-walled (15 mils or less) parts can be formed with reasonable strength, and preprinted sheet can be used, avoiding some secondary operations. The drawbacks of the process, assuming acceptable sheet is available, center around design limitations that restrict the detail in parts (unless expensive tooling is used), limit the sharpness of corners, and require relatively uniform wall thicknesses in finished parts. Trimming of brittle plastics can be a problem, and the appearance of a trimmed edge may not be acceptable. Trimmed

scrap must be recycled into sheet for best economics, and hot sheet
is easily marred when touched by tooling during the thermoforming.

One segment of the thermoforming market is primarily devoted to
packaging and disposable containers, utilizing thin sheet made into
thousands of small parts. The other segment of the market is involved
with larger functional parts made from relatively thicker sheet. Typical of the large-area parts are refrigerator doors and liners, seating,
and tote bins. Competitive plastic processes are structural foam injection molding and, to some extent, blow-molding. The thinner sheet
and film applications are in packaging, fast food restaurant serving
containers, and vending machine cups. Here, the competition is
from parts made by injection molding or from paper.

9.2 THE THERMOFORMING PROCESS

The thermoforming process is relatively simple to describe. It involves the heating of a thermoplastic sheet to a point of softness at
which it can be formed, without rupture, by external forces while the
sheet is held in a clamping frame. The sheet is cooled, and the parts
are trimmed from the web.

The object of the processing is to distribute the plastic sheet over
the mold dimensions without weakening it by excessive thinning. The
selection of the sheet thickness to be used is a compromise between
having acceptably strong finished parts and leaving a heavy weight
of scrap for recycling or sale.

Sheet temperatures for forming range from 225 to 425°F for most
thermoplastic materials. The heating of the sheet can be accomplished
by conduction, convection, or radiation. Conduction is used for very
thin sheet, which is quickly heated, since hot plastic sheet is easily
marked by direct contact with metal heaters. Convection is a slow
method, usually done in a circulating air oven, and sheet sagging can
cause indexing problems. Radiation is the most commonly used heating
technique, employing electrical radiant heaters that are either bare,
glass covered, or ceramic covered. Reflectors are used behind the
heaters to concentrate the radiant energy. Since heating from one
side is inefficient for thick sheet, due to plastics' slow conduction, two
heating banks are often used. Some convection heating occurs as
well. Spacing between the sheet and the heaters must allow for sag of
the sheet and the sensitivity of the particular sheet to overheating.

The time required for heating is dependent on the thermal characteristics of the sheet and the forming temperature that must be
achieved. This is a controllable variable. The output of the heater
bank, or banks, is also controllable by the design of the placement of
heaters, their type and intensity, and the use of reflectors.

The sheet is shaped by a variety of methods to achieve the objective of acceptably strong finished parts. After forming, the sheet is

cooled by ambient air, forced drafts, or a water spray unit so it can be safely removed from the mold. Trimming of the formed parts from the surrounding web can take place either while the sheet is still in the clamping device, or the sheet, with parts, can be removed to separate trimming equipment. The web, or trim scrap, is collected, granulated, and returned for sheet extrusion, in most cases, or sold.

The basic thermoforming machine is a single-station machine as seen in Fig. 9.1 This equipment uses individual sheets of plastics. Single-

Figure 9.1 Single-station machine. (Courtesy of Brown Machine Co., Beaverton, Michigan.)

Figure 9.2 Rotary thermoformer. (Courtesy of TMI Division, Thorne United, El Segundo, California.)

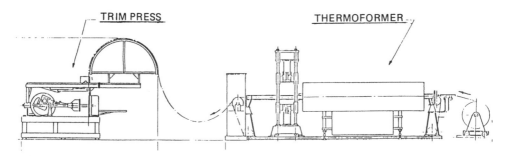

Figure 9.3 Typical continuous thermoforming line.

station equipment is mainly used for large-area part production. The requirements of the refrigeration industry for faster production led to multiple station thermoformers. An example is seen in Fig. 9.2. Here, the sheet is loaded at one station, moved to one or more stations for heating and forming, returned to the starting station for unloading, and removed for trimming. The assembly rotates, giving this equipment the name of rotary thermoformer. Four-station machines have two heating stations, and forming is carried out at the second heating station. However, three-station machines are more common.

Packaging applications require even faster speeds. Probably the first high-speed machine was an automatic thermoformer developed by C. B. Strauch in the late 1930s and purchased by the Plax Corporation in 1938 [1]. This all-mechanical, cam-operated unit was continuously fed a roll of sheet plastic that was forwarded by chain-mounted grippers into a radiant heating oven. A forming and trimming station followed. Forming was by air pressure only. The Strauch machine was ahead of its time because the only available thermoformable sheet was cellulose acetate, and markets for parts made from this plastic were limited. Nevertheless, its operating features were very much like modern, high-speed vacuum and pressure formers, as seen in Fig. 9.3.

The basic elements of the thermoforming process are (1) a plastic sheet thick enough to produce parts of acceptable strength after the drawing process, (2) a heating chamber, (3) a clamping device that holds the sheet so that it can resist forming pressures, (4) a mold with some form of cooling to prevent overheating due to constant contact with heated plastic, and (5) some sort of force to cause the sheet to assume the shape of the desired part. Heated sheet moves from the oven and is shaped by the mold, either male or female, by a variety of forces, to be discussed in the next section. Once formed, the part remains in contact with the mold unit and is rigid enough to retain its shape. It is then blown out of the mold as the mold is removed. Additional cooling can be provided by fans or water mist spray to prevent distortion of the part in subsequent indexing and to make trimming easier [2].

9.3 THERMOFORMING TECHNIQUES

When heated sheet plastic touches cooler metal, it cools rapidly and the amount of drawing subsequently done is limited. There is little drawing from the sheet adjacent to the clamps. The aim is to arrange for the drawing to be done in such a manner that the wall thicknesses in the finished part provide enough strength for the intended use. Heat patterning can be done to provide nonuniform sheet heating, so that differential drawing will occur. This is not an accurate procedure; therefore, a forming plug is often used to prestretch the hot sheet before it enters the mold. Air or vacuum pressure then completes the forming.

Vacuum and drape forming are common simple techniques, not using plug assist, which are seen in Figs. 9.4 and 9.5. Both show the forming done with a sheet held in the lower clamping device, or platen. The platen is lowered over the mold, and vacuum is applied through the mold to bring the sheet into conformance with the mold. The vacuum forming is done in a female mold and could also be done by forcing the sheet down into the mold with air pressure. This technique is used for relatively shallow draws. Parts have good detail on the mold side, have thin sections in bottom corners, and shrink away from the mold on cooling for easy removal.

Drape forming is done with male molds and is used where draw depth is greater than can be done by vacuum forming. The platen is moved down quickly over the form (or the mold moved up) so that the sheet is stretched by the mold before vacuum is applied. Parts have good detail on the mold side and the bottom of the part is thick, since the mold touches the sheet there and stretching is retarded. Drag lines can occur between the chilled sheet and sheet that has not touched metal, but rapid drape speeds and warm molds can reduce this problem. Webbing between parts may have to be corrected by metal assists.

Where the depth of the draw greatly exceeds the width of the part, the term "high draw ratio" is used to describe this condition. Severe thinning of the part at the bottom of the draw occurs. More prior stretching of the sheet is needed than can be provided by drape forming, and the plug assist is used. The plug, located usually above the sheet, is brought in contact with the sheet. The sheet touching the plug is thicker than that not touched as it moves into the mold. When vacuum is applied, the bottom of the part is less thinned out than would be the case with other techniques. This can be seen in Fig. 9.6.

Variants of plug assist forming are many to achieve a sheet distribution most advantageous to the finished part. The sheet may be blown

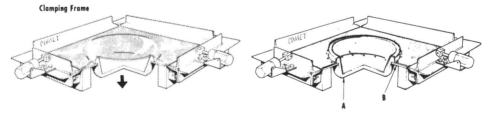

Figure 9.4 Vacuum form lower platen negative mold. (Courtesy of Comet Industries, Richmond, Indiana.)

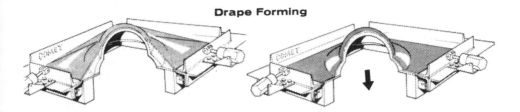

Figure 9.5 Drape forming positive mold. (Courtesy of Comet Industries, Richmond, Indiana.)

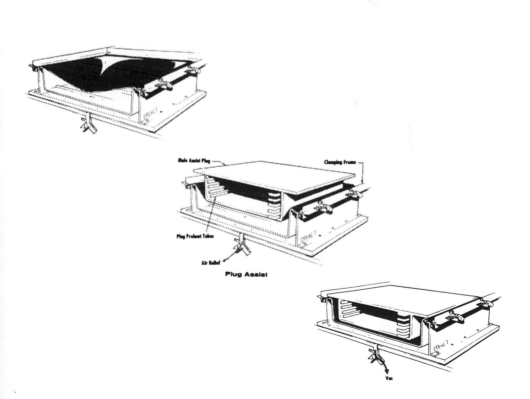

Figure 9.6 Plug assist lower platen. (Courtesy of Comet Industries, Richmond, Indiana.)

by air pressure into a bubble, or billowed, until it touches the plug.
The sheet is thus stretched before it meets the plug. The sheet,
loosely conforming to the plug, moves into the mold where vacuum is
applied. This can be seen in Fig. 9.7. The sheet may be pulled into
a chamber, billowed ahead of an advancing plug by air pressure, and
pulled back on the plug (which is also the male mold) by vacuum.
This can be seen in Fig. 9.8. The sheet may be billowed up from the
platen, billowed down by air pressure through the plug, and pulled
into the mold by vacuum, as seen in Fig. 9.9. More variants are in
use. All these techniques have advantages and disadvantages depending on the application.

   In practice, a sheet-fed single-station machine is often used to
determine the sheet thickness and thermoforming technique to use.
Prototype molds may be used at this point and approximate cycles established. Tooling for multiple molds, if such are used, is made. Roll-
fed or multiple station thermoformers may then be used. Experimental runs are carried out, changing the variables of heat and heat distribution, heating time, timing of air and/or vacuum application, and

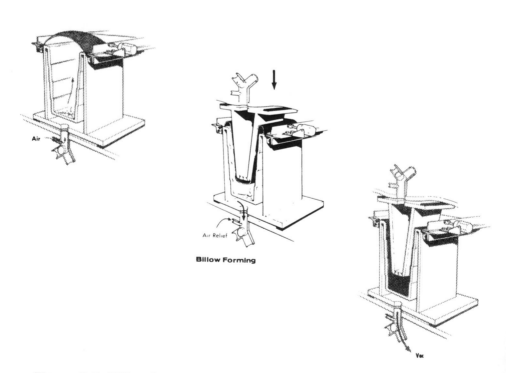

Billow Forming

Figure 9.7 Billow forming negative mold lower platen with plug assist.
(Courtesy of Comet Industries, Richmond, Indiana.)

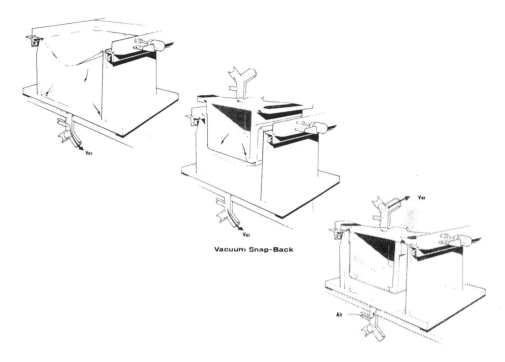

Figure 9.8 Snap-back with air cushion top platen. (Courtesy of
Comet Industries, Richmond, Indiana.)

plug movement and plug speed, but not the effects on part dimensions
and wall thicknesses until the optimum conditions occur. Thermoforming is still more of an art than a science.

Mold temperatures, plug temperatures, and even the temperature
of the pressuring air are important. Ambient air temperature and
drafts in the production area can cause changes in the product. For
this reason, quality control is difficult.

Shrinkage is considerable during cooling, and there is some postforming shrinkage that must be allowed for. Trimming forms the part
edge, and tools must be kept sharp for best appearance. Trimming
in the mold is an advantage with multiple molded parts since it avoids
double handling.

Molds are generally made of aluminum, cored for water cooling and
for better heat conduction rates. Extended production runs can afford to use metal molds. Wood or epoxy molds can be used for prototype products, where heat buildup is less of a problem. Aluminum
molds are polished or vapor blasted, depending on the surface re-

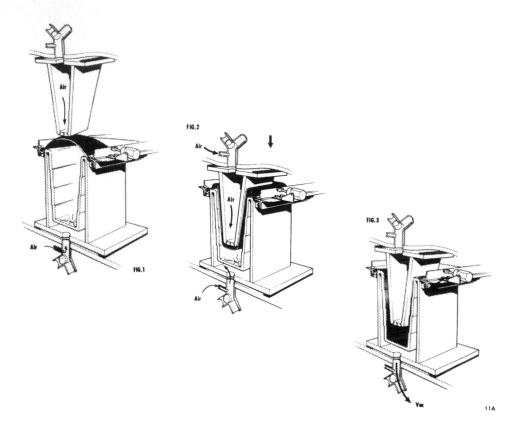

Figure 9.9 Plug assist with air cushion lower platen. (Courtesy of Comet Industries, Richmond, Indiana.)

quired in the part. Plug assists may be of aluminum, with water cooling, or may be of materials selected to prevent surface marking of the sheet where the cooler plug contacts it [3].

The sheet source, in addition to the material characteristics, is a consideration in thermoforming. Stresses that are built in during sheet production can be relieved during heating, which can change forming thicknesses. This causes problems unless the sheet is tested for shrinkage prior to use and adjustments made in the process. Stresses can also be produced in the thermoforming process by differential stretching or strains during demolding. These can affect shrinkage. With all these factors, thermoforming is a relatively sensitive process that can produce reproducible finished parts only with considerable expertise on the part of the operators, who must maintain consistent conditions of manufacture with relatively limited controls.

9.4 RIGID PVC THERMOFORMING

Thermoforming of PVC progressed slowly after World War II in the United States. It was not until the late 1940s that it was being done, although it was a common practice in Europe earlier in the decade. The lack of satisfactory sheet was the major problem. Applications requiring heat and chemical resistance required high-molecular-weight grades, and these had the poorest processability in both extrusion and thermoforming. Heat stability was the major problem.

Increased usage of PVC in pipe and profile extrusion led to improvements in formulations, and sheet extrusion benefited from these. Lead stabilizers were replaced in compounds by organic tin stabilizers, which improved overall processability, heat stability, and transparency. Fillers and lubricants were added, and processing aids were incorporated. Impact modifiers, lower molecular weight PVC grades, copolymers with vinyl acetate, and new plasticizers were evaluated in sheet. Many of these developments tended to improve processability at the sacrifice of hardness, modulus, heat resistance, and chemical resistance. The major thermoforming product was a thick, opaque vinyl acetate copolymer sheet, mostly used for chemical-resistant linings. Calendered sheet was less thick and was satisfactory for thermoforming relief maps, signs and displays, toys and games, and milk bottle caps. The 1957 production of copolymer sheet was 13 to 15 million pounds, much of which was used in thermoforming, although some sheet went for credit cards, playing cards, book bindings, and templates.

Extruded copolymer sheet was experimentally made in widths up to 24 in. in 1957 and to 36 in. in 1959. Calendered sheet was available up to 48 in., but lamination was required to achieve sheet thick enough for parts for the chemical process industry. In 1960, packaging became a major use of rigid PVC, and 2 million pounds was used in that year as thermoformed parts. Extruded rigid vinyl sheet in thicknesses from 2 to 30 mils became available from Flex-O-Glass in Chicago and Nixon-Baldwin (now part of Tenneco Chemicals, Inc.) in Nixon, New Jersey. Thermoformable clear sheet for blister packaging was also available. Reynolds Metals made both cast and extruded thin film for packaging, as did Borden, Atlantic Refining, Clopay of Cincinnati, Ohio, and Filmco of Aurora, Ohio. Goodyear Tire and Rubber, a pioneer in the extrusion of minimally plasticized PVC, had Vitafilm, which competed with cellophane. Union Carbide had Bakelite VGA 0403, an unplasticized extruded PVC film of high clarity, good printability, and acceptability to the FDA.

Nixon-Baldwin introduced Vynex, a clear, unplasticized calendered rigid vinyl sheet in 24 in. wide rolls of 2 to 80 mils in thickness at the 1961 Packaging Show. The problem of fish-eyes in the film, often seen in earlier clear sheet, appeared to have been eliminated. This FDA-accepted film was 30% cheaper than cast vinyl film and is be-

lieved to have been an English import made under a BASF license.
The German PVC fabricators, with much more experience in rigid PVC,
were then extruding PVC homopolymer up to a meter in width in
Europe.

By 1963, between 5 and 10 million pounds of imported unplasticized
homopolymer calendered film was being used in the United States, but
less than half went into thermoformed products. Luncheon meat packaging was developed by American Can for Hormel Packing. Nixon-
Baldwin built a plant to make their imported product domestically.
Stauffer Chemical and Hoechst built a joint-ventured PVC and calendered rigid film plant. Union Carbide and Siberling Rubber of New-
comerstown, Ohio, also became producers of rigid PVC film and sheet.
Growth of rigid PVC for thermoforming, using both extruded and calendered film and sheet, was rapid after this time, and by 1973, 63
million pounds were used in packaging alone.

The improvements in formulations and processing techniques that
allowed the production of high-quality transparent rigid PVC sheet increased the interest in thermoformed products. Low-gel resins, impact modifiers [such as terpolymer of methyl methacrylate-butadiene-styrene (MBS)] with good dispersibility, and heat stabilizers
that provided excellent early color protection, high transparency, and
the ability to perform at melt temperatures of 410 to 430°F, were important developments. [di-n-butyltin and di-n-octyltin-S, S'-bis(iso-
octylmercaptoacetates) are recommended as the most effective stabilizers
in this application.] The use of copolymer PVC was essentially eliminated by the use of an impact-modified homopolymer formulation that
had equivalent deep-draw properties [4].

Lubricants and processing aids must be considered together because
of their often opposite effects on the fusion characteristics of powder
PVC compounds. Very low levels of processing aid are required with
low molecular weight PVC since such resin fuses well and requires
little help. Indeed, the fusion of the compound really needs to be retarded for this process to prevent temperature rise driving the product off-color. This is achieved by reducing the level of processing
aid and by choosing lubricants which allow appropriate fusion rates.
Another critical function of lubricants is to decrease the tendency of
the very soft melt to adhere to metal surfaces [5].

Clean recycle, moisture removal, and clean, smooth, and well-
chromed surfaces plus careful control of gage were found to be critical
in providing a good sheet for thermoforming. Gels and thick spots
that were flattened in processing show up on vacuum forming, and
good polish cannot be achieved with uneven gage [6]. With high-
quality sheet, thermoforming is not considered difficult. The lower
thermoforming temperature limit, below which unacceptable stresses
occur, is 200°F. The upper limit is believed to be 300°F. The normal
forming temperature is 250°F, and the temperature at which parts can
be removed from the mold, without warpage, is 150°F.

Grain direction in calendered sheet must be considered, since longitudinal orientation is present, which will be released on heating. Dimensional changes can affect preprinted patterns and wall thicknesses. With reasonable care of the amount of scrap recycled, 20 to 30% is acceptable, and an awareness that rigid PVC is a heat-sensitive material, no serious thermoforming problems should be encountered if operations are conducted within recommended temperature limits.

9.5 MARKETS FOR THERMOFORMED RIGID PVC

The early uses of rigid PVC in thermoforming were in chemical-resistant applications of shallow draw such as photographic development trays. Calendered sheet, available in the early 1950s, was thermoformed into small individual serving packs for jelly by Kraft. Vinyl milk bottle caps and fruit trays were also made. The total volume of thermoformed rigid PVC, of the type available in 1951, was estimated at 9 to 10 million pounds. The largest volume was in nonpackaging applications [7].

As formulations improved and clear, gel-free, sheet became available in the early 1960s, packaging usage increased rapidly until the 1974 volume in thermoformed packaging was 65 million pounds and nonpackaging was 12 million pounds. In 1975, the possible migration of vinyl chloride monomers (VCM) from PVC used in the packaging of foods caused much of the food-packaging industry to change to other products, and an estimated 24 million pounds of volume was lost by 1980 [8].

Drug packagers added a barrier coating, or lamination, and so were able to satisfy FDA regulations. Although no prohibition of the use of rigid PVC in food packaging occurred, the uncertainty regarding continued acceptability limited growth. The total 1980 consumption of rigid extruded and calendered film and sheet was estimated at 123 million pounds. This was about 5% of the total market for thermoformed plastics. Table 9.1 shows a distribution of this usage.

Tool and hardware packaging has long been a market for tough, clear PVC sheet, but medical supplies have become a newer usage. These products do not fall under FDA regulations. High gloss and clarity are important, and applications are for test kit containers, blister packs for disposable instruments, and boxes for syringes and thermometers, for example. The sheet used is normally 20 to 25 mils.

Drug packaging has grown rapidly, spurred by the increase in individual dosage packaging. The sheet is usually laminated with a barrier film for increased moisture transmission resistance.

The packaging for pens, lighters, razors and blades, tools, and hardware generally uses a $7\frac{1}{2}$ mil sheet for blister packaging. The air freshener container application uses a 15 mil, glossy, opaque-colored sheet of PVC. The major user is Johnson Wax of Racine, Wisconsin, whose product appears to be losing its market share.

Table 9.1 Rigid PVC Used in Thermoforming

Application	1980 Volume, million pounds
Packaging	
Tools and hardware	20
Medical supplies	17.
Food packaging	14
Drug packaging	13
Pens and lighters	7
Razors and blades	4
Air fresheners	4
Cosmetics packaging	3
Subtotal	82
Non packaging	
Cooling tower trays	16
Fire vent skylights	2
Novelties, masks, other	23
Subtotal	41
Total	123

Source: From Ref. 9.

The cooling tower application is a thermoformed tray for cascade and trickle-type water-cooling towers, where the water flows from tray to tray. The trays replace baffles of asbestos-cement sheet. The PVC trays are used in larger cooling towers since the asbestos-cement sheets are very heavy. Cascade trays use a 20 to 25 mil sheet, and trickle trays use a 10 mil corrugated sheet. Both calendered and extruded sheet are used.

Fire vent skylights use heavy, 50 to 125 mil, extruded clear sheet. These units have a price advantage over acrylic and drop out of the frame at relatively low temperatures, venting hot gases from a fire.

Novelties, masks, and other products are part of the multitude of minor applications into which PVC sheet is thermoformed. Wall plaques, shutters, and signs are a significant portion of this usage. Imported and reprocessed PVC sheet plays a large part in this market.

PVC has managed to ward off threats to its blister packaging market up to this time. A new polypropylene grade of sheet has been developed targeted for pharmaceutical blister packaging that could prove to be major competition. This polyolefin sheet is reported to have improved clarity and stiffness over conventional polypropylene, and economics are reported to be 25% more favorable. If vapor barrier properties are adequate, or can be improved by lamination or coating, it could be a threat to PVC. Thermoformability could favor PVC.

REFERENCES

1. *Thermoforming, 1978-1982*, Springborn Laboratories, December 1978, p. 71.
2. *Thermoforming, 1978-1982*, Springborn Laboratories, December 1978, p. 80.
3. J. L. Throne, *Plastics Process Engineering*, Marcel Dekker, Inc., 1979, p. 636.
4. R. C. Neuman, Extrusion of thin gauge transparent rigid vinyl sheet Part II. Raw materials and operating requirements, BF Goodrich, *ANTEC 82*, May 10-13, 1982, p. 519.
5. *Ibid.*, p. 520.
6. *Ibid.*, p. 521.
7. Modern Plastics *June*:75, 1966.
8. Modern Plastics *January*:63, 1981.
9. *A World Analysis of the Vinyl Plastics Industry & Markets*, Springborn Laboratories, January 1982, p. 81.

Index

A

Acetone immersion test, 18, 86, 236
Acrawax C, 36, 38, 296 (*see also* Ethylene bissteara-mide)
Acrylic elastomers, 41, 212, 289, 342
Acrylonitrile-butadiene-styrene (ABS) impact modifier, 4, 40, 395, 407
Acrylonitrile-methylmethacry-late-butadiene-styrene (AMBS), 40 – 41, 342, 394, 395, 407
Acrylonitrile-styrene copoly-mers (SAN), 54, 339, 400, 402
Activation energy of flow, 113, 124-125, 141-143, 148
Additive master-batch, 45, 49, 53
Adiabatic energy economy, 163-164
Adiabatic process, 80, 84, 163, 164, 165
Adjuvant heat stabilizers (See also costabilizer), 16, 24, 71, 73, 76, 297-299

ADS extrusion blow-molding machines, 258, 263, 270, 274, 287-288
Alkyl-aryl phosphite, 24, 25, 27, 31, 44, 49, 71, 76
All-acrylic impact modifiers, 4, 41, 342 (*see also* Acrylic elasto-mers)
All-acrylic processing aids, 34, 49, 67, 71, 76, 296, 339, 394
Alphaphenylindol, 24, 31, 292 (*see also* α - Phenylindole)
American Society for Testing and Materials (ASTM), 22, 23, 49, 236, 399, 403
β-Aminocrotonate esters, 24, 31
Andouart two-stage extruders, 77, 251, 252, 253
Anger twin-screw extruders, 69, 71, 230, 247-249
Antimony heat stabilizers, 24, 32-33
Antimony-S,S' S"-tris (isooctyl-mercaptoacetate), 24, 32-33, 36
Antioxidants, 56, 71, 76, 292, 297
Antiplasticization, 409
Apparent melt viscosity, 33, 112, 118, 122, 124, 220, 428, 430

[Apparent melt viscosity]
  versus temperature, 120-122
Associated bulk properties, dry
  blend, 55
Automa extrusion blow-molding
  machines, 286
Average degree of polymerization ($\bar{P}$), 22-23, 303-305

                        B

Baker-Perkins MPC/V compounding extruders, 79
Banbury mixers, 12, 40, 43, 46,
  84-86, 290, 411 (see also
  Internal fluxing mixers)
Banbury operation, 89
Barium-cadmium heat stabilizers,
  16, 24-26, 32, 36, 49, 71,
  73, 334 (see also Cadmium
  heat stabilizers)
Barium stearate, 26
Barrier properties, 9, 12, 56,
  279, 398, 400
Barrier-type screws, 8, 156,
  187-190, 258-259
Barus effect, 7, 20, 99, 118,
  129-130 (see also postextrusion swelling, percentage puff-up, percentage
  memory)
Bekum extrusion blow-molding
  machines, 246, 257-258,
  260, 263, 265-269, 283-
  285, 369
Bell extrusion blow-molding machines, 249, 260, 263,
  269
Benzophenone UV absorbers, 44
Benzotriazole UV absorbers, 44,
  71, 72
Biaxial orientation, 11, 278-280,
  287
Blow-molding
  via extrusion, 8, 9, 132, 246-
  300 (see also Extrusion
  blow-molding)

[Blow-molding]
  via injection, 9, 10, 363-383
  (see also Injection blow-
  molding)
Blown, rigid PVC film, 396, 433
Bottles, blowing, 246
Brabender torque rheometer, 141,
  325-332, 409
BTA III MBS impact modifier, 41,
  298
Bueche relation, 119
Bulk density, 17, 46, 53, 55, 57
Bulk polymerization, 17-19, 35,
  53 (see also mass poly-
  merization)
Bundesgesundheitsamt (BGA),
  29, 30, 291, 293
Buss-Kneader, 13, 77, 80-83, 411
Butyltin maleates, 24, 30, 52, 72
Butyltin mercaptoesters, 24, 28-
  29, 52, 76, 211, 394-395

                        C

Cadmium heat stabilizer, 16, 24-
  26, 32, 36, 49, 71, 73, 334
  (see also Barium-cadmium
  heat stabilizer)
Calcium carbonate, precipitated,
  34, 42, 70, 207, 212, 343
Calcium stearate, 26, 34, 36, 37,
  49, 73
Calcium-zinc heat stabilizer, 24,
  31, 32, 76, 291, 297-299,
  334, 401
Calendered rigid PVC products,
  12-13
  cooling tower baffles, 404
  credit card overlay film, 403-
  404
  credit card stock, 402
  floppy disk cartridge film,
  401-402
  insulation facing for pipe, 405
  light-gage films, 404
  orange mask film, 404
  packaging film, drug, 401

[Calendered rigid PVC products]
  packaging film, food, 399
  packaging film, industrial, 402
Calenderettes, 13, 387
Calendering lines for rigid PVC
  automated control systems,
    431
  capital investment, 386
  companies, 387
  roll arrangements, 387, 388
Calendering process, factors
    controlling web thickness
  drawdown, 391
  pick-off rolls, 391
  roll-separating forces, 389,
    390, 426-430
  roll speeds, 387
  tentering, 393
Calendering rigid PVC, 5, 19,
    31, 78, 385-436
  formulations, 393, 394, 395
  manufacturing operations
    controlling finish, 432-433
    drawdown and cooling, 433-
      434
    feeding the calender, 412
    fluxing mixers, 411, 412
    raw materials, 406-410
    relaxing, tempering, 434
    tentering, 393, 434
    trimming, slitting, and wind-
      up, 434, 435
    use of accumulators, 435
    use of rework, 436
    weighing and mixing, 410,
      411
  market volume statistics, 393,
    396
  processing and product prob-
    lems, applying theory to
    solve, 413--432
    belly, 417—418
    gauge problems, 414, 415
    race-tracking, 415-417
    roll deflection, 418, 425
  relating rheology to film quali-
    ty, viscous heating, 430

[Calendering rigid PVC]
  roll deflection, compensation
    use of cross axis, 389
    use of roll-bending, 390,
      423
    use of roll crown, 390, 420,
      421
Capillary flow measurement, 104,
    105
Capillary rheometers, 105, 106,
    107, 135, 136, 146, 409
Carbon black, 46, 395
Cascade extruders, 251, 256
  single-screw extruders arrange-
    ment, 256, 257
  twin-screw and single-screw ex-
    truders arrangement, 253
Characteristics curves
  head die, 182-185
  screw, 178-182, 185-187
Check rings, 350, 352, 382
Chemical resistance, 1, 12, 102,
    306, 309, 314, 316, 318,
    342
Chlorinated polyethylene (CPE),
    4, 35, 39, 48, 71, 72, 73,
    342, 407
Chlorinated PVC (CPVC), 140
Coefficient of linear thermal ex-
    pansion, 235
Coefficient of viscosity, 103
Coextrusion, 241-242
Cohesive energy density, 117,
    138, 146
Cold-runner injection mold, 67
    353-354, 382
Color compounding, 5, 46, 343
Color propagation curves, 138
Compatibility, 36
Complex modulus, 126, 127,
    128
Compound forms, 94, 95
Compounding, equipment classi-
    fication and techniques,
    42-95
  batch-type internal, high-in-
    tensity mixers, 85-93

[Compounding, equipment classi-
    fication and techniques]
  high-shear vertical non-
    fluxing mixers, 47-53
  low-shear and high-shear
    mixing in horizontal,
    cylindrical blenders, 55-
    56
  low-shear horizontal non-
    fluxing mixer; ribbon
    blenders, 43-47, 49, 86
  multiscrew extruders,
    twin-screw extruders, 63-
    77
    two-stage compounding
      extruders, 77-84
    two-stage continuous, high-
      intensity fluxing mixers,
      84-85
  single-screw extruders, 58-
    62
Condition injection blow-mold-
    ing (CIBM), 11, 364
Conduits, 8, 17, 34, 66, 67
Conical twin-screw extruders,
    62, 71-77, 247
  processing, 210-212, 214-215
Conversion operations, 1, 5
Cooling bath requirements in
    single-screw pipe or hol-
    low profile extrusion, 221-
    228
Copolymers, vinyl chloride-
    vinyl acetate, 139, 394-
    395, 403 (see also VC/VA
    copolymers)
Corotating, intermeshing,
    parallel twin-screw extru-
    ders, 62-68, 248
Costabilizers, 16, 24, 71, 73,
    76, 297-299 (see also Ad-
    juvants heat stabilizers)
Cost performance of rigid PVC,
    16, 304, 308, 316-320,
    323, 336, 343

Counterrotating, intermeshing,
    conical twin-screw extru-
    ders, 62, 71-77, 247 (see
    also Conical twin-screw ex-
    truders)
Counterrotating, intermeshing,
    cylindrical, twin-screw
    extruders, 62, 68, 248,
    250
Credit card core stock calender-
    ed, 402-403
Credit card overlay film,
    calendered, 403-404
Critical pressure point at which
    melt fracture occurs, 131,
    135
Critical shear rate at which melt
    fracture occurs, 133, 134
Cross-linking, 106, 135, 136,
    327
Crystalline melting point, 101,
    130, 140-141, 227
Crystallinity, 124, 125, 129,
    144, 304, 332, 333
C.W. Brabender plasticorder,
    141, 325-332, 409 (see
    also Brabender torque
    rheometer)

                    D

Decompression zone in extrusion,
    58
Deflection temperature under
    load (DTUL), 5, 16, 19,
    29, 33-34, 309, 321-322,
    332-333 (see also Heat
    distortion temperature)
Degree of compounding, 67, 86-
    89
Degree of fusion, 18, 56, 86-87,
    125, 215, 289 (see also
    gelation)
Degree of polymerization, aver-
    age ($\bar{P}$), 22-23

Densification of the dry blend, 48, 53, 55
Devolatilization, 251
Dibasic lead phosphite, 16, 26, 27
Dibasic lead stearate, 16, 26, 37, 66
Dibutyltin maleate, 30, 52
Dicing, 92
Die adaptor, 157, 192, 215, 229, 231, 238
Die design for pipe extrusion, 215-220
Die flow-restricting devices, 200-201
Die wall axial temperature profile, 200
Diffusion, 125, 191, 202, 216
Diglycol stearate, 49
β-Diketone, 24, 293
Dimensional stability, 303, 309, 348, 350, 358
Dimethyltin-S,S'-bis (isooctyl mercaptoacetate), 24, 52, 292
Di-n-butyl-S,S'-bis(isooctyl mercaptoacetate), 28, 52, 394-395, 450
Di-n-octyltin maleate, 28, 30, 52, 56, 76
Di-n-octyltin-S,S'-bis(isooctyl mercaptoacetate), 28-29, 76, 292, 394, 408
Diosna high-shear vertical mixers, 47, 48
Discoloration streaks, 158
Dispersibility, 44, 52, 55
Dispersing agent, 46-47
Distributive blending, 43, 55
Double-batch operation in dry blending, 53
Double-conical twin-screw design, 62, 75, 249-250
Double die heads, 263, 266 (see also Twin screw extrusion heads)
Double-platen bottle-blowing machines, 266-269
Double-wave screw design, 259

Drag capacity of the metering section of a single screw, 166, 171
Drag flow rate, 157, 161, 164, 171, 173
Drain, waste, and vent (DWV), 26, 34, 42, 43
Dry blend
   compactor, 177
   coolers, 45, 49, 56
   extrusion, 2, 8, 246
   extrusion blow-molding, 246-259
   fundamental properties, 54-55
   preparation, 17, 43, 47-48
   procedures, 43-46, 48-53, 52
   properties, 16, 54-55
   stratification, 46
Davis-Standard barrier (DSB) screw, 190
Dynamic mechanical properties, 126-129

E

Elastic memory, 130
Elastic modulus, 4, 397
Electric insulation boosters, 16
Embossing, 232-234, 401, 403
Emulsion resins, 20, 66, 78, 130, 131, 387
Energy economy, 8, 59-60, 68, 79, 83-84, 155, 162-164, 167, 169, (see also Power economy, Energy factor)
Enthalpy, 162-164
Entry angles, adaptors and dies, 133-135
Epoxidized soybean oil (ESO), 25, 30-31, 49, 67, 71, 73, 76, 297-299
Epoxy aliphatic resins, 24
Epoxy plasticizers, 16, 30, 44, 67, 70
Epoxy tallate, 95
Equations for shear rate calculations, 109-113
Estertin stabilizers, 28, 30
2-Ethyl-hexyl acrylate, 41

N,N-Ethylene bisstearamide, 36,
    38, 296 (*see also* Acrawax C)
Ethylene vinyl acetate copolymer
    (EVA), 4, 39, 71, 342,
    395
European Economic Community
    (EEC), 291
Experimentally determined screw
    performance data, analy-
    sis, 172-176
External lubricants, 16, 29, 35-
    38, 71-73, 76, 210-211,
    293-294, 337-338 (*see
    also* Lubricants, external)
Extrudate
  quality, 157, 158
  swelling, 7, 20, 29, 99, 118,
    129, 130 (*see also*
    Percentage memory)
  temperature, 157-159
  yellowness, 45, 138, 158, 162,
    212, 230
Extruder
  barrel wear, 29, 206-207
  flushing, 70, 261
  instrumentation, 157-160 (*see
    also* Instrumentation
    to monitor operating
    variables)
  screw wear, 27, 29-30, 206-
    207
  shutdowns, 68, 70, 261
  start-ups, 68-69
  temperature control, 194-196
Extrusion blow-molding of rigid
    PVC containers, 1, 8, 19,
    132, 246-300
  biaxially oriented bottles
    compounds, 295
    equipment, 280-287
    process, 278-279
  bottle-blowing machines, 264-
    278
    platen-type, 264-270
    rotary wheels, 271-278
  compounds and formulations,

[Extrusion blow-molding of rigid
    PVC containers]
    288-299
  extruders
    single-screw, 257-259
    twin-screw, 246, 248-250
  extrusion heads
    double-die heads, 263
    parison wall thickness con-
      trol, 260-261, 263
  extrusion systems, two-stage,
    251-257
    single-screw extruders com-
      bination, 251-252, 256
    twin-screw and single-screw
      extruders combination,
      247, 251, 253-254
Extrusion stability, 159-160
Extrusion surging, 8, 54, 155-
    156, 159, 177

                    F

Farrel continuous mixer (FCM),
    84-85, 411
Feed zone of extruder screws,
    60, 162, 167, 351, 372
Fillers, 121, 129, 343, 407
Film blown, 31, 396, 399
Film calendered, 397-406
Fines, 54
Fischer Johann extrusion blow-
    molding machines, 246,
    257, 286, 369
Flame retardance, 304-306, 308-
    311, 315, 323, 409
Flammability, 1, 308-309, 311,
    313
Flaws of the semigeometric scal-
    ing rules, 170-171, 173
Floppy disk cartridge film, 12,
    395, 401
Flow function for screw, 157,
    160-161
Flow index, flow-behavior index,
    (*see also* Power low-flow index),

[Flow index, flow-behavior index]
    99, 105, 112-114, 171-180, 182, 430
Flow mechanism on capillary rheometers, 106-109
Flow properties, dry blend, 55
Flow-restricting devices in extrusion heads, 200-201
Fluxed compounding, 43, 56-91
    (see also Melt compounding)
Food and Drug Administration (USFDA), 26, 29, 31, 36, 38, 292, 314, 323, 334, 400, 405, 408, 451
Food-grade rigid PVC compounds, 7, 57, 75, 76, 296, 297, 394
Food-packaging film, 12, 399-401
Force feeding, 77, 162, 193, 213, 218
Forming techniques, 2, 5
Formulations, 66-67, 71-73, 76, 211-212, 296-299
Free-radical initiation, 18, 20
Frohn-Alfing extrusion blow-molding machines, 250, 263-264
Fumed silica, 45, 394
Furniture, 314-315
Fusion, degree of, 18, 56, 67, 86-87, 125, 289 (see also Gelation)

G

Gas permeation resistance, 1, 400
Gel permeation chromatography, 126-129, 144
Gelation, 18, 56, 67, 86-87, 125, 215, 289 (see also Degree of fusion)
Glass transition temperature,

[Glass transition temperature]
    35, 40-41, 99, 101-102, 138
Glyceryl monoesters, 28, 30, 38, 76, 296
Glyceryl monooleate, 28, 36, 38, 76
Graft copolymers, 39-40
Grafting PVC, 39-40, 122, 212
Graham blow-molding wheels, 258
Grain, shape and size, 55
Grooved feed zone, 164, 256
Gutters, 8, 166

H

Hagen-Poiseuille equation, 109
Head pressure variation versus output rate, 188-189
Heat distortion
    improvers, 16, 140
    temperature, 5, 16, 29, 33-34, 309, 321-322, 332-333 (see also Deflection temperature under load)
Heat sealing, 409
Heat stabilizers, 16, 22-23, 49, 52, 66-67, 71-73, 76, 211, 296-299, 394-395, 408
Heat transfer
    coefficient, 169
    rate, 169-170
Heisler temperature charts, 223, 225
Henschel high intensity mixer, 48, 408
High density polyethylene (HDPE), 36, 39
High-impact rigid PVC, 309-310, 314
High-intensity fluxing mixers, 12, 40, 43, 46, 84-86, 290, 411 (see also Banbury and internal fluxing mixers)
High molecular weight PVC resins, 78, 116, 141, 166, 303, 341

High nitrile copolymers, 54, 137-138

High shear
  choppers, 55
  vertical nonfluxing mixers,
      43, 45-47, 411
    blending procedures, 48-49,
      52, 411

Homogenization, dry blend, 17, 48

Horizontal rotary table blow-
      molding machines, 271,
      277-278

Horizontal L-shaped two-stage
      extrusion, 252-255

Hostalit Z, 39

Hot runner molds, 382

Hydrogen chloride, 23-26, 333

Hydroxyphenyl chlorobenzotria-
      zole, 72

I

Ideal fluids, 102-104, 109 (see
      also Newtonian fluids)

Impact resistance (or impact
      strength), 1, 66, 309-
      311, 314, 317, 322, 325,
      332, 339, 343

Inherent viscosity (IV), 21,
      333 (see also logarithmic
      viscosity number)

Injection blow-molding of rigid
      PVC containers
  biaxial orientation, 379
  parison blowing, 368
  parison process
    cavity filling, 366-367
    injection pressure, 366
    temperature control, 367-368
  process and variables
    blow pressure, 365, 368
    blow rate, 365, 368
    manifold and nozzle design,
      376-378
    mold design and materials,
      373-375
    parison temperature, 364
    process elements, 371-373

[Injection blow-molding]
  process machinery, 370-371
  resin-compound picture, 369-
      370

Injection molding
  computer-aided design (CAD),
      348
  computer-aided manufacturing
      (CAM), 348
  cycle, 346, 356
  economics, 316-321, 324
  flow restrictions, 346-347
  formulations, 67, 324-345
  market trends and opportuni-
      ties, 306-316
  mold construction, 347-348
  polymer preparation
    colorants, 343
    fillers, 343
    heat stabilizers, 334-337
    impact modifiers, 339-343
    lubricants, 337-338
    processing aids, 338-339
    resin, 332-333
  processing recommendations
    cold slug wells, 354
    equipment, 349
    gates, 347, 354, 355
    machine size, 349-350
    nozzle, 352-353
    runners, 353-354
    screw design, 350-352
    sprue bushing, 353
  processing techniques
    drying, 356
    heater band setting, 357,
      360
    injection and molding,
    injection and molding pres-
      sure, 358
    injection rpm, 357
    injection speed, 358
    machine start-up techniques,
      359
    mold temperature, 356
    purging and shutdown tech-
      niques, 359, 360

[Injection molding]
screw back pressure, 357
stock temperature, 356, 357
product design considerations
flow restrictions, 346, 347
gate dimensions, 347
gate location, 347
wall thickness, 346
reheat stretch blow, 380
testing, 325-332
Instrumentation to monitor
operating variables, 157-
160
Intermeshing twin-screw extruders, 62-75, 203-208
Internal fluxing
mixers, 40, 43, 46, 84-86,
290, 411 (see also high-
intensity fluxing mixers
and Banbury mixers)
mixing, operation, 89-90
Internal lubricants, 28-29, 35-
37, 71-73, 76, 210-211,
293-294, 337-338 (see
also Lubricants, internal)
Intrinsic viscosity, 21 (see also
Limiting viscosity number)
Izod impact, 16, 66, 309, 322,
340-341, 370
versus impact modifier level,
340-341

K

Kane Ace impact modifiers, 41, 292
Kautex extrusion blow-molding
machines, 246, 257, 286,
369
Kneaders, 77, 80-83, 411
Knife die face pelletizers, 61,
64, 69, 79
Kombiplast, 63, 67, 78
K value, 2, 9-10, 18-20, 22, 48,
82, 280, 303, 333, 370, 450

L

Labile chlorine, 24
Laminar flow, 101, 107-108, 135,
197
Lead heat stabilizers, 24, 26,
334-335, 395
Lead stearate, normal, 27, 66-67,
71
Lead sulfur staining, 28
Leakage flow, 162, 171
Length-to-diameter ratios (L/D),
159, 188-189, 252-253,
257-258, 350, 372
Lewis acid, 25
Light stability, 16, 24, 39, 409
Limiting viscosity number, 21
(see also intrinsic viscosity)
Linseed oil, 31
Littleford-Lödige high-shear vertical non-fluxing mixers,
48, 408
Logarithmic viscosity number,
21, 22 (see also Inherent
viscosity)
Low molecular weight partially
oxidized polyethylene, 25,
27, 37-38, 49, 73, 293 (see
also partially oxidized low-
molecular weight PE)
Low molecular weight resins, 20,
48, 114, 119, 304, 306,
333, 341
Low shear limiting viscosity, 129
L-shape combination of extruders,
247, 251, 257, 280
Lubricants, 16, 35-38, 337-338,
408-409, 450
external, 16, 29, 35-38, 71-73,
76, 210-211, 293-294, 337-
338
internal, 28, 35-37, 71-73, 76,
210-211, 293-294, 337-338
selection, 37-38, 210, 241

Luvitherm process, 78, 387-388,
    404

M

Maleates, tin, 29, 52, 72, 76,
    334
Marrick cold-parison blow-mold-
    ing, 246, 279
Mass extrusion rate, 161-163,
    168
Mass polymerization, 17-19, 35,
    53 (*see also* Bulk poly-
    merization)
Matching formulation to the ex-
    trusion process and to the
    finished products, 208-
    212
Mechanical power, 161, 169
Melt
    compounding, 6, 17, 56-93
    extensiometer, 135
    flow, 215, 302-305, 310-311,
        320-322, 325-332, 339,
        342, 345-346, 348-349
      index, 290
    fracture, 7, 100, 131-135
    homogeneity, 34, 90, 256
    pressure, 159-160
      stability, 189-190
    processing, 2, 5, 99, 125
    strength, 34, 213, 294, 339
    temperature, 37, 59, 180
    thermal stability, 106, 135-
        136, 138, 325
Melting rate, 177, 180
Memory percentage, 7, 20, 99,
    118, 129-130 (*see also*
    Postextrusion swelling,
    Barus effect, puff-up)
Methylmethacrylate-butadiene-
    styrene, 40, 76, 296, 297,
    342, 394, 400
Methylmethacrylate-polybutyla-
    crylate copolymers, 41
Methacrylonitrile-styrene co-

polymer, 75, 117
α-Methyl styrene, 34, 339
Methyltin heat stabilizers, 29, 52,
    211, 292
Mills blow-molding wheels, 251,
    258, 261, 271-272
Mill-rolls operation, 90-91
Mixing
    impellers, 51
    mechanism
      in twin-screw extruders,
        corotating and counter-
        rotating, 65, 68, 72
      in two-stage single-screw
        vented extruders, 59, 64,
        209-210
    plows, 55
Modern single-screw machines,
    58, 155-156
Modulus
    in flexure, 309, 322
    in tension, 397-398
Moisture vapor permeability, 398,
    400
Molecular weight, 20-23
    average, 20, 119-120, 126-129,
        142
Moment of inertia, 60-61, 419, 424
Montan-wax acid esters, 25, 37,
    70-72, 78
Multiple-head die, 32, 263
Multiscrew extruders, one-stage,
    57, 62-77

N

National Sanitation Foundation
    (NSF), 26, 30, 32, 212
Newtonian
    flow behavior, 104
    fluid concept of viscosity, 102-
        103
    fluids, 102, 104-105
    viscosity coefficient, 103
Nonflammability, 1
Nonnewtonian flow behavior, 102,

[Nonnewtonian flow behavior]
104, 109
Nontoxic stabilizers, 30-32, 76,
292-293, 296-299, 370,
394, 399
Normal lead stearate, 25, 37
Number average molecular
weight, 21
Nylon, 36, 317

O

n-Octyltin derivatives, 24, 28-
30, 52, 76, 292, 296,
334, 370, 401, 408 (see
also Di-n-octyltins)
Oil canning, siding, 5, 232 (see
also Surface distortion,
siding)
One-pack stabilizer-lubricant
system, 26, 38, 335
Orange peel, 17, 216
Organotin carboxylates, 30, 52,
72, 76 (see also Tin
maleates)
Organotin heat stabilizers, 24,
28-30, 52, 76, 394, 408
Organotin mercaptide heat stabilizers, 24, 28-30, 52,
76, 211
Orientation, 4, 9, 11, 126, 278-
280, 287, 379-380, 404,
410
Outdoor applications, 27, 66,
405
Over-the-counter (OTC) pharmaceutical, 369, 401
Overworked compound, 17, 87,
95
Oxygen permeation, 398, 400

P

Pack-out in Banbury lines, 92-
93
Paraffin waxes, 25, 37, 70, 202,

[Paraffin waxes]
211, 293
Parison, 8-9, 260-264, 364-369
die, 199-200
wall thickness control, 260-261
Partially oxidized low-molecular-
weight polyethylene, 25,
27, 37-38, 49, 71, 73,
293
Pelletizing, 6, 61, 64, 69, 73, 75,
87, 91
Pellet post lubrication, 95
Pendant chlorine atoms, 6, 102,
138, 337
Pentaerythritol, 25, 27, 32, 394,
408
Percentage memory, 7, 20, 99,
118, 129-130 (see also Postextrusion swelling, Puff-up
percentage, Barus effect)
Permeability, 102, 400
α-Phenylindole, 24, 31, 292
Phthalocyanine blue, 46, 75
Phthalocyanine green, 46
Pigments, 44, 46-47, 75, 343-344
Pipe die design, matching the die
to single-and twin-screw
extrusion, 215-220
Pipe and electrical conduits
extrusion takeoff equipment,
221-228
formulations, 66-67, 211
Pipe extrusion, 125, 212-215
processing and quality problems
that can be remedied with
temperature adjustments,
228-230
Pipe sizing, 125
Planetary-gear extruders, 13, 16,
33, 63, 77, 122, 410
Plasticizers, 19, 33, 122
Platen-type blow-molding machines, 278, 286
Plate-out, 20, 29, 32, 34, 42,
201-203, 295
Plow-shape mixing tools, 55

Polarity, 102, 117, 138, 210
Polyacrylic esters (PAE), 4, 41
Polybutylacrylates, 4, 41, 414
Polyethylene (PE), 36, 120, 218,
    227, 257, 318, 400
Polyethylene terephthalate
    (PET), 8, 259, 369
Polyethylene waxes, 37, 49,
    211, 229, 293, 298
Polymer
    degradation, 23, 58, 65, 303-
        304, 327, 333-338, 359-360
    preparation, 1, 5-6, 16, 99
    processability, 2, 5, 16, 68,
        302-306, 317, 319
    residence times, 156
    rheology, 6, 99-149
Polymerization processes, 10, 17
Poly(methylmethacrylate) PMMA,
    140, 339
Poly-α-methylstyrene, 34, 339
Polyols, 25, 27, 203
Polypropylene, 36, 141, 318
Polystyrene, 120, 318, 360, 400
Polytetrafluoroethylene (PTFE),
    26
Postextrusion swelling, 7, 20, 99,
    118, 120-130 (see also Bar-
    us effect, Puff-up percent-
    age, Percentage memory)
Powder
    conveying, 54
    flow, 54
Power economy, 8, 60, 61, 68, 79,
    83-84, 155, 162-164, 167,
    169
Power law fluids, 117, 180, 429
Power low-flow index, 99, 105,
    112-114, 171-180, 182,
    430 (see also flow-be-
    havior index)
Preforms, 11, 279, 281, 286-
    287, 380-381
Pressure-building capabilities,
    second-stage two-stage
    screw, 185-187

Pressure
    dependence of viscosity on, 99,
        113, 145-146
    drop, 100, 107-108, 135, 197,
        216, 218-220
    flow, 157, 161, 164, 171, 173
    profile, 164
    stability in extruders, 159-
        160, 178
Primary PVC resin particles, 18,
    124
Processability, 1, 66, 302-308,
    315, 317, 320, 323, 339,
    349-359
Processing aids, 16, 33, 294-
    295, 338-339, 450
Puff-up percentage, 7, 20, 99,
    118, 129-130 (see also Bar-
    us effect, memory, Post-
    extrusion swelling)
    vs shear rate, 123, 129-130
Pumping rate, 177, 179
Purging and shutdown tech-
    niques, 359-360
PVC-PP copolymers, 333

Q

Quality control for pellets, 86-
    89

R

Rabinowitsch correction, 112
Radial temperature gradient,
    100, 157-159
Radial velocity gradient, 100-
    101
Radiant heaters, 13, 440, 443
Refractive index, 39
Relative viscosity, 21-22
Residential siding, 3, 34, 44,
    56, 193, 218, 230-236
Resins, 17-20
Reworked material or rework,
    34, 49, 188, 194, 218,

[Reworked material or rework]
    260, 281, 410, 441, 450
Rheology, 99-149
Rhone-Poulenc, 19
Ribbon blenders, 43-47, 49, 86
Ribbon extrusion test, 89
Rigid PVC compound forms, 94
Rotary blow-molding machines,
    271-278

S

Saturated polyester polymers
    (PET), 36, 259, 369
Scale-up rules for single screws,
    166-174
Screws, single
  characteristics curves, 178-
    182, 185-187 (see also
    Characteristics curves,
    screw)
  combined screw and die per-
    formance, 196-199
  cooling and heating, 81
  feed zone depth, 161-162, 167
  helix angle, 161-162, 178-
    179, 181
  metering zone depth, 6, 59,
    101, 167, 176, 185
  nose design, 192-193
  temperature control, 193
  torque, 34, 59, 161, 168, 172
Shearing forces, 107
Shear rate, 59, 64, 109-110, 170
  calculations, 109-110
Shear rate-shear stress curves,
    131-132, 220
Shrinkage, 309, 398, 447
  index of rigid PVC pellets,
    88-89, 220
Sidel blow-molding wheels, 274-
    275, 277
Siding and accessories
  processing, 230-235
  troubleshooting, 235-236
Single conical twin-screw

[Single conical twin-screw de-
    sign]
    250
Single die platen-type bottle-
    blowing machines, 264-267
Single-screw extruders, com-
    pounding, 58-62
Single-screw extrusion, 6, 8,
    29, 32, 5, 57-58, 64, 127,
    155-203, 257-259
  analysis, 161
Specific viscosity, 21, 23
Spider extrusion head, 261-263
  spider lines, 202, 216-217,
    219, 261
Spiral flow, 304-305, 322
Solvay blow-molding wheels,
    252
Sorbitan monooleate, 36
Sorbitol, 25
Standard-reference state, 115-
    117
Static charges, 20
Static mixers, 154
Stearoylbenzoylmethane, 31, 297
Storage modulus, 126-128
Strand dies for pelletizing, 61,
    64, 69
Stretch blow-molding, 286-287
Styrene-acrylic copolymers, 34
Styrene-acrylonitrile copolymers
Sulfide staining, 25, 27, 32, 343
Surface distortion, siding, 5,
    232 (see also Oil canning)
Surfaceskin, PVC resin particle,
    18
Suspension polymerization, 17-18,
    35, 48, 53, 82, 120, 289
Syndiotacticity, 139-140

T

Taper flow path, 100, 134-135,
    197, 216
Temperature controllers
  extrusion, 194-196

[Temperature controllers]
  injection, 352, 357
Temperature dependence of viscosity, 99, 115, 141-143
Temperature gradient, 6, 57,
  100, 197, 213
Tensile
  properties, 100
  strength, 100
  stress, 100, 135
Thermal degradation, 303-304,
  327, 333-338, 359-360
Thermal properties
  heat capacity, 222
  thermal conductivity, 222
  thermal diffusivity, 222-224
Thermal stability, 10, 304-305,
  327, 330, 332, 334-337,
  342, 348
Thermoforming, 1, 13, 439-440,
  449-450
  markets, 440, 451
  process, 440-443
  rework usage, 450
  techniques, 443
Tin mercaptides (see also Organotin mercaptides)
Titanium dioxide, 24, 27, 29,
  44, 46, 49, 73, 166
Toxicity, 26, 29-31
Translucent impact-modified
  rigid PVC formulations,
  71-72
Traversing thermocouples, 6,
  100, 157
Tribasic lead sulfate, 26, 66
Triple-batch operation in dry
  blending, 53
Twin die platen bottle-blowing
  machines, 268-270
Twin extrusion heads, 263, 266
Twin-screw extrusion process,
  203-208, 246, 248-250
  analysis, 208
  blow-molding, 9, 246-247, 249
  compounding, 63-77
  extruders, 63-77, 203-208,

[Twin-screw extrusion process]
  246-251
  flushing procedure, 70
  hoppers, preheated, 250
  opposite-end driven extruders,
    247, 253
  pelletizing heads, 69
  power economy, 79, 214, 215
  screw-tip effect, 215, 248
  shutdown, purging, 70
  start-up procedure, 69
  starve-feeding mode, 69
Two-stage compounding extruders, 77, 79-80, 84
Two-stage continuous high-
  intensity, fluxing mixers,
  Farrel continuous mixers
  (FCM) 8, 84-85
Two-stage extruders systems in
  blow-molding, 247, 251-257
Two-stage single-screw extrusion in a vented barrel
  extruder, 58, 156
  analysis, 174-182
  processing, 182-187, 209-210
  second-stage optimization,
    198-200
Two-stage twin screw extruders,
  62, 203-205 (see also Four-
  screw extruders)

                  U

UV stability, 24, 133, 214, 342,
  405
UV stabilizers, 16, 24, 71-72

                  V

Vacuum forming, 13, 444, 446
Van der Waals forces, 101
Velocity gradient, 107, 197
Velocity profile, 107
Venting, 58, 251, 256
  on multiextruders combination,
    256
Vent zone operation, 58, 191-192

Vertical blow-molding wheels,
    271, 273, 281
Vertical single-screw extruders,
    251, 256, 270
Vinyl acetate copolymers (VC/
    VA), 139, 333, 394-395,
    403, 449, 451
Vinyl chloride monomer (VCM),
    3, 52, 54, 365, 400, 410, 451
Viscosity calculations in capil-
    lary extrusion rheometers,
    111
Viscous dissipation, 197

                    W

Wall slip-stick, 113-114
Wax E, 37, 71

Wax OP, 37, 72
Weatherability, 1, 16, 25, 56,
    314, 316, 342, 344
Weight average molecular
    weight (MW), 2, 116-117
Weissenberg rheogoniometer,
    126-128, 409
Weld lines, 202, 216-217, 219,
    261 (*see also* Spider ex-
    trusion head lines)
White, opaque impact-modified
    formulations, 72, 211-212
Window lineals, 8, 212, 236-
    241 (*see also* Window pro-
    files)
Window profiles, 8, 212, 236-
    241
    processing, 236-242

Printed and bound by CPI Group (UK) Ltd, Croydon, CR0 4YY

23/10/2024

01778224-0018